Multifunctional Polymer Nanocomposites

Multifunctional
Polymer
Nanocomposites

Edited by
Jinsong Leng
Alan kin-tak Lau

CRC Press
Taylor & Francis Group
Boca Raton London New York

CRC Press is an imprint of the
Taylor & Francis Group, an **informa** business

CRC Press
Taylor & Francis Group
6000 Broken Sound Parkway NW, Suite 300
Boca Raton, FL 33487-2742

First issued in paperback 2017

© 2011 by Taylor and Francis Group, LLC
CRC Press is an imprint of Taylor & Francis Group, an Informa business

No claim to original U.S. Government works

ISBN-13: 978-1-4398-1682-0 (hbk)
ISBN-13: 978-1-138-11180-6 (pbk)

Visit the Taylor & Francis Web site at
http://www.taylorandfrancis.com

and the CRC Press Web site at
http://www.crcpress.com

Contents

Preface

A nanocomposite is a multiphase solid material in which one of the phases has one, two, or three dimensions smaller than 100 nanometers, or structures having nano-scale repeat distances between the different phases that make up the material (*Nanocomposite Science and Technology*, Wiley, 2003). Nano inorganic or organic powders or films with special physical properties are combined with polymers to form polymer nanocomposites whose physical properties and mechanical performance differ from those of the component materials significantly. A primary purpose of producing polymer nanocomposites is to impart the composites with multifunctional properties. There has been a rapid development of multifunctional polymer nanocomposites and a number of achievements have been reported. Due to their novel properties, multifunctional polymer nanocomposites can be used in a broad range of applications from outer space to automobiles, and to address challenges in organic solar cells, and biological technologies.

The book is aimed at audiences at different levels to provide a comprehensive discussion of multifunctional polymer nanocomposites. Both theoretical work and experimental results on the relationships between the effective properties of polymer composites and the properties of polymer matrices and reinforcements are discussed. Chapter 1 presents the overview of the development from bulk to nano for academics and industries, the importance of understanding the role of the nano and multifunctional polymer composites in definition, fabrication, design, nanotechnology, and nano products, and the increasing trend of using nano and multifunctional polymer composites.

Chapter 2 gives a comprehensive review on the structural properties of nanotubes and their related polymer composites. Many key factors such as dispersion, interfacial bonding characteristics, novel types of nanotubes in relation to the resultant mechanical, electrical, and thermal properties of the composites are discussed and analyzed through theoretical and computational studies. Chapter 3 provides some recent advances in multifunctional polymeric smart materials, including electroactive polymers and shape memory polymers and their composites. A comprehensive discussion is devoted to the definitions, architectures, applications, and fundamental principles of various functions of multifunctional polymeric smart materials.

Chapter 4 reviews the development of magnetic polymer nanocomposites. Their fabrication, processing, and physicochemical property analyses are taken into account. The effect of magnetic oxide nanoparticles on the chemical polymerization of polymer matrices is investigated. The morphology (size and shape) and other physicochemical properties of the polymer matrix are significantly influenced by the magnetic oxide nanoparticles.

Chapter 5 focuses on the nanomechanics of carbon nanotubes and modeling of carbon nanotube-based composites. This chapter introduces the atomistic modeling technique, the modeling of electromechanical coupling behaviors of carbon nanotubes, and the modeling of electrical conductivity of nanotube-based composites. Furthermore, the relevant experimental studies are also introduced. Chapter 6 presents the development of conventional biomaterials to the state-of-the-art biocomposites for biomedical and bioengineering applications. The advantages and disadvantages of different types of biomaterials, their material properties, structures, biodegradability and biocompatibility to the host body, and applications of the biocomposites are given. Chapter 7 provides the advances in synthesis, processing and testing of condensed phase flame retardant polymer nanocomposites. The morphologies, thermal stability, flammability, and char formation of polymer composites coated with carbon nanopaper are studied. The fire retardant mechanism of carbon nanopaper is discussed.

Chapter 8 follows the nanomaterials through engineering to applications and focuses on the recent development of polymer nanocomposites coatings for aeronautical applications. This provides an informative account of the challenges and opportunities for nanocomposite coatings in aeronautical operations under conditions such as lightning strike, erosion, ice accretion, and environmental corrosion. Chapter 9 describes the surface modification of carbon nanotubes for composites, which includes chemical modification, substitution reaction, electrochemical modification, and photochemical modification. Chapter 10 describes the ocean engineering application of nanocomposites, dealing with deformation, damage initiation, damage growth, and failure and corrosion in nanopaticles.

Multifunctional Polymer Nanocomposites provides the reader with the latest thinking on polymer nanocomposites by the scientists and researchers actually involved in their development. This book will be a useful reference not only for engineering researchers, but also for senior and graduate students in their relevant fields.

We would like to take this opportunity to express our sincere gratitude to all the contributors for their hard work in preparing and revising the chapters. We also wish to thank and formally acknowledge all the members of our team, as well those who helped with the preparation of this book. Finally, we are indebted to our families and friends for all their patience and support.

Jinsong Leng
Cheung Kong Scholars Professor, SPIE Fellow
Editor-in-Cheif: International Journal of Smart and Nano Materials
Centre for Composite Materials and Structures
Harbin Institute of Technology, PR China

Alan K. T. Lau
Professor and Executive Director
Centre of Excellence in Engineered Fibre Composites
Faculty of Engineering and Surveying
University of Southern Queensland
Australia
Department of Mechanical Engineering
The Hong Kong Polytechnic University
Hong Kong SAR China

About the Editors

Jinsong Leng is a Cheung Kong Chair Professor at the Centre for Composite Materials and Structures of Harbin Institute of Technology, China. His research interests include smart materials and structures, sensors and actuators, fiber-optic sensors, shape-memory polymers, electroactive polymers, structural health monitoring, morphing aircrafts, and multifunctional nanocomposites. He has authored or coauthored over 180 scientific papers, 2 books, 12 issued patents, and delivered more than 18 invited talks around the world. He also serves as the chairman and member of the scientific committees of international conferences. He served as the editor-in-chief of the *International Journal of Smart and Nano Materials* (Taylor & Francis Group) and as the associate editor of *Smart Materials and Structures* (IOP Publishing Ltd.). He is the chairman of the Asia-Pacific Committee on Smart and Nano Materials. Prof. Leng was elected an SPIE Fellow in 2010.

Alan K. T. Lau is professor and executive director of the Centre for Excellence in Engineered Fibre Composites, University of Southern Queensland, Australia. His research directions are mainly focused on smart composites, bio-nano-composites, and FRP for infrastructure applications. Due to his significant contribution to the field of science and engineering, he was elected as a member of European Academy of Sciences with the citation "for profound contributions to materials science and fundamental developments in the field of composite materials" in 2007. Dr. Lau has published over 190 scientific and engineering articles, and his publications have been cited over 1250 times (with an h-index of 17, over 950 times for non-self-cited articles) since 2002. He has also successfully converted his research findings into real-life practical tools, and therefore a total of six patents have been granted to him. Currently, he has been serving more than 40 local and international professional bodies as chairman, committee member, editor, and key officer to promote the engineering profession to the public. He is also the chairman of the 1st International Conference on Multi-Functional Materials and Structures, 2008.

Contributors

Xueting Chang
Institute of Marine Materials
  Science and Engineering
Shanghai Maritime University
Shanghai, China

Karen Hoi-yan Cheung
Department of Mechanical
  Engineering
Hong Kong Polytechnic University
Hong Kong, SAR China

Tsu-Wei Chou
Department of Mechanical
  Engineering and Center for
  Composite Materials
University of Delaware
Newark, Delaware

Shanyi Du
Center for Composites and
  Structures
Harbin Institute of Technology
Harbin, China

Jihua Gou
Composite Materials and
  Structures Laboratory
Department of Mechanical,
  Materials, and Aerospace
  Engineering
University of Central Florida
Orlando, Florida

Zhanhu Guo
Integrated Composites Laboratory
Dan F. Smith Department of
  Chemical Engineering
Lamar University
Beaumont, Texas

Nam Hoon Kim
Department of Hydrogen and Fuel
  Cell Engineering
Chonbuk National University
Jeonbuk, South Korea

N. Satheesh Kumar
Faculty of Chemical and Process
  Engineering
National University of Malaysia
Selangor, Malaysia

Xin Lan
Center for Composites and
  Structures
Harbin Institute of Technology
Harbin, China

Alan kin-tak Lau
Centre for Excellence in Engineered
  Fibre Composites
Faculty of Engineering and
  Surveying
University of Southern Queensland
Queensland, Australia, and
Department of Mechanical
  Engineering
Hong Kong Polytechnic University
Hong Kong, SAR China

Joong Hee Lee
BIN Fusion Research Team
Department of Polymer and Nano
  Engineering
Chonbuk National University
Jeonbuk, South Korea

Jinsong Leng
Center for Composites and
  Structures
Harbin Institute of Technology
Harbin, China

Chunyu Li
Department of Mechanical
  Engineering and Center for
  Composite Materials
University of Delaware
Newark, Delaware

Jianjun Li
Center for Composites and
  Structures
Harbin Institute of Technology
Harbin, China

Yanju Liu
Center for Composites and
  Structures
Harbin Institute of Technology
Harbin, China

Pallavi Mavinakuli
Department of Chemical
  Engineering
Lamar University
Beaumont, Texas

Hua-xin Peng
Advanced Composites Centre for
  Innovation and Science
Department of Aerospace
  Engineering
University of Bristol
Bristol, United Kingdom

Basavarajaiah Siddaramaiah
Department of Polymer Science and
  Technology
Sri Jayachamarajendra College of
  Engineering
Mysore, India

Yong Tang
Composite Materials and
  Structures Laboratory
Department of Mechanical,
  Materials and Aerospace
  Engineering
University of Central Florida
Orlando, Florida

Erik T. Thostenson
Department of Mechanical
  Engineering and Center for
  Composite Materials
University of Delaware
Newark, Delaware

Suying Wei
Department of Chemistry and
  Physics
Lamar University
Beaumont, Texas

Yansheng Yin
Institute of Marine Materials
  Science and Engineering
Shanghai Maritime University
Shanghai, China

D. Zhang
Integrated Composites Laboratory
Lamar University
Beaumont, Texas

Jiahua Zhu
Integrated Composites Laboratory
Lamar University
Beaumont, Texas

1

Introduction

Jinsong Leng and Jianjun Li
Harbin Institute of Technology

Alan kin-tak Lau
University of Southern Queensland
Hong Kong Polytechnic University

CONTENTS

1.1 Overview

The nano era, similar to the mid-industrial steel era, not only stands for great technical innovations but also indicates the future trend of existing technologies. It is believed that this period will dominate and transform people's daily lives. "Nano" is a unit of length defined as 10^{-9} m. To give you an idea of how small it is, the width of a human hair is 10^6 nm, and the size of an atom is 0.1 nm.

In recent decades, the development of microscopes has enabled scientists to observe the structures of the materials at nanoscale and investigate their novel properties. In the early 1980s, IBM (Zurich) invented the scanning tunneling microscope, which was the first instrument that could "see" atoms. In order to expand the types of materials that could be studied, scientists invented the atomic force microscope. Now, these instruments can be used to observe the structures and different properties of materials at nanometer scale. Physics reveals big differences at the nanometer scale, and the properties observed on a microscopic scale are novel and important. For example, quantum mechanical and thermodynamic properties have pushed forward the development of science and technology in the 20th century.

Nanotechnology means the study and application of materials with structures between 1 and 100 nm in size. Unlike bulk materials, one can work with individual atoms and molecules and learn about an individual molecule's properties. Also, we can arrange atoms and molecules together in well-defined ways to produce new materials with amazing characteristics. For example, nanotechnology has produced huge increases in computer speed and storage capacity. That is why "nano" has attracted attention in the research fields of physics, chemistry, biology, and even engineering. This word has entered the popular culture and can be found in television, movie, and commercial advertisements. Politicians and leaders around the world have realized the importance and urgency of developing nanoscience and nanotechnology, so countries have promoted research in nanoscience and nanotechnology in their universities and laboratories. With the huge increase in funding, scientists are pursuing nano research intensively, and the rate of discovery is increasing dramatically.

1.2 Classification of Nanomaterials and Nanostructure

The main classification of nanomaterials can be described as the following: carbon-based materials, nanocomposites, metals and alloys, nanopolymers, and nanoceramics. Carbon-based materials refer to carbon black, fullerenes, single-walled or multiwalled carbon nanotubes, and other carbides. Carbon nanotubes (shown in Figure 1.1), discovered in 1991 by S. Iijima, are hollow cylinders made of sheets of graphite [1]. The dimensions are variable, and one nanotube can also exist within another nanotube, which leads to the formation of multiwalled carbon nanotubes. Carbon nanotubes have amazing mechanical properties due to strength of the sp^2 carbon–carbon bonds. Young's modulus and the rate of change of stress with applied strain represent the stiffness of a material. The Young's modulus of the best nanotubes can reach 1000 GPa, which is approximately five times higher than the Gpa of steel. The tensile strength can be as high as 63 Gpa, and this value is around 50 times higher than steel. Depending on the graphite arrangement around the tube, carbon nanotubes exhibit varying electrical properties and can be insulating, semiconducting, or conducting. Also, carbon nanotubes are interesting media for electrical energy storage due to their large surface area, and they are still being investigated as a hydrogen storage medium.

Organic–inorganic nanocomposite is the fast-growing area of current materials research. Significant effort is directed at developing synthetic approaches and controlling their nanoscale structures. The properties of nanocomposite materials are determined not only by the properties of their individual

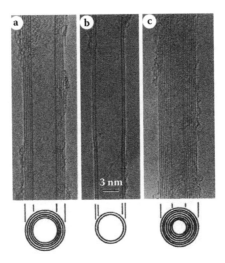

FIGURE 1.1
Electron micrographs of microtubules of graphitic carbon. Parallel dark lines correspond to the (002) lattice images of graphite. A cross section of each tubule is illustrated. (a) Tube consisting of five graphitic sheets, diameter 6.7 nm. (b) Two-sheet tube, diameter 5.5 nm. (c) Seven-sheet tube, diameter 6.5 nm, which has the smallest hollow diameter (2.2 nm). (Reprinted with permission from Iijima, S., *Nature*, 1991, 354, 56–58. Copyright 1991 Nature Publishing Group.)

components but also depend on their morphology and interfacial characteristics. The rapid research in nanocomposites has already generated many exciting new materials showing novel properties. It is also possible to discover new properties that are still unknown in the parent constituent materials.

Metal and alloy nanomaterials generally include gold, silver, magnetic iron-based alloys, and magnesium-based alloys. Gold and silver nanoparticles can be easily prepared, and they are promising probes for biomedical applications. Unlike other fluorescent probes such as organic dyes, gold and silver nanoparticles do not burn out after long exposure to light. Gold nanoparticles have already been used as ultrasensitive fluorescent probes to detect cancer biomarkers in human blood. Iron, cobalt, and their alloys are classes of magnetic nanoparticles whose magnetic performance can be modified by controlling synthesis method and chemical structure of the materials. In most cases, the magnetic particles ranging from 1 to 100 nm in size may display superparamagnetism.

Polymers are large molecules (macromolecules) composed of repeating structural units typically connected by covalent chemical bonds. They are widely used in our lives and play an important role in industry. The Nobel Prize in Chemistry in 2000 was awarded for the discovery and development of conductive polymers. In future, one can use such new exciting materials based on conductive polymer technology. In nanostructured polymers, the attractive force between polymer chains plays an important role in determining their properties. When inorganic or organic nanomaterials are dispersed in the polymers, the nanostructures of polymers can be modified and the desired properties can be obtained.

Nanoceramics considered in the study are oxide and non-oxide ceramic materials. Since nanocrystalline materials contain a very large fraction of atoms at the grain boundaries, they can exhibit novel properties. One important class of nanoceramics is semiconducting materials such as ZnO, ZnS, and CdS; they are synthesized by different methods, and the scientist can control their size and shape easily. They show quantum confinement behavior in the 1–20 nm size range. For such materials, the focus is on the production and application of ultrathin layers, fabrication, and molecular architecture.

Nanostructure is defined as an object of intermediate size between molecular and microscopic (micrometer-sized) structures. Based on the different shapes, generally they can be classified into nanoparticle, nanofiber, nanoflake, nanorod, nanofilm, and nanocluster types, and the typical photos are shown in Figure 1.2 [2–4]. Materials with different nanostructures can have obviously different properties, so one of the tasks scientists face is to find the relationship between properties and structure. We know that the structure of materials will determine their properties, and the properties of materials can reveal their structure. Thus, it is necessary to focus on exploring size-controllable and shape-controllable nanomaterials.

1.3 Nanomaterials from Academia to Industry

The general meaning of synthesis and assembly of nanomaterials is engineering materials with novel properties through the preparation of material at the nanoscale level. In fact, nanomaterials already existed before the advanced microscope was invented. The problem is that scientists cannot observe their nanometer structures directly at the moment. Research in nanomaterials and their novel properties is motivated by understanding how to control the building blocks and enhance the properties at the macroscale. For example, scientists can increase the magnetic storage ability, catalytic enhancement, electronic or optical performance, hardness, and ductility by controlling the size and method of assembly of the building blocks.

The most frequent techniques used in the laboratory to synthesize nanomaterials include chemical vapor deposition, physical vapor deposition, sol–gel technique, and precipitation from the vapor and supersaturated liquids. These techniques have been applied in the industry for the preparation of nano products ranging from electronics to drug delivery systems. There are several reviews on the synthesis and assembly of nanomaterials [5–8].

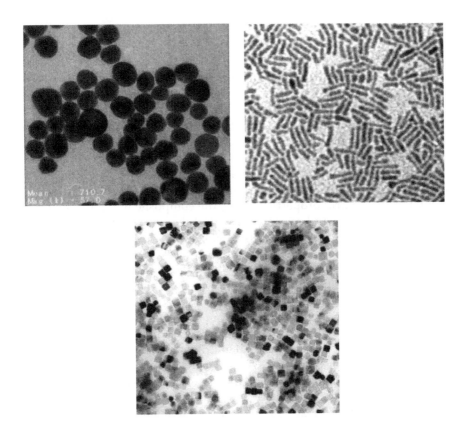

FIGURE 1.2
(Top left) TEM image of 20 nm Au nanoparticles produced by the Turkevich method. (Reprinted with permission from Hodak et al., *J. Phys. Chem. B,* 2000, 104, 9954. Copyright 2000 American Chemical Society); (Top right) CdSe quantum rods grown from the dots by a secondary injection and subsequent growth for 23 h. (Reprinted with permission from Peng, Z. A. and Peng, X., *J. Am. Chem. Soc.,* 2001, 123, 1389. Copyright 2001 American Chemical Society); (Bottom) TEM image of nanocrystalline KMnF₃. (Reprinted with permission from Carpenter, E. E., Ph.D. thesis, University of New Orleans, New Orleans, LA, 1999. Copyright 1999 Everett E. Carpenter.)

In the global market for nanomaterials, conventional materials such as SiO_2, TiO_2, Fe_2O_3, and ZnO are the main products that make the greatest initial commercial impact. The reason is that these nanoceramics can be easily synthesized, and the cost of production is lower. In future, the possibility of incorporating "smart" features in nanomaterials should be explored. Smart materials can also be termed *intelligent* materials. Such materials have one or more properties that can be significantly changed in a controlled fashion by external stimuli, such as stress, light, magnetic fields, or electric fields. Smart materials, including piezoelectrics, electrostrictors, magnetostrictors, and shape memory alloys, can perform sensing and actuating functions. It is believed that the smart materials with nanostructures will dominate our lives.

1.4 Nanocomposites

Composite materials (or composites for short) are combined from two or more constituent materials that have significantly different physical or chemical properties. The constituent materials will remain separate and distinct at a macroscopic level within the finished structure. Generally, two categories of constituent materials, matrix and reinforcement, exist in the composite. The matrix materials maintain the relative positions of the reinforcement materials by surrounding and supporting them, and conversely the reinforcements impart their special mechanical or physical properties to enhance the matrix properties. Thus, the composite will have the properties of both matrix and reinforcement, but the properties of a composite are distinct from those of the constituent materials. Thousands of years ago, people used straw to reinforce mud in brick making to increase the strength of the brick.

A nanocomposite is defined such that the size of the matrix or reinforcement falls within the nanoscale. The physical properties and performance of the nanocomposite will greatly differ from those of the component materials. According to the type of matrix, nanocomposites can be classified into ceramic matrix nanocomposites, metal matrix nanocomposites, and polymer matrix nanocomposites.

In ceramic matrix nanocomposites, the main volume is occupied by ceramics including oxides, nitrides, borides, and silicides. In most cases, a metal as the second component is combined into ceramic matrix nanocomposites. Ideally, the metal and the ceramic matrix are finely dispersed in each other to form a nanocomposite that has improved nanoscopic properties, including optical, electrical, and magnetic properties.

In metal matrix nanocomposites, ceramics are often used as reinforcement and matrices are based on most engineering metals, including aluminum, magnesium, zinc, copper, titanium, nickel, cobalt, and iron. Depending on the properties of the matrix metal or alloy and of the reinforcing phase, the metal matrix nanocomposites can have the features of low density, increased specific strength and stiffness, increased high-temperature performance limits, and improved wear-abrasion resistance. Compared with polymer matrix composites, metal matrix composites can offer higher modulus of elasticity, ductility, and resistance to elevated temperature. However, they are more difficult to process and are more expensive.

Polymer composites are generally made of fiber and matrix. Usually glass, but sometimes Kevlar, carbon fiber, or polyethylene, is used as the fiber. The matrix usually refers to a thermoset such as an epoxy resin, polydicyclopentadiene, or a polyimide. The fiber is embedded in the matrix so as to increase the strength of matrix. Such fiber-reinforced composites are strong and light, and they are even stronger than steel, but weigh much less. This means that composites can be widely used in industry for their high strength-to-weight ratio. For example, reinforced polymer composite is used in the automotive industry to make automobiles lighter.

1.5 Multifunctional Polymer Nanocomposites

The purpose of producing polymer nanocomposites is to give the composite multifunctional properties, so nano inorganic or organic powders or films with special physical properties are combined with polymers to form polymer nanocomposites. Long ago, people living in South and Central America fabricated polymer composites. They used natural rubber latex, polyisoprene, to make gloves, boots, and raincoats, but one will feel uncomfortable if he wears a raincoat just made from the single polymer. Then, a young man Charles Macintosh used two layers of cotton fabric and embedded them in natural rubber to fabricate a composite with three-layered sandwich structure. Thus, the raincoats made from composites have the advantages of the two components: waterproofing of rubber and comfort of the cotton layers. From this story, we know that the polymer composites with designed structures have the properties of a polymer matrix and inorganic or organic reinforcement. In multifunctional polymer composites, the reinforcements impart their special mechanical, optical, electrical, and magnetic properties to the composite. The polymer matrix holds the reinforcements and retains the properties of polymer; for example, polymer matrix can absorb energy by deforming under stress, thus overcoming the disadvantages of the brittleness of the reinforcement.

The current research emphasis and theoretical work on polymer nanocomposites is to find relationships between the effective properties of polymer composites (such as Young's modulus, tensile strength, and thermomechanical parameters) and the properties of constituents (polymer matrix and reinforcement), volume fraction of components, shape and arrangement of reinforcement, and matrix–reinforcement interaction. These results can help predict and control the properties of polymer nanocomposites. Experimental results have revealed that the reinforcement size and morphology of polymer and reinforcement play a very important role in determining the performance of polymer nanocomposites. Also, the properties of polymer nanocomposites will depend on the nature of dispersion and aggregation of the reinforcements. In the polymer composite, there exists an interphase with a layer of high-density polymer around the particle. The strength of the reinforcement–matrix interaction will affect the thickness and density of the interphase. The inter-reinforcement distance and the arrangement of reinforcement are important factors to be investigated. Generally, as their size decreases and number increases, the reinforcements become closer, and the bulk properties can be modified significantly. Due to the different physical and chemical systems for different researchers, different processing conditions are needed for different polymer systems to be formed. And it is difficult to provide one universal technique for producing polymer nanocomposites. Methods including melt mixing and in situ polymerization have been applied in the laboratory.

Compared with other conventional composites, the unique characteristics of polymer composite material with optimized structures are corrosion resistance, high strength-to-weight ratio, and more design flexibility. Large panels can be easily fabricated to reduce labor costs and assembly time by using polymer composites. Polymer composites can replace traditional materials, precluding the use of heavy-duty installation equipment. Unlike the traditional materials such as concrete and steel, polymer composites provide a robust alternative to the highly corrosive properties of concrete and steel. Another advantage of polymer composites is their efficient manufacturing processes; people can integrate the unique advantages of polymer matrix and reinforcement filler in the manufacturing process. According to the requirement of customers, the engineers and designers can provide high-quality, innovative, fashionable products from concept to production. The advantages of polymer composites make them highly competitive in the range of services and products for the industrial parts market, medical, architectural, building, construction, and food packaging industries.

Polymer matrix-based nanocomposites with exfoliated clay, one of the key modifications, have been reviewed [9]. In this review, the author compares properties of nanoscale dimensions to those of larger-scale dimensions. In order to get the optimized resultant nanocomposite, it is necessary to have a better understanding of the property changes as the powder (or fiber) dimensions decrease to the nanoscale level. Unlike the in situ polymerization and solution and latex methods, the melt processing technique is usually considered prior as an alternative to clays and organoclays for its economical, more flexible characteristics. As shown in Figure 1.3, immiscible (conventional or microcomposite), intercalated, and miscible or exfoliated, three different states of dispersion of organoclays in polymers, have been proposed from WAXS and TEM results. The mechanism of organoclay dispersion and exfoliation during melt processing is shown in Figure 1.4 [10].

Carbon nanotubes are ideal as advanced filler materials in polymer composites because of their good mechanical, thermal, and electronic properties. A review of the mechanical properties of carbon nanotube–polymer composites has been conducted by Coleman [11]. The progress to date in the field of mechanical reinforcement of polymers using nanotubes has been reviewed. Large aspect ratio, good dispersion, alignment, and interfacial stress transfer are important factors that affect the properties of such polymer composites. For example, the author showed that the difference between random orientation and perfect alignment is a factor of five in composite modulus. The aligned composites, especially, fiber-reinforced polymers have very anisotropic mechanical properties. If one wants to avoid the

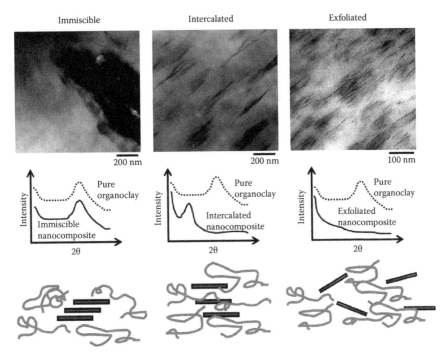

FIGURE 1.3
Illustration of different states of dispersion of organoclays in polymers with corresponding WAXS and TEM results. (Reprinted with permission from Paul, D. R. and Robeson, L. M., *Polymer*, 2008, 49, 3187–3204. Copyright 2008 Elsevier B.V.)

anisotropic properties in bulk samples, one can align the fibers randomly. In carbon nanotube–polymer composites, only the external stresses applied to the composite are transferred to the nanotube; the composite can show its excellent loadability. The interaction between polymer and nanotube in the vicinity of the interface is an important issue to investigate. Research results have indicated that interfacial interactions with nanotubes lead to an interfacial region of polymer with morphology and properties different from the bulk. The SEM images of composite fibers containing carbon nanotubes are shown in Figure 1.5 [12].

Conductive polymer is one important class of organic polymers that can conduct electricity. Inorganic nanoparticles of different nature and size were embedded into conducting polymers to endow such nanocomposites with further interesting physical properties and important application potential. A review of conducting polymer nanocomposites is given by Amitabha De [13]. Depending on the preparation methods and the nature of the inorganic materials, the properties of the resulting composite can be controlled. The TEM of a dilute dispersion of a PPy-silica colloidal nanocomposite is displayed in Figure 1.6 [14].

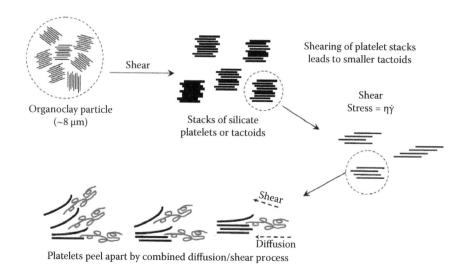

FIGURE 1.4

Mechanism of organoclay dispersion and exfoliation during melt processing. (Reprinted with permission from Fornes, T. D. et al., *Polymer* 2001, 42, 9929. Copyright 2001 Elsevier B.V.)

Poly (ethylene oxide) (PEO)-based composite polymer electrolytes have been widely investigated, and a review of PEO-based composite polymer electrolytes can be found in Reference [15]. In this paper, the author not only focuses on the experimental work but also summarizes the theoretical models for ion transport in such composite polymer electrolytes. Two different systems called *blend-based* and *mixed-phase* composite electrolytes are proposed. Blend-based systems are obtained from homogeneous solutions of two components in an appropriate common solvent. Mixed-phase systems mean that the polymer and inorganic or organic additives, not dissolved in a common solvent, are mixed inhomogeneously. In PEO-based composite polymer electrolytes, inorganic nanoparticles ZnO, $LiAlO_2$, or zeolites are combined with poly (ethylene oxide) to improve the mechanical properties, conductivity, and the interfacial stability. Also, the organic entities are added in the system to modify the structure of PEO, and this idea focuses on the two following research subjects: (1) synthesis of PEO-based flexible networks and (2) synthesis of composites by the addition of organic fillers [15]. Polystyrenes, polymethylmethacrylate, poly-acryloamides, and polyacrylates are added to the PEO-based composite polymer electrolytes, and their effect on the conductivity is studied. Poly(ethylene glycol methylether) (PEGME) molecules modified on SnO_2 nanoparticle surfaces through exchange reactions are managed to obtain PEGME–SnO_2 stable colloids, and the PEGME–SnO_2 composites themselves are able to dissolve lithium salts to form a new type of solid polymer electrolyte. PEGME–SnO_2 stable colloids also can be used as fillers for prototypical PEO-based electrolytes, where they exhibit advantages of both organic plasticizers and inorganic

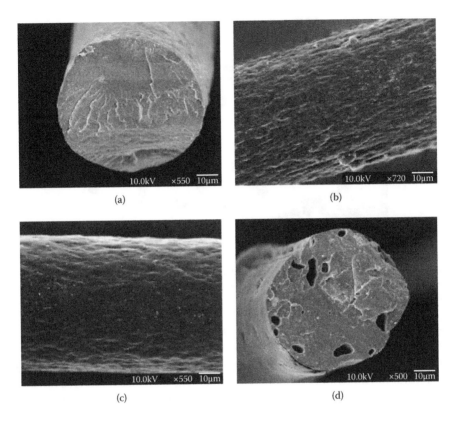

10.0kV ×550 10μm

(a)

10.0kV ×720 10μm

(b)

10.0kV ×550 10μm

(c)

10.0kV ×500 10μm

(d)

FIGURE 1.5
SEM images of composite fibers containing (a) carbon nanofibers, (b) entangled MWNT, (c) aligned MWNT, and (d) arc-MWNT. (Reprinted with permission from Sandler, J. K. et al. *Polymer* 2004, 45, 2001–2015. Copyright 2004 Elsevier B.V.)

fillers [16]. Transmission electron micrographs of the PEGME–SnO_2 nanoparticles are shown in Figure 1.7.

Polymer materials can be used in biological research and applications [17,18]. Biodegradable polymers and bioactive ceramics can be combined to form composite materials for tissue engineering scaffolds. Bioactive ceramics refer to alumina, titania, zirconia, bioglass, hydroxyapatite, etc. Also, some metals and alloys, including gold, stainless steel, and NiTi shape memory alloys can be used as biomaterials. Low biological compatibility and corrosion of metal, brittleness, high density, and low fracture strength of ceramics make polymers such as polyethylene, polyacetal, and the corresponding polymer composites excellent candidates for use in biological applications. Figure 1.8 lists the various applications of different polymer composite materials. In living systems, the responses of the host to the polymer materials can make a big difference; some polymer materials can be accepted by the body and others cannot. Great work has led to considerable progress in understanding

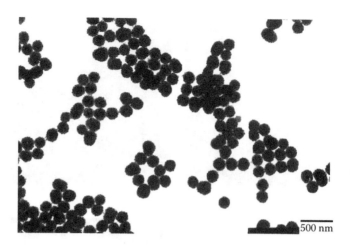

FIGURE 1.6
Transmission electron micrograph of a dilute dispersion of a PPy-silica colloidal nanocomposite. (Reproduced with permission from Butterworth, M. D. et al., *J. Colloid Interface Sci.* 1995, 174, 510. Copyright 1995 Academic Press.)

the interactions between the tissues and the materials. For the application of polymer biocomposites, the first thing to be considered is their safety and compatibility with living systems. Compatibility implies surface and structural harmony between the tissues and the polymer materials. Surface compatibility implies that researchers must take surface morphology into account, as also the physical and chemical suitability of the surface of the implant to the tissue. Structural compatibility reflects optical adaptation of the mechanical performance of the implant, including strength and effective loadability. Thus, only the ideal interfacial interaction is reached, and the

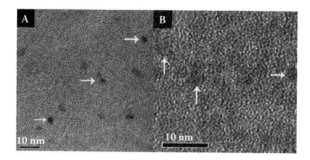

FIGURE 1.7
Transmission electron micrographs of the PEGME–SnO$_2$ nanoparticles (marked by arrows). A shows that nanoparticles are uniform, monodispersed and about 3 nm in diameter, while B, with a higher magnification, illustrates the crystalline phases of the nanoparticles. (Reproduced with permission from Xiong, H. et al., *J. Mater. Chem.*, 2004, 14, 2775–2780. Copyright 2004 Royal Society of Chemistry.)

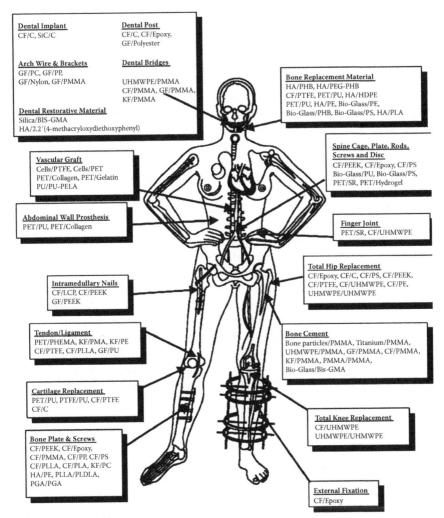

Dental Implant
CF/C, SiC/C

Dental Post
CF/C, CF/Epoxy,
GF/Polyester

Arch Wire & Brackets
GF/PC, GF/PP,
GF/Nylon, GF/PMMA

Dental Bridges

UHMWPE/PMMA
CF/PMMA, GF/PMMA,
KF/PMMA

Dental Restorative Material
Silica/BIS-GMA
HA/2.2´(4-methacryloxydiethoxyphenyl)

Bone Replacement Material
HA/PHB, HA/PEG-PHB
CF/PTFE, PET/PU, HA/HDPE
PET/PU, HA/PE, Bio-Glass/PE,
Bio-Glass/PHB, Bio-Glass/PS, HA/PLA

Vascular Graft
Cells/PTFE, Cells/PET
PET/Collagen, PET/Gelatin
PU/PU-PELA

Spine Cage, Plate, Rods,
Screws and Disc
CF/PEEK, CF/Epoxy, CF/PS
Bio-Glass/PU, Bio-Glass/PS,
PET/SR, PET/Hydrogel

Abdominal Wall Prosthesis
PET/PU, PET/Collagen

Finger Joint
PET/SR, CF/UHMWPE

Total Hip Replacement
CF/Epoxy, CF/C, CF/PS, CF/PEEK,
CF/PTFE, CF/UHMWPE, CF/PE,
UHMWPE/UHMWPE

Intramedullary Nails
CF/LCP, CF/PEEK
GF/PEEK

Tendon/Ligament
PET/PHEMA, KF/PMA, KF/PE
CF/PTFE, CF/PLLA, GF/PU

Bone Cement
Bone particles/PMMA, Titanium/PMMA,
UHMWPE/PMMA, GF/PMMA, CF/PMMA,
KF/PMMA, PMMA/PMMA,
Bio-Glass/Bis-GMA

Cartilage Replacement
PET/PU, PTFE/PU, CF/PTFE
CF/C

Total Knee Replacement
CF/UHMWPE
UHMWPE/UHMWPE

Bone Plate & Screws
CF/PEEK, CF/Epoxy,
CF/PMMA, CF/PP, CF/PS
CF/PLLA, CF/PLA, KF/PC
HA/PE, PLLA/PLDLA,
PGA/PGA

External Fixation
CF/Epoxy

CF: carbon fibers, **C:** carbon, **GF:** glass fibers, **KF:** kevlar fibers, **PMMA:** polymethylmethacrylate, **PS:** polysulfone, **PP:** polypropylene, **UHMWPE:** ultra-high-molecular weight polyethylene, **PLDLA:** poly(L-DL-lactide), **PLLA:** poly (L-lactic acid), **PGA:** polyglycolic acid, **PC:** polycarbonate, **PEEK:** polyetheretherketone, **HA:** hydroxyapatite, **PMA:** polymethylacrylate, **BIS-GMA:** bis-phenol A glycidyl methacrylate, **PU:** polyurethane, **PTFE:** polytetrafluoroethylene, **PET:** polyethyleneterephthalate, **PEA:** polyethylacrylate, **SR:** silicone rubber, PELA: block co-polymer of lactic acid and polyethylene glycol, **LCP:** liquid crystalline polymer, **PHB:** polyhydroxybutyrate, **PEG:** polyethyleneglycol, **PHEMA:** poly(20-hydroxyethyl methacrylate)

FIGURE 1.8

Various applications of different polymer composite materials. (Reproduced with permission from Ramakrishna, S. et al., *Compos. Sci. Technol*, 2001, 61, 1189–1224. Copyright 2001 Elsevier B.V.)

harmony between the tissues and implants made from the polymer materials can be satisfied.

The phrase *low carbon economy* now becomes synonymous with "warm gas," and solar energy is the best renewable energy. The solar cell is a device that can convert the energy of sunlight directly into electricity by the photovoltaic effect. Assemblies of solar cells are used to make solar panels, solar modules, and photovoltaic arrays that can be widely used in industry. Organic solar cells, known as *excitonic solar cells*, belong to the class of photovoltaic cells. Recently, a review of organic photovoltaics based on polymer–fullerene composite solar cells was published. The schematic illustration of a polymer–fullerene BHJ solar cell is presented in Figure 1.9. Such polymer-based photovoltaics can be processed in solution, and they have the features of low cost, low weight, and greater flexibility. Scientists now have a much better understanding of the complex interplay between the electronic and physical interactions of the polymer, and the fullerene component can assist in the design of the next generation of optimized organic solar cells [19].

Shape memory polymers (SMPs) are able to recover their original shape upon exposure to an external stimulus [20,21]. SMPs can be activated not only by heat and magnetism (similar to shape memory alloys [SMAs]) but also by light and moisture and even by a change of pH value. Thermoresponsive, light-responsive, and chemical-responsive SMPs were researched recently, although with a focus on thermoresponsive SMPs because they are better developed at present. In addition, based on the shape memory effect, some novel multi-functional SMPs or nano SMP composites have also been proposed.

1.6 Fabrication of Polymer Nanocomposite

It is a challenge to fabricate polymer matrix nanocomposites with good performance. Several processing techniques including melt mixing and in situ polymerization have been applied in the laboratory. Melt mixing as a method of dispersing carbon nanotubes into thermoplastic polymers is discussed in the reference [22]. The author provides two ways of introducing nanotubes into polymer matrices. In the first case, commercially available master batches of nanotube or polymer composites are diluted by the pure polymer in a subsequent melt mixing process (master batch method), which means that the starting materials are already preformed polymer composites. However, in the other case, nanotubes as reinforcement and polymer as matrix are used as the starting materials, then nanotube reinforcement is directly incorporated into the polymer matrix. The results show that the master batch method is applicable and easily accessible for small- and medium-sized enterprises. And the percolation only slightly depends upon the mixing equipment and the processing conditions. In the direct incorporation method, the electrical

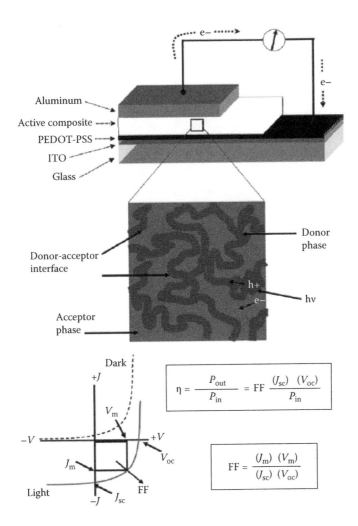

FIGURE 1.9

Schematic illustration of a polymer–fullerene BHJ solar cell, with a magnified area showing the bicontinuous morphology of the active layer. ITO is indium tin oxide, and PEDOT-PSS is poly(3,4-ethylenedioxythiophene)-polystyrene sulfonate. The typical current–voltage characteristics for dark and light current in a solar cell illustrate the important parameters for such devices: J_{SC} is the short-circuit current density, V_{OC} is the open circuit voltage, J_m and V_m are the current and voltage at the maximum power point, and FF is the fill factor. The efficiency (h) is defined, both simplistically as the ratio of power out (P_{out}) to power in (P_{in}), as well as in terms of the relevant parameters derived from the current–voltage relationship. (Reproduced with permission from Thompson, B. C. and Frechet, J. M. J., *Angew. Chem. Int. Ed.* 2008, 47, 58–77. Copyright 2008 Wiley-VCH Verlag GmbH & Co. KGaA.)

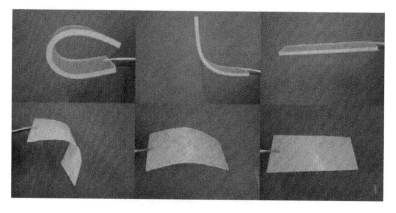

FIGURE 1.10
Time series photographs that show the recovery of a pure SMP (top row) and a glass fiber-reinforced SMP (bottom row).

properties of composites prepared will be determined by the kind of the nanotubes, different sizes, purity, and defect levels, even the purification and modification steps [22]. In polymer chemistry, in situ means "in the polymerization mixture." The in situ polymerization is an effective method applied to prepare polymer nanocomposite material. In this method, the nanoreinforcement is fixed with a structure that is impregnated with a monomer and the monomer polymerized. Usually, the monomer is very fluid and will not disturb the nanoreinforcement arrangement. After polymerization, the polymer nanocomposite material is obtained. The in situ polymerization method has the advantage of dispersing nanofillers effectively in monomeric liquids and consequently in the polymer nanocomposite.

Open (closed) mold processes; filament winding, and pultrusion processes are the main methods of fabricating polymer nanocomposites in industry. Molding is the process of manufacturing the pliable raw material using a rigid frame or model. In open-mold processes, a single positive or negative mold surface is used to produce laminated fiber-reinforced polymer structures, and the starting materials (resins, fibers, etc.) are applied to the mold in layers, building up to the desired thickness. Then, curing and part removal follow until the final product is finished. Compared with the open-mold processes, the tooling cost of closed-mold processes is more than double due to the more complex equipment required. However, closed-mold processes have the following advantages: good finish on surfaces, higher production rates, closer control over tolerances, and more complex three-dimensional shapes. In the filament winding technique, the mandrel rotates, while a carriage moves horizontally, laying down nanofibers in the desired pattern. For example, the glass or carbon nanofibers are coated with resin as they are wound. After reaching the desired thickness, the mandrel is put in an oven for resin solidification. Finally, the mandrel is removed, and we get the hollow final product. Pultrusion is a manufacturing process for producing continuous lengths of reinforced polymer structural shapes with constant cross sections. The starting materials can be liquid resin mixtures and flexible reinforcement fibers. This process involves pulling the resin matrix and the reinforcement fiber through a heated steel-forming die, where the resin undergoes polymerization. The reinforcement materials are in continuous forms, and the cured profile is formed based on the shape of the die.

While the polymer matrix and the reinforcement are being combined, the starting materials may be kept as separate entities when they arrive at the fabrication operation. Then, the polymer matrix and the reinforcement are combined into the composite during shaping. Also, the polymer matrix and the reinforcement can be combined into some starting form that is convenient for use in the shaping process. In future, research activity in industry will focus on the optimization of polymer composite formulations, optimization of polymer composite forming processes, improvement of polymer composites performance, and reduction of the cost.

1.7 Future Trends

There have been many achievements in the field of polymer nanocomposite technology. We can easily see the potential for applications of polymer nanocomposites in the civil and military fields. For example, polymer nanocomposites have emerged as an attractive material in construction, aerospace, biomedical, marine, electronics, and recreation industries for their excellent properties. Great efforts have been undertaken to modify these materials at the nanoscale to optimize their performance. Carbon nanotube-reinforced nanocomposites, multifunctional polymeric smart materials, and new functional polymer nanocomposites are being explored for unusual behaviors in order to fulfill more than one task in automotive, biological, aerospace, marine, manufacturing, and defense technologies. It is believed that multifunctional polymer nanocomposites will dominate our lives in future.

References

1. Iijima, S., *Nature*, 1991, 354, 56–58.
2. Hodak, J. H., Henglein, A., Hartland, G. V., *J. Phys. Chem. B*, 2000, 104, 9954.
3. Peng, Z. A., Peng, X., *J. Am. Chem. Soc.*, 2001, 123, 1389.
4. Carpenter, E. E., Ph.D. thesis, University of New Orleans, New Orleans, LA, 1999.
5. Katz, E., Willner, I., *Angew. Chem. Int. Ed.*, 2004, 43, 6042–6108.

6. Trindade, T., O'Brien, P., Pickett, N. L., *Chem. Mater.*, 2001, 13, 3843–3858.
7. Fendler, J. H., *Chem. Mater.*, 1996, 8, 1616–1624.
8. Cushing, B. L., Kolesnichenko, V. L., O'Connor, C. J., *Chem. Rev.*, 2004, 104, 3893–3946.
9. Paul, D. R., Robeson, L. M., *Polymer*, 2008, 49, 3187–3204.
10. Fornes, T. D., Yoon, P. J., Keskkula, H., Paul, D. R., *Polymer* 2001, 42, 9929.
11. Coleman, J. N., Khan, U., Blau, W. J., Gun'ko, Y. K., *Carbon* 2006, 44, 1624–1652.
12. Sandler, J. K. W., Pegel, S., Cadek, M., Gojny, F., van Es, M., Lohmar, J. et al. *Polymer* 2004, 45, 2001–2015.
13. Gangopadhyay, R., De, A., *Chem. Mater.* 2000, 12, 608.
14. Butterworth, M. D., Corradi, R., Johal, J., Lascelles, S. F., Maeda, S., Armes, S. P., *J. Colloid Interface Sci.* 1995, 174, 510.
15. Quartarone, E., Mustarelli, P., Magistris, A., *Solid State Ionics* 1998, 110, 1–14.
16. Xiong, H., Liu, D., Zhang, H., Chen, J., *J. Mater. Chem.*, 2004, 14, 2775–2780.
17. Ramakrishna, S., Mayer, J., Wintermantel, E., Leong, K. W., *Compos. Sci. Technol,* 2001, 61, 1189–1224.
18. Rezwan, K., Chen, Q. Z., Blaker, J. J., Aldo Roberto Boccaccini, *Biomaterials* 2006, 27, 3413–3431.
19. Thompson, B. C., Frechet, J. M. J., *Angew. Chem. Int. Ed.* 2008, 47, 58–77.
20. Behl, M., Lendlein, A., *Mater. Today* 2007, 10, 20–28.
21. Leng, J. S., Lv, H. B., Liu, Y. J., Huang, W. M., Du, S. Y., *MRS Bulletin.* 2009, 34, 848–855.
22. Pötschke, P., Bhattacharyya, A. R., Janke, A., Pegel, S., Leonhardt, A., Täschner, C., Ritschel, M., Roth, S., Hornbostel, B., Cech, J., *Fullerenes, Nanotubes, Carbon Nanostruc.*, 2005, 13, 211–224.

2

Carbon Nanotube-Reinforced Nanocomposites

Alan kin-tak Lau

University of Southern Queensland

Hong Kong Polytechnic University

CONTENTS

2.1 Structure of Carbon Nanotubes (CNTs)

A single-walled nanotube (SWNT), similar to a sheet of graphene, rolls over to form a tube with the longest ends joined together to form a cylinder, and theoretically possesses superior mechanical, electrical, and thermal properties. Such properties are governed by perfect carbon–carbon bonds with a bond length of about 0.142 nm, as shown in Figure 2.1 [1]. In this regard, graphene has been referred to as an infinite alternant (only a six-member carbon ring). Mechanical property tests showed that graphene has a breaking strength 200 times greater than steel. However, the process of separating it from graphite, where it occurs naturally, will require some technological developments before it becomes economical enough to be used in industrial processes [2]. For a multiwalled type, a stack of sheets are rolled together, with the separation of each sheet being 0.034 nm. All sheets (hereafter called "the layers" for multiwalled types) attract together by only a weak van der Waals interaction, with this force being very small compared with the in-plane carbon–carbon bond. One mole of C–C bonds has a bond energy of 347 kJ/mol. The energy required to break a single C–C bond is 3.74 nN. From Cumings and Zettle's analysis [3], they found that the movement of inner

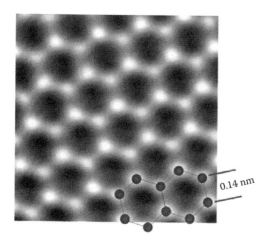

FIGURE 2.1
Graphene sheet captured by TEM (http://en.wikipedia.org/wiki/File:Real_graphene.jpg).

layers of multiwalled nanotubes (MWNTs) behaves telescopically due to very low friction between the layers. The static and dynamic friction forces of the nanotubes were $< 2.3 \times 10^{-14}$ N per atom (6.6×10^{-15} N/Å^2) and $< 1.5 \times 10^{-14}$ N per atom (4.3×10^{-15} N/Å^2), respectively. Comparing with the C–C bonds in the layers and attractive force between the layers, it is almost on the order of 100,000. Therefore, the use of MWNTs as reinforcement for polymer-based or metallic composites may not be appropriate as only their outer layer can take loads (this will be discussed in Section 2.3).

The structure of carbon nanotubes has been well described elsewhere. Simply speaking, nanotubes, can be categorized into three structural forms: (1) armchair, (2) zigzag, and (3) chiral nanotubes. Apart from their mechanical properties, the major difference between these nanotubes is their electrical conductivity, and possibly thermal conductivity due to their atomic arrangement. This property is highly sensitive to a light change of their chiral parameters (n, m) [4], which causes the change of a nanotube from metallic (armchair) to semiconducting status. When $n - m = 3p$ (where p is any integer), such as (3,0), (7,1), (8,5), etc. the nanotubes are expected to be metallic. In the remaining cases, $n - m \neq 3p$, and the nanotubes are predicted to be semiconductors, with an energy gap of the order of ~0.5 eV. This gap is highly dependent on the nanotube diameter and can be evaluated by

$$E_{gap} = \frac{2\gamma a_{c-c}}{d} \tag{2.1}$$

where g, a_{c-c}, and d denote the C–C tight binding overlap energy (2.45 eV), the nearest-neighbor C–C distance (~ 1.42 Å), and the diameter of the nanotubes,

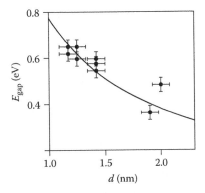

FIGURE 2.2
Energy gap E_{gap} versus diameter d of the nanotubes with a conducting chirality. The solid line denotes the theoretical prediction calculated from Equation 2.1.

respectively. In Figure 2.2, a plot of the band gap energy (E_{gap}) versus the diameter of nanotubes (in nanometers) is shown.

The diameter of the nanotubes is determined by

$$d = \frac{\sqrt{n^2 + m^2 + nm}}{\pi} a \tag{2.2}$$

The preceding equation can only be valid for a SWNT or outer shells of MWNTs. By substituting the values of m and n into it, it is easy to see that the size of an armchair nanotube is a bit bigger than the zigzag type as n is equal to zero. The theoretical model based on the properties of a graphene sheet, in which all carbon-carbon bonds are in perfect condition, can be used to estimate the tensile modulus and strength of the SWNT to reach 1 TPa and 200 GPa, respectively. These values are almost five times the value of steel (~200 GPa). Early studies mainly employed Young's modulus to predict the mechanical properties of nanotube-related composites. In the early research, the nanotubes were modeled, like a multishell cylinder, for investigating their bending stiffness, torsion behavior, and even local buckling while they were subjected to bending. Those studies normally assumed that the nanotubes are defect free. The formation of Stone–Wales defects (5-7-7-5) during the stretching process was not taken into account. In reality, at the nanoscale level, it does exist when the energy applied to the system over E_{SW}:

$$E_{SW} = E_{SW}^o + ae^{-br} + \frac{\sigma}{r^2} \tag{2.3}$$

where E_{SW}^o is the energy required to produce a Stone–Wales defect in a flat graphene sheet; r is the radius of the nanotube; and a, b, and c are the fitting parameters. Such energy is relatively small and can be easily transmitted to the nanotube during the loading process. Such a result may cause a localized defect or damage of the nanotube and, thus, the previous prediction on the mechanical properties may be overestimated.

According to the foregoing equations, it is not difficult to see that traditional linear elastic theories may be inadequate to predict the mechanical properties of nanotubes and their related composites as the physical and mechanical deformations are in a nonlinear pattern. In general, molecular dynamics (MD) simulations have been widely adopted for anticipating localized properties of the nanotube at nanoscale and then further using the estimated properties or a block macroscale mechanics analysis, such as continuum models and/or finite element analysis (Section 2.5).

As mentioned earlier, MWNTs are formed by many layers of nanotubes, and these layers are bound by a very weak van der Waals interaction. Therefore, in terms of conducting the mechanical property tests, such as tensile and pullout for nanotube–polymer samples, a strong bond between the outer layer of the nanotube and surrounding matrix exists; sliding between all inner layers does happen which highly affects the overall mechanical properties of the nanotubes, for use in real life. This argument may be reflected in inconsistent results from tensile tests conducted under the same conditions but for different types of nanotubes [5].

2.2 Dispersion Properties of CNT Nanocomposites

To manufacture high-strength biocompatible nanotube-related polymer composites with good thermal and electrical conductivity, the control of their homogeneity at nanoscale is the key factor during the whole manufacturing process. Nanotubes tend to form bundles, and their solubility in common solvents is very limited. The key challenge is how to break the cohesion of aggregated nanotubes in order to obtain fiber dispersion in the selected solution or matrices. Many studies attempted to use mechanical stirring with the high-energy sonication process to separate entangled nanotubes. However, another key factor, viscosity of resin, was somehow ignored in the foregoing procedures.

Typically, the preparation of nanotube–polymer composites starts from the preliminary mixing process, which is designed to reduce the size of nanotube aggregates in order to avoid blocking of the circulating system. Subsequently, the increase of temperature, and thus decrease of the viscosity of resin, will then be applied followed by the sonication process. Many studies were conducted to investigate the sonication effect on the properties of nanotubes. It was found that, in excessive vibration motion at high frequency, the nanotube may be damaged [6–7]. It was reported that by sonicating nanotubes for over 4 h may destroy most of their graphene layers and possibly cause the formation of junctions among the nanotubes. Such distortion may be due to the violent motions of the nanotubes, which damage the carbon–carbon bonds. However, this effect may vary with the environment. Tarawneh et al. [6] have reported that a high mechanical strength of nanotube–rubber composites was achieved by sonicating MWNTs for 1 h. This result was due to the viscosity effect, which retards the collision of nanotubes during vibration motion.

Functionalization of nanotubes is another method of producing uniformly dispersed nanotube–polymer composites. This can enhance the bonding efficiency of the nanotubes with other matrices, and their biocompatibility (with biocompatible amylose and cellulose) and fluorescence (ability to form nanoscale photoactive materials). In general, there are two approaches currently developed for functionalization: (1) direct covalent attachment of functional groups (COOH, OH, CH, NH$_2$, NO$_2$, and NO) with the nanotubes and (2) noncovalent adsorption of various functional molecules (such as surfactants, sodium bromide (DTAB), octyl phenol ethoxylate (Triton X-100), and sodium dodecyl sulfate (SDS) onto the surfaces of nanotubes. The molecule layer should be in just sufficient concentration to coat the nanotube surface.

The direct covalent sidewall functionalization is associated with a change in hybridization from sp^2 to sp^3 and a simultaneous loss of conjugation. The indirect covalent functionalization takes advantage of chemical transformations of carboxylic groups at the open ends and holes in the sidewalls of the nanotubes. In order to increase the reactivity of nanotubes, the carboxylic acid groups usually need to be converted into acid chloride and then undergo an esterification or amidation reaction. However, the drawback of the covalent functionalization is the need to break carbon–carbon bonds, thus resulting in significant changes in the mechanical and physical properties of the nanotubes. Noncovalent functionalization is mainly based on supramolecular complexation using various adsorption forces, such as van der Waals interaction, hydrogen bonds, electrostatic force, and p-stacking interactions [8]. In terms of biocompatibility, Zheng et al. [9] have reported that bundled nanotubes could be effectively dispersed in water by sonication in the presence of single-stranded DNA (ssDNA).

Solvent effects during the dispersion process are other key factors that may influence the resultant properties of nanotube–polymer composites. Lau et al. [10] studied the mechanical properties of nanotube–epoxy composites by using three common types of solvent, namely, *N*-dimethylformamide (DMF), ethanol, and acetone for dispersion of nanotubes into epoxy resin. These solvents possess different boiling points, which are 153°C, 78.4°C, and 56.53°C, respectively, and all are miscible. The results from hardness and flexural strength tests as shown in Table 2.1 show that the hardness and flexural strength of a sample dispersed by acetone are higher compared with

TABLE 2.1

Mechanical Properties of Epoxy Composites Mixed by Using Different Dispersed Solvents

Sample	Boiling Temperature	Microhardness (Hv)	Flexural Strength (MPa)
Pure epoxy	—	17.8 (0%)	74.3 (0%)
Dispersed with acetone	56.53°C	18.0 (+1%)	75.6 (+2%)
Dispersed with ethanol	78.4°C	14.4 (−19%)	63.3 (−15%)
Dispersed with DMF	153°C	7.70 (−56%)	6.80 (−90%)

the use of other solvents. However, DMF influenced the properties adversely. It was suspected that due to the high boiling point of DMF, a large amount of nonvaporized DMF still remained inside the resin and thus affected its chemical reaction with the hardener.

Fujigaya et al. [11] have reported that UV-curable monomers are able to disperse into nanotubes without the need for any solvent if they are simply mixed at warm temperature (50°C) under sonication. The mixture is a transparent viscous liquid and forms a stable dispersion for more than 6 months.

As the dispersion properties of nanotubes inside the polymer matrix do substantially affect the thermal and electrical conductivities of the resultant composites, many studies have attempted to find ways to monitor these properties by dividing them into two categories: (1) destructive and (2) nondestructive methods. For the destructive method, fractured samples are required for microscopic examination with x-ray diffraction (XRD), scanning electron microscopy (SEM), and transmission electron microscopy (TEM) to capture nanotube distribution images. However, because only a small number of nanotubes are used in general for strength enhancement for polymer-based materials, they are difficult to track during the scanning process. Besides, the accuracy of producing a complete picture of the nanotube distribution cannot be ensured as the image can only reflect a small portion of the samples. Figure 2.3 shows images of nanotube–polymer composites using SEM and TEM, respectively.

Recently, Battisti et al. and Ilcham et al. [12,13] measured the electrical resistivity and thermal conductivity of mixtures of nanotube–resin in relation to their homogeneity. The resistivity of the mixtures was measured by using a tailor-made flow-through resistivity cell, which is shown in Figure 2.4.

The resistivity of the mixtures is calculated by

$$r = (2pLR)/\text{In}(t_o/t_i) \tag{2.4}$$

where R is the measured resistance, L denotes the sensing length, r_o is the internal radius of the outer cylinder, and r_i the radius of the inner rod. In their work, it was found that the dispersion quality change due to the sonication process was corroborated by the accompanying change in the DC

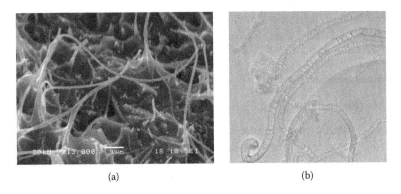

(a)   (b)

FIGURE 2.3
Nanotube images captured by (a) SEM (after fracture of MWNT/epoxy sample) and (b) TEM (a bamboo-type nanotube).

electrical conductivity of samples. The results were validated with SEM imaging. Similarly, for samples subjected to different sonication times, the resistance for a nanotube–poly(p-phenylene) mixture varied. The dispersion of the mixture was dependent on the sonication time and amount of binding initiator used (Figure 2.4). However, the monitoring of dispersion quality of nanotubes in the processing of curing has not yet been studied comprehensively. The measurement of their dispersion status was discussed in the previous paragraph; note that the nanotubes normally sink to the bottom inside a mold or get entangled again after sonication. The viscosity of the resin would increase to hold the nanotubes in place only after hardener is added. However, sonication would not be introduced into the mixture

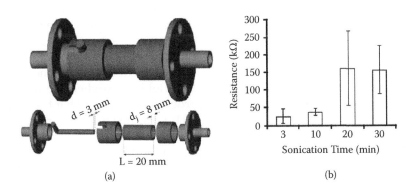

FIGURE 2.4
(a) A tailor-made flow-through electrical resistivity cell (From Battisti et al. *Comp. Sci. Technol.* 2009; 69: 1516–1520) and (b) dependence of MWNT/PP composites mixed at different sonication times (From Ilcham, A. et al. *Particuology.* 2009; 7: 403–407).

(nanotube–resin/hardener), and within the first 30 min, such sediment and entanglement actions may occur.

2.3 Interfacial Bonding Properties

To ensure that high-conductive and high-strength nanotube–polymer composites are produced, the fundamental mechanism that governs the stress and electron transfers is their interfacial bonding properties. Many experimental and theoretical approaches have been found in the past decade. As nanotubes can exist in many forms, such as single-walled or multiwalled structural arrangements with any of armchair, zigzag, or chiral types, these atomic configurations will surely alter the properties of nanotubes and their related polymer composites.

Lau et al. [14] have reported that the interfacial bonding strength between the nanotube and surrounding matrix is highly dependent on the type of nanotube used for a composite. The Young's modulus in tension of nanotubes is dependent on its outer layer diameter (for MWNTs). The inner layers would not be able to take any load as only a weak van der Waals interaction exists, and this binding force is not large enough for transferring stress from the outer to inner layers. However, if the compression due to the load is applied directly on both ends of the nanotubes (outer and inner), the modulus is higher than if the number of layers increases. The stiffness of nanotubes is on the order of zigzag > chiral > armchair, which means that high interfacial shear stress occurs for a single-walled zigzag nanotube.

Experimental studies [15,16] also indicated that pullout of an outer layer of nanotubes did exist; when they were embedded inside the polymer matrix, they demonstrated telescopic behavior. However, as mentioned in Chapter 1, Section 1.2, functionalized nanotubes can have better bonding characteristics. The bridging effect for enhancing the fracture toughness of nanotube–polymer composites under the crack growth process has also been recorded in the literature. Such bridges can stop crack growth due to the energy for deformation of the nanotubes. Zheng et al. [16] have studied the bonding effect between the nanotubes and the surrounding matrix, using functional groups ($-COOH$, $-CONH_2$, $-C_6H_5$ and $-C_6H_{11}$) attached to their surfaces (see Figure 2.5).

In their study, simulations of pullout of a nanotube in a polymer environment were conducted by using the force field technique. In the past, quantum mechanical techniques could accurately simulate a system of atomic particles, but due to the need for powerful computers for data processing and the large number of atoms needed to determine the microscale properties of materials, simulations are costly and time consuming. Generally, the

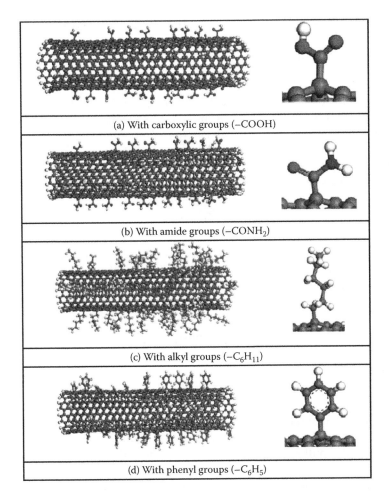

FIGURE 2.5
A (10, 10) SWNT with different functional groups. (From Zheng et al. *Appl. Surf. Sci.* 2009; 255: 3534–3543.)

drawbacks of using quantum mechanics to predict the mechanical properties of nanotubes are [17]

- Limitation of short/small time/length scale on the order of picoseconds/nanometers
- Lack of economic viability for a model with a large number of atoms
- Huge amount of computational effort, thus increasing cost
- Complex formulation

The force field technique is based on the energy induced by both attractive and repulsive forces, and such energy can be converted to show force-replacement quantities:

$$E_{total} = V_{alence} + E_{cross-term} + E_{non-bond} \qquad (2.5)$$

$E_{valance}$ and $E_{cross-term}$ denote the valence energy and cross-term interacting energy. The nonbond interaction term accounts for the interactions between nonbonded atoms and includes the van der Waals energy, the Coulomb electrostatic energy, and hydrogen bond energy. To determine their energy level under loading, molecular mechanics (MM) and MD were conducted, and the expression of these energies in relation to structural properties [18] is

$$E = \frac{1}{2} \sum_{k=i} (DR_i)^2 \qquad (2.6)$$

where dR_i is the elongation of the bond identified by the label i and K_i is the force constant associated with the stretching of the i bond. The detailed expressions are shown in Reference [13a]. The pullout energy can be calculated by using simple shear lag theory:

$$E_{pullout} = \int_{x=0}^{x=L} 2\pi r(L-x)\tau_i dx \qquad (2.7)$$

$$\tau_i = \frac{E_{pullout}}{\pi r L^2} \qquad (2.8)$$

In the MD simulation, the pullout energy was calculated by the difference between potential energy of the composites (nanotubes and matrix) after and before the pullout action. Their results show that a PE molecule could wrap around or fill a (10, 10) SWNT due to the attractive van der Waals interactions. The interfacial bonding characteristics between the nanotubes, on which –COOH, –CONH$_2$ and –C$_6$H$_{11}$, or –C$_6$H$_5$ groups have been chemically attached, proved that appropriate functionalization of nanotubes at low densities of functionalized carbon atoms drastically increases their interfacial bonding and shear stress between the nanotube and the polymer matrix. Figure 2.5 shows the pullout action in MD simulation and the bonding stress at the interface between the nanotubes and the matrix (PE) with different functional groups (carboxylic group –COOH, amide group –CONH$_2$, alkyl group –C$_6$H$_{11}$, and phenyl group –C$_6$H$_5$) attached to the surfaces of the nanotubes.

Recently, some studies in the literature have started using macroscale modeling techniques, such as finite element analysis to estimate the interfacial bonding properties of nanotubes–polymer composites. However, many parameters and factors have not been well understood to date, so the reliability of the results estimated by different models is difficult to assess. Traditional elastic theory assumes that the relationship between the displacement (deformation) and energy (force) is linear and the application of the rule of mixture for macroscale structural analysis was used in many theoretical studies. In reality, by simply looking at the Lennard–Jones potential function, the energy between two atoms is no longer linear in respect to the distance between the atoms. Atomic arrangement of nanotubes and Stone–Wales defects are always the key factors that influence the energy parameters (force and displacement) of nanotubes. Shokrieh and Rafiee [18] introduced the use of a 3D multiscale finite element model, under a nonbond interphase circumstance between the nanotubes and surrounding matrix, to determine their interfacial bonding properties. In their work, nanotubes were modeled at nanoscale while the surrounding matrix was at microscale. A nonlinear finite element model based on the nonlinear behavior of van der Waals interactions (Newton–Raphson iterative method) was constructed. It was found that the rule of mixture, under the assumption of the linear elastic principle, overestimates the properties of nanotubes–polymer composites. Micromechanics equations cannot precisely predict the properties of nanoscale structures because of the different governing assumptions and structural behaviors.

Zhang and Wang [16] theoretically analyzed the thermal effect of stress transferability of nanotube–polymer composites under different temperature conditions. A shear-lag model modified from Lau [18] was used to analyze the stress induced due to the differential thermal expansion of nanotubes and matrix. It was found that stress could be induced on the composites under a temperature change due to differential thermal expansions of nanotubes and the matrix. However, the result was based on the linear elastic deformation regime, and the local deformation based on energy differentiation of atoms was not considered. Twisting of nanotubes may occur under temperature, which was predicted by MD [19]. In such case, localized transverse and longitudinal stresses may exist at the same time.

As a summary, the interfacial bonding properties of nanotubes with the surrounding matrix are highly dependent on a few factors, which are listed below:

- Type of nanotubes (single-walled or multiwalled)
- Structure of nanotubes (zigzag, armchair, or chiral)
- Degree of functionalization (covalent or noncovalent)
- Degree of dispersion (fully, partially, or entangled to form aggregates)

As mentioned previously, functionalization of nanotubes may cause the distortion of their carbon–carbon bond and thus affect their resultant mechanical properties as well as the efficiency of thermal and electrical conductivities. Recently, several studies have been conducted on developing different types of nanotubes, in which surface functionalization is not necessary when the bonding between the nanotubes and surrounding can be enhanced.

To enhance the bonding efficiency (or increase the friction) between the nanotubes and the matrix, many studies changed the surface contour or configuration of the nanotubes. Cheng et al. [20] developed bamboo-like nanotubes synthesized

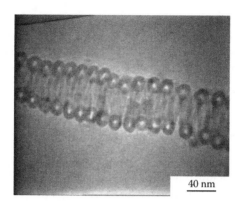

FIGURE 2.6
Coiled nanotube produced by CCVD.

by the ethanol catalytic combustion (ECC) technique. A corrugated section was formed between short sections of nonfunctionalized nanotubes. Coiled multi-walled nanotubes (Figure 2.6) were produced by using the catalytic chemical vapor deposition (CCVD) method on finely divided Co nanoparticles supported in silica gel under reduced pressure and relatively low gas flow rates [21]. It was observed that a regular pitch of 67 nm and coil diameter of 40 nm of coiled nanotubes were obtained. The average length is up to 5.5 mm (aspect ratio ~140). In terms of the mechanical properties of coiled nanotubes–polymer composites as compared with the straight type, it was found that the hardness of a coiled nanotube (2 wt%)/epoxy composite was much better than a pure epoxy sample and a straight SWNT (2 wt%)–epoxy composite. The improvements in hardness of single-walled and coiled nanotube–epoxy composites compared with pure epoxy were 19% and 54%, respectively [22].

A study by Lau et al. [23] also investigated the heat absorbability of single-walled (straight), coiled, and bamboo-type nanotubes. It was found that the endothermic ability of the nanotubes was in the order of straight type > bamboo type > coiled type, which was consistent with the result of T_g (Table 2.2). In conclusion, straight-type nanotubes can act as a heat sink to accelerate

TABLE 2.2

Initial Temperature, Peak Temperature, Final Temperature, Glass Transition Temperature, and Transition Enthalpies of Different Types of Nanotube–Epoxy Composites

Sample	$T_i(°C)$	$T_p(°C)$	$T_e(°C)$	$T_g(°C)$	ΔH (J g^{-1})
SWNT–Epoxy	52.7	57.3	61.9	7.3	7.9
Bamboo NT–epoxy	50.2	55.7	60.3	55.3	6.8
Coiled NT–epoxy	46.6	50.8	55.3	51.0	0.7
Pure epoxy	50.0	54.8	59.0	54.5	7.3

the heat absorption of the matrix, while the coiled types can act as heat-shielding fillers and prevent the matrix from exchanging energy with outside systems.

As nanotubes serve as high-cost nano-reinforcements for polymer-based composites, incorporating nanotubes with other low-cost reinforcements can compensate for this disadvantage in polymers and their related composite industries, Lu et al. [24] produced nanoclay support nanotubes to reinforce epoxy mechanically.

2.4 Mechanical Properties and Conductivity of Nanotube–Polymer Composites

Since nanotubes possess superior mechanical, thermal, and electrical properties, mixing these nano-structural materials into soft polymers as nanoreinforcements alters the overall properties of the polymers and their related composite materials. Numerous researchers have reported that the enhancements of mechanical properties and electrical conductivity of polymer-based composites were achieved by adding a small number of nanotubes [25–28]. The maximum tensile modulus and strength measured by the conventional tensile property test of thin films increased more than 50% as compared with host materials. A recent experimental study, through the use of the nano-indentation technique (see Figure 2.7), also revealed that an increase of the nanotube content resulted in increasing the Young's modulus of nanotube–polystyrene (PS) composites [29].

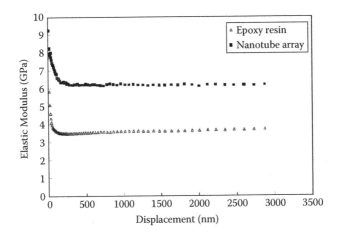

FIGURE 2.7
Comparison of the elastic moduli of patterned nanotube array reinforced epoxy and neat epoxy resin, measured by a nano-indentation technique. (From Lee, C. et al. *Comp. Pt. B. Eng.* 2007; 38: 58–65.)

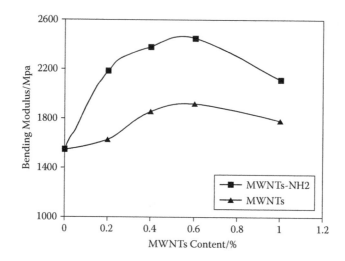

FIGURE 2.8
The effect on bending modulus of MWNTs and MWNTs-NH2 content. (From Zheng et al. *Mater. Sci. Eng. A.* 2006; 435–436: 145–149.)

Several key parameters that govern the production of high-strength nanotube–polymer based composites are

- Uniformly dispersed properties of the nanotubes inside the matrix
- Good interfacial bonding properties between the nanotubes and the surrounding matrix
- Alignment of the nanotubes inside the composites
- The types of nanotubes used

Uniformly dispersed nanotube–polymer composites are difficult to achieve due to an agglomeration of high aspect ratio nanotubes in the resin-curing process. In-situ polymerization can effectively disperse nanotube bundles into a polymer matrix [30]. Functionalization of nanotubes subjected to treatment with triethylenetetramine to form function groups on the nanotube surfaces can also prevent the agglomeration of the nanotubes inside an epoxy matrix [31]. It was found that the bending modulus of the composites was enhanced (see Figure 2.8). Mukhopadhyay et al. [32] have reported that the sonication time is also a key parameter in controlling the integrity of the nanotubes inside the composites. Sonication for more than 4 h may destroy most of the graphitic layers, and thus the superior mechanical properties of the nanotubes cannot be maintained. Layer-by-layer assembly of nanotube–polymer films is able to produce better dispersion properties in nanocomposites [33]. This technique can avoid production of a nanotube- rich film at the bottom and resin-rich at the top.

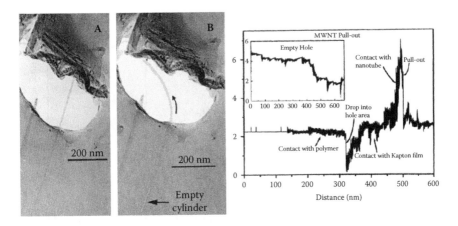

FIGURE 2.9
An indirect pullout test was conducted. (From Qian, D. et al. *Appl. Phys. Lett.* 2000; 76(20): 2868–2870.)

To produce high-strength and electrically conductive nanotube–polymer composites, the properties at a bonding interface are crucial to ensure good stress transfer and charge transportation between the nanotubes and surrounding matrix. The interfacial shear stress between the nanotubes and the matrix was measured through different experimental and computational approaches. At an early stage, stress transferability inside the nanotube–polymer composites was microscopically observed [34–36]. A maximum shear strength of 376 MPa was measured, depending on the length and diameter of the nanotubes based on a nanotube pullout test (see Figure 2.9). Fractographic diagrams show that the nanotubes were bonded well inside the polystyrene matrix. Barber et al. [37] conducted a nanotube pullout test by AFM. A nanotube was embedded into a polymer film and then pulled out from the film. A maximum stress of 47 MPa was measured (see Figure 2.10).

On the other hand, some researchers reported that poor adhesion was found at the interface of nanotube–polymer composites because pullout of nanotubes was observed. Until recently, a coiled-type multiwalled carbon nanotube (hereafter called "coiled nanotube") was developed by using a reduced-pressure catalytic chemical vapor deposition (CCVD) method to produce a high-strength nanotube–polymer composite [38] as shown in Figure 2.6. Due to their coiled configuration, sliding inside inner layers of the nanotubes become less significant to the mechanical properties of their related composites. The highest Young's modulus of the coiled nanotubes was 0.7 TPa. Experimental results proved that the hardness and bending stiffness of epoxy composites were enhanced by adding a small amount of coiled nanotubes to them. The strengthening mechanism was due to mechanical interlocking between the

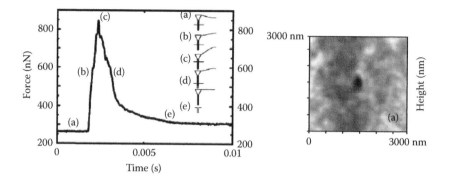

FIGURE 2.10
A force–time plot of the nanotube pullout test from a polymer film. (From Wong, M. et al. *Polymer.* 2003; 44: 7757–7764.)

coiled-shape nanostructures and polymer chains even with no direct chemical bonding between the surfaces of the nanotubes and the surrounding matrix.

2.5 Modeling of CNT Nanocomposites

As nanotubes are too small for their mechanical and physical properties to be tested and measured through mechanical property tests, many modeling techniques have been applied in the past ten years to estimate their properties from atomic to micro-scale levels. In regard to nanotube-related nanocomposites, a large number of studies have been conducted to date. In these studies, three focus areas were (1) the basic properties of nanotubes, (2) their interface bonding properties, and (3) the overall mechanical properties of a nanotube–polymer nanocomposite, based on the traditional cylindrical model.

For conventional fiber-reinforced polymer-based composites, single-fiber pull-out and push-in tests have always been adopted in many industries to identify bonding properties between the fiber and matrix. However, because the size of the nanotubes is very small, these tests are not feasible. Because the diameter of the nanotubes is at nanometer scale, the interaction at the nanotube–polymer interface is highly dependent on the local molecular structure and bonding, and any single atom–atom bond defect may alter the result substantially. Molecular mechanics and molecular dynamics simulations have been used to investigate molecular interactions at the nanotube–polymer interface. Lordi and Yao [39] first used force field-based molecular mechanics to calculate the binding energies and sliding friction between the nanotubes using a surface functional group with a helical polymer conformation and surrounding matrix. They found that the binding energies and frictional force (assuming there is no chemical bonding between the nanotube surface and matrix) play a minor role in determining the strength of the interface. Gou et al. [40] examined the molecular interactions between the nanotubes and thermosetting matrix. A pullout simulation of a single-walled nanotube surrounded by epoxy matrix was conducted. In the simulation, a periodic model of a composite system was constructed as a unit cell. The following relationship for the nanotube, epoxy, and their composite must be satisfied:

$$\frac{n_{NT}}{N_A} \cdot m_{NT} + \frac{n_{polymer}}{N_A} \cdot m_{polymer} = a \cdot b \cdot c \cdot \rho_{composite}$$

where n_{NT} and $n_{polymer}$ are the number of the SWNT fragments and cured epoxy resin molecules, respectively; m_{NT} and $m_{polymer}$ are the molecular weights of the SWNT fragment and the cured epoxy resin molecule, respectively; N_A is Avogadro's number, 6.023×10^{23} formula units/mol; a, b, and c are parameters of the simulation cell; and $\rho_{composite}$ is the density of the resulting nanotube composite. The two terms on the left side of the equation are the masses of the polymer matrix and the nanotube in the composite, which are calculated by using the molecular weight and Avogadro's number. The term on the right side of the equation is the mass of the single-walled polymer composite system, which is calculated by the volume and density of the simulation cell. The density of the composite could be conserved during molecular dynamics simulations.

The initial density can be determined by the rule of mixture if the volume fraction of the nanotubes is known. In their study, the composite was made of a fragment of (10,10) single-walled nanotube totally embedded inside the amorphous polymer matrix of cured epoxy resin. The (10,10) nanotube has a diameter of 13.56 Å. A model of the composite with 21,288 atoms is shown in Figure 2.11, which consisted of a supercell in the range of $50 \times 50 \times 100$ Å. The configuration was initiated by randomly generating the epoxy resins surrounding the nanotube using an initial density of 1.2 g/cm³. The matrix

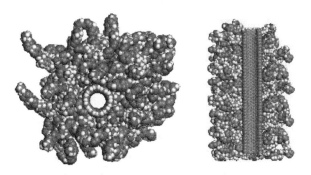

FIGURE 2.11
Molecular model of SWNT–epoxy resin composite system.

polymer was then equilibrated for approximately 20 ps with a time step of 0.2 fs while holding the nanotube rigid. The system was further equilibrated for 60 ps at a time step of 2 fs with a nonrigid nanotube to create a zero initial stress state. The energy of the nanotube–epoxy resin composite system was minimized during the calculations to achieve strongest bonding between the SWNT and epoxy resin.

In nanotube–polymer composites, the bonding strength between the nanotube and epoxy resin can be evaluated by interfacial bonding energy that comes from the electrostatic and van der Waals forces in the molecular system. Generally, the interaction energy is estimated from the energy difference, E, between the total energy of the composite and the sum of the energies of individual molecules as follows:

$$E = E_{total} - (E_{nanotube} + E_{polymers}) \tag{2.9}$$

where E_{total} is the total potential energy of the composite system, $E_{nanotube}$ is the potential energy of the nanotube without the polymer, and $E_{polymer}$ is the potential energy of the polymer without the nanotube. In other words, the interaction energy can be calculated as the difference between the minimum energy and the energy at an infinite separation of the nanotube and polymer matrix. The interaction energy of the nanotube–epoxy resin composite system was calculated using molecular mechanics after energy minimization was applied, as shown in Table 2.3. In this case, the interaction energy came from both van der Waals and electrostatic interactions between the nanotube and epoxy resin. The negative values of interaction energies were assumed to be attractive forces between the nanotube and epoxy resin. The electrostatic forces resulted from coulombic attraction between the positive hydrogen on both ends of the nanotube and epoxy resin. However, carbon atoms of the nanotube were neutrally charged. Therefore, the electrostatic forces were much smaller than the van der Waals forces. The total interaction energy, E, is −900 kcal/mol, resulting in attractive forces between the nanotube and epoxy resin. The total interaction energy, E, is twice the interfacial bonding energy γ scaled by the contact area A [41]:

$$\gamma = \frac{E}{2A} \tag{2.10}$$

TABLE 2.3

Interaction Energies of SWNT–Epoxy Resin Composite System

	Electrostatic		van der Waals		Nonbonded Interaction Energy	
	Resin	SWNT	Resin	SWNT	Resin	SWNT
Resin	7851.58	−4.202	7308.72	−445.554	15160.3	−449.756
SWNT	−4.202	72.350	−445.554	1861.82	−449.75	1934.17

For the nanotube–epoxy composite system, the interfacial bonding energy was 0.1 kcal/mol × Å², calculated from the preceding equation.

The mechanical properties of nanotube composites are known to strongly depend on the magnitude of interfacial shear stress from the polymer matrix to the nanotubes. It is also clear that the level of physical–chemical interactions established at their interface plays an important role in determining the magnitude of load transfer. To characterize the interfacial shear strength of the composites, pullout simulations of a nanotube were performed. The nanotube was pulled out of the epoxy resin along the tube axis direction. The results indicated that the potential energy of the SWNT–epoxy resin composite system increased as the nanotube was pulled out of the epoxy resin. In fact, the increase of potential energy resulted from a decrease of the interaction energy and increase of the energies of both the nanotube and epoxy resin. The interaction energy increased due to the decrease of contact area during the pullout. In the pullout simulation, the nanotube and epoxy resin were not held fixed. The potential energy of the nanotube and epoxy resin increased due to changes in their configurations during the pullout. The deformation of the nanotube and epoxy resin during the pullout influences the pullout energy. During the pullout, the interaction energy changed linearly with the displacement. This is due to the stable interfacial binding interaction between the SWNT and epoxy resin. The interfacial binding energy was kept constant with a value of 0.1 kcal/mol × Å² during the pullout. After the nanotube was completely pulled out of the epoxy resin, the potential energy of the system leveled off because there was no interaction between the nanotube and epoxy resin. The interaction energy was then kept at zero, and there was no change in the potential energies of the nanotube and epoxy resin.

The pullout energy, E_{pullout}, is defined as the energy difference between the fully embedded nanotube and the complete pullout configuration. The pullout energy was divided into three terms, which included the energy change in the nanotube, polymer, and their interaction as follows:

$$E_{\text{Pullout}} = E_2 - E_1 = (E_{NT2} - E_{NT1}) + (E_{\text{resin2}} - E_{\text{resin1}}) + (E_2 - E_1) \qquad (2.11)$$

where E_{NT}, E_{resin} is the potential energy of the nanotube and epoxy resin, respectively, and E is the interaction energy between the nanotube and epoxy resin. The pullout energy can be related to the interfacial shear stress, τ_i, by the following relation:

$$E_{\text{pullout}} = \int_0^L 2\pi r(L-x)\tau_i\,dx = \pi r\tau_i L^2$$

$$\tau_i = \frac{E_{\text{pullout}}}{\pi r L^2} \qquad (2.12)$$

where r and L are the radius and length of the nanotube, respectively, and x is the displacement of the nanotube. At the initial stage of the pullout, the potential energy of the composite was 152,665 kcal/mol. After the pullout, the potential energy increased to 154,951 kcal/mol. From the pullout simulations, the interfacial shear strength between the nanotube and epoxy resin was about 75 MPa.

Liao and Li [42] studied the mechanical interlocking due to the mismatch in the coefficient of thermal expansion (CTE) between the nanotube and polystyrene (PS) in the nanocomposites. The thermal residual radial stress was estimated to be about −40MPa/K from the three-phase concentric cylinder model of elasticity for a unidirectional composite. In addition, they performed the pullout simulations for both single-walled and double-walled nanotubes from the PS matrix based on molecular mechanics. In the absence of atomic bonding between the nanotube and PS matrix, the physical attraction came from the nonbonded interactions, which consisted of electrostatic and van der Waals interactions and deformation induced by these forces. The pullout energies for single-walled and double-walled nanotubes are 158.7 and 154.4 kcal/mol, respectively. Based on Equation (2.12), the interfacial shear stress between the nanotube and the PS matrix, τ_i, estimated from the molecular simulation was calculated to be 160 MPa, which is significantly higher than for most carbon fiber-reinforced polymer composite systems.

A second approach to modeling the interfacial bonding behavior between the nanotube and polymer matrix is based on molecular dynamics simulations. Molecular dynamics simulations can generate information at the molecular level, including atomic positions and velocities. Therefore, they can be performed to analyze the interfacial sliding between the nanotube and polymer matrix in the composites during the pullout process. At the molecular level, the interfacial sliding is closely linked with the fundamental origins of the sliding friction. The physical principles of sliding friction can found in Reference [43].

Frankland et al. [44,45] characterized the nanotube–polyethylene interfacial sliding during the entire pullout process of nonfunctionalized and functionalized nanotube–polyethylene composite systems through molecular dynamics simulations. For both composite systems, the initial configuration for molecular dynamics simulation was the polyethylene containing a (10,10) nanotube, as indicated in Figure 2.11. In the functionalized composite system, a total of six cross-links containing two methylene units each were created between the nanotube and polyethylene. In their study, a unidirectional force was applied to each atom of the nanotube along the nanotube axis. The applied force was increased incrementally over time. After the MD simulations, the velocity and displacement of the nanotube were recorded to characterize the interfacial interactions during the pullout.

Li and Chou [46] developed a new approach to modeling of nanotube-related composites by using a multiscale modeling approach. A single-walled nanotube is viewed as a space frame, where covalent bonds are represented as connecting beams and the carbon atoms as joints. This model has been further utilized by many researchers to simplify the process of estimating the properties of nanotube–polymer composites. Some criticisms have been made of the results due to nonlinear deformation between atoms, which may not be able to be modeled by using elastic frame-like structures based on the principle of solid mechanics. However, nonlinear relationships can be overcome by the coupling of tensile resistance, flexural rigidity, and torsional stiffness.

Namilae and Chandra [47] have used molecular dynamics simulation to study the load transfer condition of a nanotube–polymer composite subject to compressive loading. Most of the studies in the literature to date have been mainly focused on tensile and flexural loadings 12-6 using Lennard–Jones potential for nonbond interactions between the polymer chains and nanotubes, while C–C and C–H chemical bonds were modeled using Tersoff–Brenner potential. In general, for multiwalled nanotubes, the severity of their damage is mainly due to the local buckling when they are subjecting to bending. This phenomenon was proved elsewhere by the linear-shell equation, finite element analysis, and molecular dynamics simulations for a nanotube alone. However, when bound by the surrounding matrix, assuming that a chemical bond between surface carbon atoms and the matrix does exist, the buckling behavior may change.

In the study by Namilae and Chandra, the deformation patterns of neat nanotubes and nanotubes with surface hydrocarbon functionalized group ($-C_4-H_8-$) are very similar with serious crushing occurring at the whole nanotubes. However, for nanotubes fully chemically bonded with the matrix, extensive deformation occurs at the end loading points, while the central region is relatively undistributed (Figure 2.12). In their force–displacement

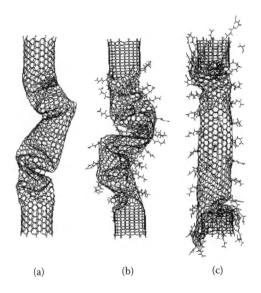

(a)   (b)   (c)

FIGURE 2.12
Simulation snapshot of (10,10) SWNT subject to compression: (a) the free SWNT, (b) functionalized SWNT, and (c) SWNT with chemically bonded interface. (From Namilae, S. and Chandra, N. *Comp. Sci. Technol.* 2006; 66: 2030–2038.)

relationship, it was also found that a relative high force was needed to cause complete failure of the nanotube, which was due to the debonding and rebonding actions of atoms before they failed completely.

Tserpes et al. [48] also simplified the process of modeling by converting properties of a nanotube, based on the modified Morse potential model to the beam model for finite element model (also called *representative volume element* [RVE]) in the model. A constant increment of displacement was applied to the nanotube, which may not be realistic since only a small part of the tension load would be transferred to the caps of the nanotubes; the deformation profile of the nanotube in its lengthwise direction may not be constant. The shear-lag model was also employed to simulate debonding of the nanotube when the stress applied exceeded the interfacial shear strength (ISS). Unfortunately, this method can only be used to provide an approximation of the whole system; as mentioned previously, the bonding between the nanotube and surrounding matrix is on the scale of atom–atom bonding, and the shear-lag model may be a bit of a rush as an average shear strength is used.

2.6 CNT Biocomposites

The geometry of nanotubes has been recognized as an excellent cell delivery system for hydrogen and drugs for fuel cells and biomedical engineering applications. Harrison and Atala [49] have reported that nanotubes can be used in tissue engineering materials for improved tracking of cells, sensing of microenvironments, delivery of transfection agents and scaffolding for incorporating with the host's body. These nanotubes can also act as reinforcement to enhance the strength of materials. Unfortunately, these nanotubes if mixed with some biodegradable or bioresorbable polymers to form biocomposites may potentially risk being trapped in the cells of the human body. The formation of a cell bridge due to the existence of nanotubes may create neural networks. Several studies have reported that carbon nanotubes are cytotoxic. For example, when nanotubes were incubated with alveolar macrophages, significant cytotoxicity was observed after 6 h of exposure. Besides, there are some limitations in adopting these materials for implant applications because of their poor biodegradability. Cytotoxic material must be totally eliminated from the human body once it is no longer needed.

Polizu et al. [50] have established that an understanding of the surface structure and related chemistry of nanotubes is a vital element in the design of biocompatible materials since adsorption and adhesion of biological components are involved. Tian et al. [51] have presented a toxicological assessments of carbon nanomaterials such as single-walled nanotubes (SWNTs), active carbon (AC), carbon black (CB), multiwalled nanotubes (MWNTs), and carbon graphite (CG) on human fibroblast cells in vitro. In their study,

TABLE 2.4

Five Carbon Materials Used in Experiment

Material	Source	Dimensions	Surface Area
SWCNT	Carbon Nanotechnologies, Inc., USA	2 nm × 500 nm	3.15 μm^2
AC	Silcarbon Aktivkohle GmbH, Kirchheim, Germany	25 nm radius	7.85 μm^2
CB	CarboTech Aktivkohle GmbH, Essen, Germany	200 nm radius	502 μm^2
MWCNT	IIJIN Diamond Co., Ltd., Korea	50 nm × 5 μm	789 μm^2
CG	Kern group at Max Planck Institute for Solid State Research, Stuttgart, Germany	500 nm radius	3.14 mm^2

Based on Tian, F., Cui, D., Schwarz, H., Estrada, G. G., and Kobayashi, H. Cytotoxicity of single-wall carbon nanotubes ion human fibroblasts. *Toxicol. In Vitro.* 2006; 20: 1202–1212.

single-walled nanotubes induced the strongest adverse effects: apoptosis and necrosis. A purified single-walled nanotube (25 mg/mL) was more toxic than an unpurified sample. They have also given the cell survival order: CG, MWNTs, CB, AC, and then SWNTs. However, it has to be noted here that their results did not take into account aspect ratios; for example, SWNTs and MWNTs were 250 and 100, respectively, while CG, CB, and AC were 1 (sphere). However, based on the shape alone one cannot come to a definitive conclusion about the toxicity effects of nanomaterials. With reference to Table 2.4, it seems that the surface is the main variable that can be best used for predicting the toxicity of carbon nanomaterials on fibroblast cells.

However, although the surface area is a factor in deciding how many atoms can react with the surrounding polymer, surface chemistry, in which the nanotubes may or may not be purified, is also key to generating toxicity in human cells. Surface chemistry modifies both the aggregation and the toxic effect of these hydrophobic nanomaterials. It was also found that unpurified nanotubes (both single-walled and multiwalled) are less harmful than dispersed purified nanotubes because of the smaller surface areas exposed to the surrounding environment. This finding is similar to the conclusion of Panessa-Warren et al. [52] that unpurified or bundles of nanotubes may reduce highly reactive surfaces available for biological interactions, and thus may reduce the rate of cytotoxicity. This result presents a new possibility that the nanotubes could be modified to neutralize the oxidized highly reactive carbon lattice surfaces in aqueous environments so as to reduce the risk of biological harm. A recent article published in *Engineers Australia* [53] reported that nanotubes could be hazardous owing to the risks of pulmonary inflammation, oxidative stress, interstitial fibrosis, and granulomas if inhaled in sufficient quantity into the human body.

To enhance their properties, in terms of strength and wear resistance, nanotubes are good nanofillers to compensate for the drawbacks of polymers, due to their low mechanical and thermal strengths. For biocomposites or natural composites engineering applications, many aspects require further study as nanotubes are neither biodegradable nor bioresorbable. However, the existence of nanotubes can create a bridge for creating neural networks [49].

In general, due to their small size, individual nanomaterials and small agglomerates may be deposited deep within the lungs when inhaled, reaching areas that are not as easily accessed by larger materials. Also, their small size may also permit them to pass directly through tissues and cell membranes, allowing them to translocate from their initial site of exposure to organs in the body. There are three generally accepted factors that determine the potential of a particle to cause harm:

- The surface area to mass ratio of the particle—A large surface area gives the particles a greater area of contact with the cellular membrane, as well as a greater capacity for absorption and transport of toxic substances.

- The particle retention time—The longer the particle stays in contact with the cellular membrane, the greater the chance of damage. This factor also incorporates the concept of particle mobility, either through clearance or migration to the surrounding tissue.

- The reactivity or inherent toxicity of the chemicals contained within the particles.

2. 7 Conclusion

This chapter gives a comprehensive review of the structural properties of nanotubes and their related polymer composites. Many key factors such as dispersion, interfacial bonding characteristics, and novel types of nanotubes in relation to the resultant mechanical, electrical, and thermal properties of the composites are discussed and analyzed through theoretical and computational studies. Recently, the hottest topic in nanotubes is how to apply these extraordinary nanomaterials to biomedical engineering and medicine fields for strength enhancement of implants, biosensor development, and drug delivery. Although many investigations have been conducted to date, mainly on structural applications, and plenty of track records and data can be found in the literature, applying nanotubes in the human body is still a big challenge. Overcoming the effects of cytotoxicity will be the main focus in the future.

References

1. Lee, C. et al. (2008). measurement of the elastic properties and intrinsic strength of monolayer graphene. *Science*. 321 (5887): 385.
2. Sanderson, B. (2008). Toughest Stuff Known to Man: Discovery Opens Door to Space Elevator. nypost.com.
3. Cumings, J. and Zettl, A. Low-friction nanoscale linear bearing realized from multiwall carbon nanotubes. *Science* July 28, 2000; 289: 602–604.
4. Lau and Hui D. The revolutionary creation of new advanced materials—carbon nanotube composites. *Comp. Pt. B: Eng.* 2002; 33: 263–177.
5. Ruoff, R. S. and Lorents, D. C. Mechanical and thermal properties of carbon nanotubes. *Carbon*. 1995; 33: 925–930.
6. Tarawneh, M. A., Ahmad, S. H., Rasid, R., Yahya, S. Y., and Lau, K. T. Sonication effect on the mechanical properties of MWNT reinforced natural rubber. *J. Comp. Mater.* 2010; submitted.
7. Mukhopadhyay, K., Dwived, C. D., and Mathur, G. N. Conversion of carbon nanotubes to carbon nanofibers by sonication. *Carbon*. 2002; 40: 1369–1383.
8. Meng, L. J., Fu, C. L., and Lu, Q. H. Advanced technology for functionalization of carbon nanotubes. *Progr. Nat. Sci.* 2009; 19: 801–810.
9. Zheng, M., Jagota, A., and Strano, M. S. Structure-based carbon nanotube sorting by sequence-dependent DNA assembly. *Science*. 2003; 302(5650): 1545–1548.
10. Lau, K. T., Lu, M., Lam, C. K., Cheung, H. Y., Sheng, F. L., and Li, H. L. Thermal and mechanical properties of single-walled carbon nanotube bundle-reinforced epoxy nanocomposites: the role of solvent for nanotube dispersion. *Comp. Sci. Technol.* 2005; 65: 9–725.
11. Fujigaya, T., Fukumaru, T., and Nakashima, N. Evaluation of dispersion state and thermal conductivity measurement of carbon nanotubes/UV-curable resin nanocomposites. *Syn. Metals*. 2009; 159: 827–830.
12. Battisti, A., Skordos, A. A., and Partridge, I. K. Monitoring dispersion of carbon nanotubes in a thermosetting polyester resin. *Comp. Sci. Technol.* 2009; 69: 1516–1520.
13. Ilcham, A., Srisurichan, A., Soottitantawat, A., and Charinpanitkul, T. Dispersion of multiwalled carbon nanotubes in poly(p-phenylene) thin films and their electrical characteristics. *Particuology*. 2009; 7: 403–407.
14. Lau, K. T., Chipara, M., Ling, H. Y., and Hui, D. On the effective elastic moduli of carbon nanotubes for nanocomposite structure. *Comp. Pt. B. Eng.* 2004; 35: 95–101.
15. Zheng, Q., Xia, D., Xue, Q. Z., Yan, K., Gao, X., and Li, Q. Computational analysis of effect of modification on the interfacial characteristics of a carbon nanotube-polyethylene composite system. *Appl. Surf. Sci.* 2009; 255: 3534–3543.
16. Zhang, Y. C. and Wang, X. Thermal effects on interfacial stress transfer characteristics of carbon nanotubes/polymer composites. *Int. J. Solid Struct.* 2005; 42: 5399–5412.
17. Shokrieh, M. M. and Rafiee, R. On the tensile behaviour of an embedded carbon nanotube in polymer matrix with non-bond interphase region. *Comp. Struct.* 2010; 92: 647–652.
18. Lau, K. T. Interfacial bonding characteristics of nanotube–polymer composites. *Chem. Phys. Lett.* 2003; 370: 399–405.

19. Tam, W. T., Lau, K. T., Liao, K., and Hui, D. Role of defects on the load transfer system in the carbon nanotube composite through the molecular simulation. *The Proceedings of the 15th International Conference on Composite Materials (ICCM15)*, June 27, 2005–July 1, 2005, Durban, South Africa.

20. Cheng, J., Zou, X. P., Li, F., Zhang, H. D., and Ren, P. F. Synthesis of bamboo-like carbon nanotubes by ethanol catalytic combustion technique. *Trans. Nonferrous Metals Soc. China*. 2006; 16: 435–437.

21. Lu, M., Lau, K. T., Xu, J. C., and Li, H. L. Coiled carbon nanotubes growth and DSC study in epoxy-based composites. *Coll. Surf. A*. 2005; 257–258: 339–343.

22. Lau, K. T., Lu, M., and Liao, K. Improved mechanical properties of coiled carbon nanotubes reinforced epoxy nanocomposites. *Comp. Pt. A. Appl. Sci. Manufact.* 2006; 37: 1837–1840.

23. Lau, K. T., Lu, M., Li, H. L., Zhou, L. M., and Hui, D. Heat absorbability of single-walled, coiled and bamboo nanotube–epoxy nanocomposites. *J. Mater. Sci*. 2004; 39: 5861–5863.

24. Lu, M., Lau, K. T., Qi, J. Q., Zhao, D. D., Wang, Z., and Li, H. L. Novel nanocomposite of carbon nanotube-nanoclay by direct growth of nanotubes on nanoclay surface. *J. Mater. Sci*. 2005; 40: 3545–3548.

25. Ramamurthy, P. C., Malshe, A. M., Harrell, W. R., Gregory, R. V., McGuire, K., and Rao, A. M. Polyaniline/single walled carbon nanotube composite electronic devices. *Solid-State Electron*. 2004; 48: 2019-2024.

26. Fisher, F. T., Bradshaw, R. D., and Brinson, L. C. Effects of nanotube waviness on the modulus of nanotube-reinforced polymers. *Appl. Phys. Lett*. 2002; 80(24): 4647–4649.

27. Andrews R and Weisenberger. Carbon nanotube polymer composites. *Curr. Opin. Solid State Mater. Sci*. 2004; 8(1): 31–37.

28. Chen, W., Tao, X. M., and Liu, Y. Y. Carbon nanotube-reinforced polyurethane composite fibers. *Comp. Sci. Technol*. 2006; 66: 3029–3034.

29. Lee, H., Mall, S., He, P., Shi, D., Narasimhadevara, S., Heung, Y. Y., Shanov, V., and Schulz, M. J. Characterization of carbon nanotube–nanofiber reinforced polymer composites using an instrumented indentation technique. *Comp. Pt. B. Eng*. 2007; 38: 58–65.

30. Park, C., Ounaies, Z., Watson, K. A., Crooks, R. E., Smith, Jr. J., Lowther, S. E., Connell, J. W., Siochi, E. J., Harrison, J. S., and Clair, T. L. Dispersion of single wall carbon nanotubes by in situ polymerization under sonication. *Chem. Phys. Lett*. 2002; 364: 303–308.

31. Zheng, Y. P., Zhang, A. B., Chen, Q. H., Zhang, J. X., and Ning, R. C. Functionalized effect on carbon nanotube–epoxy nano-composites. *Mater. Sci. Eng. A*. 2006; 435–436: 145–149.

32. Mukhopadhyay, K., Dwivedi, C. D., and Mathur, G. N. Conversion of carbon nanotubes to carbon nanofibers by sonication. *Carbon*. 2002; 40: 1369–1383.

33. Yu, H. H., Cao, T., Zhou, L., Gu, E., Yu, D. S., and Jiang, D. Layer-by-layer assembly and humidity sensitive behaviour of poly(ethyleneimine)/multiwall carbon nanotube composite films. 2006; 119: 512–515.

34. Qian, D., Dickey, E. C., Andrews, R., and Rantell, T. Load transfer and deformation mechanisms in carbon nanotube–polystyrene composites. *Appl. Phys. Lett*. 2000; 76(20): 2868–2870.

35. Cooper, C. A., Cohen, S. R., Barber, A. H., and Wagner, H. D. Detachment of nanotubes from a polymer matrix. *Appl. Phys. Lett*. 2002; 81(20): 3873–3875.

36. Wong, M., Paramsothy, M., Xu, X. J., Ren, Y., Li, S., and Liao, K. Physical interactions at carbon nanotube–polymer interface. *Polymer.* 2003; 44: 7757–7764.
37. Barber, A. H., Cohen, S. R., and Wagner, H. D., Measurement of carbon nanotube-polymer interfacial strength. *Appl. Phys. Lett.* 2003; 82(23): 4140–4142.
38. Lu, M., Li, H. L., and Lau, K. T. Formation and growth mechanism if dissimilar coiled carbon nanotubes by reduced pressure catalytic chemical vapour deposition. *J. Phys. Chem. B.* 2004; 108: 6186–6192.
39. Lordi, V. and Yao, N. Molecular mechanics of binding in carbon-nanotube–polymer composites. *J. Mater. Res.* 2000; 15: 2770-9.
40. Gou, J., Minaie, B., Wang, B., Liang, Z. Y., and Zhang, C. Computational and experimental study of interfacial bonding of single-walled nanotube reinforced composites. 2004; 31: 225–236.
41. Gou, J. H. and Lau, K. T. *Handbook of Theoretical and Computational Nanotechnology.* Edited by Michael Rieth and Wolfram Schommers, Vol. 1, American Scientific, Valencia, CA, 2005.
42. Liao K and Li S, Interfacial characteristics of a carbon nanotube–polystyrene composite system, *Appl. Phys. Lett.* 2001; 79, 4225.
43. Gao, G. H., Cagin, T., and Goddard, W. A., Energetics, structure, mechanical and vibrational properties of single-walled carbon nanotubes, *Nanotechnology.* 1998; 9, 184.
44. Nardelli, M. B., Fattebert, J. L., Orlikowski O, Roland C, Zhao Q, and Bernhold J. Mechanical properties, defects and electronic behavior of carbon nanotubes, *Carbon.* 2000; 38, 1703.
45. Lennard–Jones, J. E. The determination of molecular fields: from the variation of the viscosity of a gas with temperature, *Proc. Royal Soc. London* 1924; A106, 441.
46. Li, C. Y. and Chou, T. W. Multiscale modeling of carbon nanotube reinforced polymer composites, *J. Nanosci. Nanotechnol.* 2003; 3, 1.
47. Namilae, S. and Chandra, N. Role of atomic scale interfaces in the compressive behaviour of carbon nanotubes in composites. *Comp. Sci. Technol.* 2006; 66: 2030–2038.
48. Tserpes, K. I. and Papanilkos, P. Continuum modeling of carbon nanotube-based super-structures. *Comp. Struct.* 2009; 91: 131–137.
49. Harrison, B. S. and AStala, A. Carbon nanotube applications for tissue engineering. *Biomaterials.* 2007; 28: 344–353.
50. Polizu, S., Maugey, M., Poulin, S., Poulin, P., and Yahia, L. Nanoscale surface of carbon nanotube fibers for medical applications. structure and chemistry revealed by TOF-SIMS analysis. In press.
51. Tian, F., Cui, D., Schwarz, H., Estrada, G. G., and Kobayashi, H. Cytotoxicity of single-wall carbon nanotubes on human fibroblasts. *Toxicol. in Vitro.* 2006; 20: 1202–1212.
52. Panessa-Warren, B. J., Maye, M. M., Warren, J. B., and Crosson, K. M. Single walled carbon nanotube reactivity and cytotoxicity following extended aqueous exposure. *Environ. Pollution.* 2009; 157: 1140–1151.
53. Nanomaterials May be Toxic. *Engineers Australia,* 2009; 18(12): 60.

3

Multifunctional Polymeric Smart Materials

Jinsong Leng, Xin Lan, Yanju Liu, and Shanyi Du

Harbin Institute of Technology

CONTENTS

3.1 Overview

Smart materials are materials that have one or more properties that can be significantly changed in a controlled fashion by external stimuli, such as electricity, magnetism, temperature, moisture, and pH. Various smart materials already exist, and are being researched extensively, including piezoelectric materials, magnetorheostatic materials, electrorheostatic materials, shape memory materials, and so on. The property that can be altered influences what types of applications the smart material can be used for. In the last decade, a wide range of novel polymeric smart materials have been developed as functional nanomaterials, sensors, actuators, biomaterials, bioinspired materials, etc. The applications cover aerospace, automobile, telecommunications, smart micro-/nanocontainers for drug delivery, actively moving polymers, neural memory devices, various biosensors, biomimetic fins, etc. This chapter presents some recent advances in polymeric smart materials, including electroactive polymers (EAPs), shape memory polymers (SMPs), and their composites.

Dielectric elastomers (DEs), a type of electroactive polymer (EAP), are used to develop lightweight, low-cost, and compliant actuators. They have attracted much attention in recent years. Their potential applications include medical devices, energy harvesters, and space robotics. The electroactive effect of dielectric elastomers depends on three factors: the electrical force between electrodes, the microstructure, and the mechanical properties of the material. Dielectric elastomers work as variable capacitors. When a potential is applied across the electrodes, the induced charge causes an electrostatic attraction between them. The electrical force between electrodes, also known as Maxwell stress, leads to a reduction in film thickness, which in turn results in elongation in the plane of the film. DE was first proposed and studied in 1991 by Stanford Research Institute (SRI) International. Later, Jet Propulsion Lab from NASA, Penn State University, and ETH Swiss, etc., started their own research on DE. With its excellent mechanical performance, DE found applications in industrial areas where strict flexibility, miniaturization, and high precision are required.

Furthermore, SMPs are stimuli-responsive smart materials with the ability to undergo a large recoverable deformation upon the application of an external stimulus. While the reversible martensitic transformation is the mechanism behind the shape memory phenomenon in shape memory alloys (SMAs), the shape memory phenomenon in SMPs stems from a dual-segment system: cross-links to determine the permanent shape and switching segments with transition temperature (T_{trans}) to fix the temporary shape. Below T_{trans}, SMPs are stiff, while they will be relatively soft upon heating above T_{trans} and, consequently, they can be easily deformed into a desired temporary shape by applying an external force. Upon cooling and subsequent removal of this external force, their temporary shape can be maintained. Upon reheating, their temporary deformed shape will automatically revert to the original permanent shape. SMPs can be activated not only by heat and magnetism, as can SMAs, but also by electricity, light, moisture, and even certain chemical stimuli (e.g., a change in pH value), etc. SMPs present many potential technical advantages that surpass those of SMAs and shape memory characteristics such as good shape recoverability (up to 400% recoverable strain), low density, ease in processing and in tailoring of properties (e.g., transition temperature, stiffness, biodegradability, and ease of functionally grading), programmability and controllability of recovery behavior, and most importantly, low cost. Based on these advantages, SMP composites and some novel multifunctional SMPs have also been developed. Furthermore, there is an increased activity in integrating SMPs with nanotechnology to develop novel materials for realistic applications. The latest developments in nanotechnology enable SMPs to better accommodate the requirements of a particular application in biomaterials, sensors, actuators, or textiles.

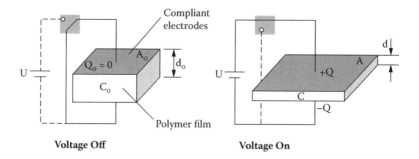

FIGURE 3.1
Working mechanism of the dielectric elastomer actuator.

3.2 Constitutive Relation and Electromechanical Stability of Dielectric Elastomers

In the past decades, electroactive materials, which are capable of elongating and bending under electric fields, have attracted much attention [1–11]. Conversely, an electroactive polymer called *dielectric elastomer* (DE) has become one of the most promising materials for actuators. Dielectric elastomers are one of the important electroactive polymers used as actuators in adaptive structures owing to their outstanding ability to generate very large deformations while subjected to an external electric field. This is due to their remarkable properties, for example, large recoverable deformation (up to 380%), high energy density (3.4 J/g), and fast response [1–9].

The electroactive effect of dielectric elastomers depends on three factors: the electrical force between electrodes, the microstructure, and the mechanical properties of the material. The electrodes are, in fact, the compliant carbon grease spread uniformly on both surfaces of the DE film [1–5,7–9]. When these electrodes are applied with a voltage, the DE film will expand in-plane and contract perpendicularly (see Figure 3.1.)

Based on this electrostatic model, one can deduce the electrostatic force developed between the electrodes. The pressure P is found to be [1]:

$$P = \varepsilon_0 \varepsilon_r E^2 = \varepsilon_0 \varepsilon_r \left(\frac{V}{t} \right)^2 \tag{3.1}$$

where ε_0 and ε_r denote the dielectric constants of the vacuum and dielectric elastomer, respectively; V is the electric field strength; E is the electric field constant; and t represents the thickness of the dielectric film.

In the above equation, P is generated by the in-plane elongation and the vertical contraction. However, the effect of the in-plane tension is predicted to be very small, and thus can be neglected. Then the vertical contraction is the only factor to be considered. The pressure induces a corresponding strain along the direction of thickness in the DE film, which is nonlinear in the case of large deformation.

The dielectric elastomers' nonlinear constitutive relation [12–22], electromechanical stability [23–28,30–42], and failure [27,31,36,38,41] have been the most popular subjects of recent investigations.

3.2.1 Constitutive Relation of Dielectric Elastomers

3.2.1.1 Constitutive Relations of Dielastic Elastomer

As a coupled system of mechanical and electric fields, the nonlinear field theory on dielectric elastomers can be used to analyze the mechanical performance of the dielectric elastomer film actuator. The free energy function can be decomposed as follows [21,23–42]:

$$W(\lambda_1,\lambda_2,\lambda_3,\tilde{D}) = W_0(\lambda_1,\lambda_2,\lambda_3) + W_1(\lambda_1,\lambda_2,\tilde{D}) \tag{3.2}$$

where $\tilde{D}$ is the nominal electric displacement and $W_0(\lambda_1,\lambda_2,\lambda_3)$ and $W_1(\lambda_1, \lambda_2,\tilde{D})$ are the elastic strain energy density function and electric energy density function, respectively. The free energy function of the dielectric elastomer electromechanical coupling system can be expressed as

$$W(\lambda_1,\lambda_2,\lambda_3,\tilde{D}) = \sum_{i=1}^{N} \frac{\mu_i}{\alpha_i}(\lambda_1^{\alpha_i} + \lambda_2^{\alpha_i} + (\lambda_1\lambda_2)^{-\alpha_i} - 3) + \frac{1}{2\varepsilon}\tilde{D}_3^2 \tag{3.3}$$

$$s_{iK}(F,\tilde{D}) = \frac{\partial W(F,\tilde{D})}{\partial F_{iK}}, \quad \tilde{E}_K(F,\tilde{D}) = \frac{\partial W(F,\tilde{D})}{\partial \tilde{D}_K} \tag{3.4}$$

$s_{iK}(F,\tilde{D})$ and $\tilde{E}_K(F,\tilde{D})$ are the nominal stress and electric field, respectively, of the electromechanical coupling system of the dielectric elastomer. F_{iK} is the tensor of deformation gradient.

$$S_{11} = \frac{\partial W}{\partial \lambda_1} = \sum_{i=1}^{N} \mu_i \left(\lambda_1^{\alpha_i-1} - \lambda_1^{-\alpha_i-1}\lambda_2^{-\alpha_i} \right) \tag{3.5a}$$

$$S_{22} = \frac{\partial W}{\partial \lambda_2} = \sum_{i=1}^{N} \mu_i \left(\lambda_2^{\alpha_i-1} - \lambda_2^{-\alpha_i-1}\lambda_1^{-\alpha_i} \right) \tag{3.5b}$$

$$S_{33} = \frac{\partial W}{\partial \lambda_3} = \sum_{i=1}^{N} \mu_i \lambda_3^{\alpha_i-1} + \frac{\tilde{D}_3^2}{\varepsilon}\lambda_3 \tag{3.5c}$$

$$\tilde{E}_3(F,\tilde{D}) = \frac{\partial W(F,\tilde{D})}{\partial \tilde{D}_K} = \frac{\lambda_3^2}{\varepsilon}\tilde{D}_3 = \lambda_3 E_3 \tag{3.5d}$$

Based on the nonlinear field theory referred to earlier, we can determine the real stress σ_{ij} of the electromechanical coupling system of the dielectric elastomer:

$$\sigma_{11} = F_{1K}S_{1K} = \sum_{i=1}^{N} \mu_i \left(\lambda_1^{\alpha_i} - \lambda_1^{-\alpha_i} \lambda_2^{-\alpha_i} \right) \tag{3.6a}$$

$$\sigma_{22} = F_{2K}S_{2K} = \sum_{i=1}^{N} \mu_i \left(\lambda_2^{\alpha_i} - \lambda_2^{-\alpha_i} \lambda_1^{-\alpha_i} \right) \tag{3.6b}$$

$$\sigma_{33} = F_{3K}S_{3K} = \lambda_3 \left(\sum_{i=1}^{N} \mu_i \lambda_3^{\alpha_i - 1} + \frac{\tilde{D}_3}{\varepsilon} \lambda_3 \right) = \sum_{i=1}^{N} \mu_i \lambda_3^{\alpha_i} + \varepsilon E_3^2 \tag{3.6c}$$

3.2.1.2 Constitutive Relations of Variable Dielectric Constants

The relationship between the variable dielectric constant $\varepsilon(\lambda_1, \lambda_2, \lambda_3)$ and the deformation for VHB 4910 has been given by Pelrine et al. [29], and cited in Figure 3.2. We obtain an expression for the dielectric constant by fitting the experimental data for VHB 4910 obtained by Pelrine [29]:

$$\varepsilon(\lambda_1, \lambda_2, \lambda_3) = \begin{cases} (c_1 \lambda_1 \lambda_2 + c_2)\varepsilon_0, & \lambda_1 \lambda_2 \leq a \\ c_3 \varepsilon_0, & \lambda_1 \lambda_2 \geq a \end{cases} \tag{3.7}$$

Here, c_1, c_2, c_3, a are constants. For acrylic VHB 4910, $c_1 = -0.016$, $c_2 = 4.716$, $c_2 = 4.48$, $a = 16$.

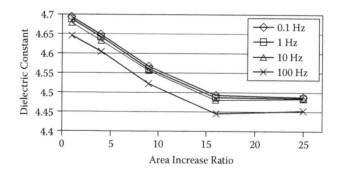

FIGURE 3.2
Dielectric constant versus area increase ratio of VHB 4910. (From Kofod, G. et al. *Journal of Intelligent Material Systems and Structures*. 2003; 14: 787.)

When $\lambda_1\lambda_2 \leq a$ due to incompressibility, we have $\lambda_3 = 1/\lambda_1\lambda_2$. The true stresses and the true electric field can be expressed as follows:

$$\sigma_{11} = \lambda_1 S_{11} = \sum_{i=1}^{N} \mu_i (\lambda_1^{\alpha_i} - \lambda_1^{-\alpha_i}\lambda_2^{-\alpha_i}) - \frac{\tilde{D}^2}{2\varepsilon_0} \frac{3c_1\lambda_1^{-1}\lambda_2^{-1} + 2c_2\lambda_1^{-2}\lambda_2^{-2}}{(c_1\lambda_1\lambda_2 + c_2)^2} \tag{3.8a}$$

$$\sigma_{22} = \lambda_2 S_{22} = \sum_{i=1}^{N} \mu_i (\lambda_2^{\alpha_i} - \lambda_2^{-\alpha_i}\lambda_1^{-\alpha_i}) - \frac{\tilde{D}^2}{2\varepsilon_0} \frac{3c_1\lambda_1^{-1}\lambda_2^{-1} + 2c_2\lambda_1^{-2}\lambda_2^{-2}}{(c_1\lambda_1\lambda_2 + c_2)^2} \tag{3.8b}$$

$$\sigma_{33} = \lambda_3 S_{33} = \sum_{i=1}^{N} \mu_i \lambda_3^{\alpha_i} + \frac{\tilde{D}^2}{\varepsilon_0(c_1\lambda_1\lambda_2 + c_2)} \lambda_3^2 \tag{3.8c}$$

3.2.1.3 Constitutive Relations of Composite Dielectric Elastomers

Figure 3.3 demonstrates that the silicone filled with barium titanate has better mechanical performance (elastic module, driving force, etc.) than pure silicone, but does not lose silicone's excellent hyperelasticity shown in Figure 3.3. To illustrate the elasticity of this new type of silicone, a developed Ogden model is given as follows:

$$W_0(\lambda_1, \lambda_2, \lambda_3, v) = \sum_{i=1}^{N} \frac{\mu_i(v)}{\alpha_i}(\lambda_1^{\alpha_i} + \lambda_2^{\alpha_i} + \lambda_3^{\alpha_i} - 3) \tag{3.9}$$

where $\mu_i(v)$ are the material constants depending on the percentage of component C.

Based on the experimental data (Figure 3.4), the dielectric constant can be expressed as follows:

$$\varepsilon(v) = \varepsilon_0' + c(v)\varepsilon_0' \tag{3.10}$$

where ε_0' is the dielectric constant at C% = 30%. $C(v)$ can be expressed as a fractional function:

$$c(v) = \begin{cases} 0.02 + 8.6v, 0 \leq v \leq 0.04 \\ -0.609 + 23.1v, 0.04 < v \leq 0.1 \end{cases} \tag{3.11}$$

We express the true stresses of the silicone composite material as follows:

$$\sigma_{11} = \sum_{i=1}^{N} \mu_i(v)\left(\lambda_1^{\alpha_i} - \lambda_1^{-\alpha_i}\lambda_2^{-\alpha_i}\right) - \frac{\tilde{D}^2}{(1+a+bv)\varepsilon_0'} \lambda_1^{-2}\lambda_2^{-2} \tag{3.12a}$$

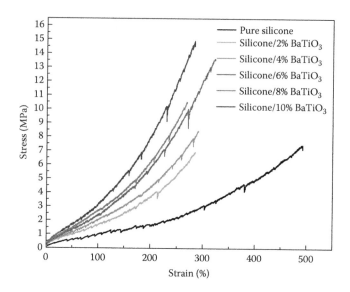

FIGURE 3.3

Elastic modulus at various barium titanate contents with 30% (percentage of component C).

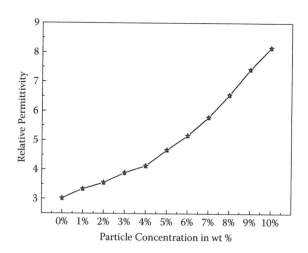

FIGURE 3.4

Permittivity of dielectric composites at varying particle concentrations.

$$\sigma_{22} = \sum_{i=1}^{N} \mu_i(v)\left(\lambda_2^{\alpha_i} - \lambda_2^{-\alpha_i}\lambda_1^{-\alpha_i}\right) - \frac{\tilde{D}^2}{(1+a+bv)\varepsilon_0'}\lambda_1^{-2}\lambda_2^{-2} \qquad (3.12b)$$

$$\sigma_{33} = \sum_{i=1}^{N} \mu_i(v)\lambda_3^{\alpha_i} - \frac{\tilde{D}^2}{(1+a+bv)\varepsilon_0'}\lambda_3^2 \qquad (3.12c)$$

3.2.2 Electromechanical Stability of Ideal Dielectric Elastomers

3.2.2.1 Method of Analyzing Electromechanical Stability of Ideal Dielectric Elastomers

When a piece of DE film is sandwiched between two compliant electrodes with a high electric field, due to the electrostatic force between the two electrodes, the film expands in-plane and contracts out-of-plane so that it becomes thinner. The thinner thickness results in a higher electric field that inversely squeezes the film again. When the electric field exceeds the critical value, the dielectric field breaks the electromechanical stability, and the actuator fails [23–42].

Referring to Figure 3.5, the elastomer has the original dimensions L_1, L_2 and L_3 in the unreformed state [23]. When subjected to the electric voltage Φ and mechanical forces P_1 and P_2, the elastomer deforms to a homogeneous state with stretches $\lambda_1, \lambda_2, \lambda_3$, as well as gaining an electric charge of magnitude Q on both electrodes. The elastomer is assumed to be incompressible, so that $\lambda_3 = 1/\lambda_1\lambda_2$.

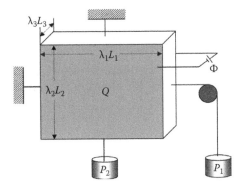

FIGURE 3.5
Layer of a dielectric elastomer coated with two compliant electrodes and loaded by a battery of voltage and by two weights. (From Zhao, X., Suo, Z. *Applied Physics Letters*. 2007; 91: 061921. With permission.)

The elastomer is assumed to be an elastic dielectric, with free energy function $W(\lambda_1, \lambda_2, D^\sim)$. The elastomer, the weights, and the battery constitute a thermodynamic system; the free energy of the system is

$$G = L_1 L_2 L_3 W(\lambda_1, \lambda_2, \lambda_3, D^\sim) - P_1 \lambda_1 L_1 - P_2 \lambda_2 L_2 - \Phi Q \tag{3.13}$$

When the generalized coordinates vary by small amounts, $\delta\lambda_1$ $\delta\lambda_2$, and $\delta D^\sim$, the free energy of the system varies by

$$\frac{\delta G}{L_1 L_2 L_3} = \left[\frac{\partial W}{\partial \lambda_1} - s_1 \right] \delta\lambda_1 + \left[\frac{\partial W}{\partial \lambda_2} - s_2 \right] \delta\lambda_2 + \left[\frac{\partial W}{\partial D^\sim} - E^\sim \right] \delta D^\sim + \frac{1}{2} \frac{\partial^2 W}{\partial \lambda_1^2} \delta\lambda_1^2$$

$$+ \frac{1}{2} \frac{\partial^2 W}{\partial \lambda_2^2} \delta\lambda_2^2 + \frac{1}{2} \frac{\partial^2 W}{\partial D^{\sim 2}} \delta D^{\sim 2} + \frac{\partial^2 W}{\partial \lambda_1 \partial \lambda_2} \delta\lambda_1 \delta\lambda_2 \tag{3.14}$$

$$+ \frac{\partial^2 W}{\partial \lambda_1 \partial D^\sim} \delta\lambda_1 \delta D^\sim + \frac{\partial^2 W}{\partial \lambda_2 \partial D^\sim} \delta\lambda_2 \delta D^\sim$$

In equilibrium, the coefficients of the first-order variations vanish:

$$s_1 = \frac{\partial W}{\partial \lambda_1} \tag{3.15a}$$

$$s_2 = \frac{\partial W}{\partial \lambda_2} \tag{3.15b}$$

$$E^\sim = \frac{\partial W}{\partial D^\sim} \tag{3.15c}$$

To ensure that this equilibrium state minimizes G, the sum of the second-order variations must be positive for an arbitrary combination of $\delta\lambda_1$ $\delta\lambda_2$, and $\delta D^\sim$, that is, the Hessian:

$$H = \begin{vmatrix} \dfrac{\partial^2 W}{\partial \lambda_1^2} & \dfrac{\partial^2 W}{\partial \lambda_1 \partial \lambda_2} & \dfrac{\partial^2 W}{\partial \lambda_1 \partial D^\sim} \\[2ex] \dfrac{\partial^2 W}{\partial \lambda_1 \partial \lambda_2} & \dfrac{\partial^2 W}{\partial \lambda_2^2} & \dfrac{\partial^2 W}{\partial \lambda_2 \partial D^\sim} \\[2ex] \dfrac{\partial^2 W}{\partial \lambda_1 \partial D^\sim} & \dfrac{\partial^2 W}{\partial \lambda_2 \partial D^\sim} & \dfrac{\partial^2 W}{\partial D^{\sim 2}} \end{vmatrix} \tag{3.16}$$

3.2.2.2 Electromechanical Stability of Neo-Hookean and Mooney–Rivlin Type Dielectric Elastomers

The electromechanical stability theory of dielectric elastomers was proposed by Suo et al. [23]. For example, the elastic strain energy function with one material constant is used to analyze the stability of an ideal elastic elastomer subjected to biaxial stress. The results revealed the relationship between the nominal electric displacement and the nominal electric field. It was proved theoretically for the first time that prestretching could enhance the dielectric elastomer's stability.

The free energy function of the neo-Hookean type dielectric elastomer is [23]

$$W(\lambda_1,\lambda_2,D^\sim)=\frac{\mu}{2}\left(\lambda_1^2+\lambda_2^2+\lambda_1^{-2}\lambda_2^{-2}-3\right)+\frac{D^{-2}}{2\varepsilon}\lambda_1^{-2}\lambda_2^{-2} \tag{3.17}$$

The first term is the elastic energy with the small strain shear modulus μ. The second term is the dielectric energy with permittivity ε. The nominal stress and nominal electric field are

$$S_1=\frac{\partial W}{\partial\lambda_1}=\mu(\lambda_1-\lambda_1^{-3}\lambda_2^{-2})-\frac{D^{-2}}{\varepsilon}\lambda_1^{-3}\lambda_2^{-2} \tag{3.18a}$$

$$S_2=\frac{\partial W}{\partial\lambda_2}=\mu\left(\lambda_2-\lambda_2^{-3}\lambda_1^{-2}\right)-\frac{D^{-2}}{\varepsilon}\lambda_2^{-3}\lambda_1^{-2} \tag{3.18b}$$

$$E^\sim=\frac{\partial W}{\partial D^\sim}=\frac{D^\sim}{\varepsilon}\lambda_1^{-2}\lambda_2^{-2} \tag{3.18c}$$

In the special case that the elastomer is under equal biaxial stresses $S_1 = S_2 = S$, the stretches are also equal biaxial, $\lambda_1 = \lambda_2 = \lambda$. The equilibrium condition (Equation 3.38) becomes

$$\frac{D^\sim}{\sqrt{\varepsilon\mu}}=\sqrt{\lambda^6-1-\frac{S}{\mu}\lambda^5} \tag{3.19a}$$

$$\frac{E^\sim}{\sqrt{\mu/\varepsilon}}=\sqrt{\lambda^{-2}-\lambda^{-8}-\frac{S}{\mu}\lambda^{-3}} \tag{3.19b}$$

Figure 3.6 shows the effects of the equal biaxial prestress. At a fixed level of the prestress S_1/μ, there is a peak in the function $E^\sim(D^\sim)$. The left-hand side of each curve corresponds to a positive-definite Hessian, the right-hand side corresponds to a non-positive-definite Hessian, and the peak is determined by $\det(H)=0$. In contrast, the true electric field is a monotonic function of $D^\sim$.

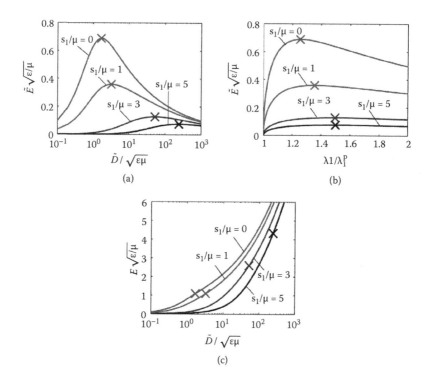

FIGURE 3.6

Behavior of a dielectric elastomer actuator under several levels of equal biaxial prestresses: (a) nominal electric field versus nominal electric displacement, (b) true electric field versus nominal electric displacement, and (c) nominal electric field versus actuation stretch. The critical points for instability are marked as crosses. (From Zhao, X., Suo, Z. *Applied Physics Letters.* 2007; 91: 061921. With permission.)

As the prestress increases, the critical nominal electric field decreases, while the critical true electric field increases.

In the absence of the prestress, maximizing $\tilde{E}$ in Equation 3.38, we have the critical stretch $\lambda^C = 1.26$, which corresponds to a reduction in thickness by 37%, and is consistent with the maximum thickness strain of 40% observed experimentally [5]. The critical nominal electric field is $E_{\tilde{C}} \approx 0.69\sqrt{\mu/\varepsilon}$.

Our group used the elastic strain energy function with two material constants to analyze the stability performance of dielectric elastomers [28]. The relationship between the nominal electric displacement and the nominal electric field of different dielectric elastomers is derived directly by using this model.

The free energy function of Mooney–Rivlin type dielectric elastomers is [28,33,34,42]

$$W(\lambda_1,\lambda_2,D^-)=\frac{\mu}{2}\left(\lambda_1^2+\lambda_2^2+\lambda_1^{-2}\lambda_2^{-2}-3\right)+\frac{G}{2}\left(\lambda_1^{-2}+\lambda_2^{-2}+\lambda_1^2\lambda_2^2-3\right)+\frac{D^{-2}}{2\varepsilon}\lambda_1^{-2}\lambda_2^{-2}$$

$$(3.20)$$

where μ and G are material constants determined by experiment. Evidently, the material constants are different for different dielectric elastomer materials (such as BJB TC-A/BC, HS3silicone, CF19-2186 silicone, VHB 4910) [9] and different structural dielectric elastomer drives (such as rolling, folding, stacking, and flattening) [9]. The nominal stress and the nominal electric field are expressed, respectively, as follows:

$$s_1 = \frac{\partial W}{\partial \lambda_1} = \mu\left(\lambda_1 - \lambda_1^{-3}\lambda_2^{-2}\right) + G\left(-\lambda_1^{-3} + \lambda_1\lambda_2^{2}\right) - \frac{D^{-2}}{\varepsilon}\lambda_1^{-3}\lambda_2^{-2} \tag{3.21a}$$

$$s_2 = \frac{\partial W}{\partial \lambda_2} = \mu\left(\lambda_2 - \lambda_2^{-3}\lambda_1^{-2}\right) + G\left(-\lambda_2^{-3} + \lambda_2\lambda_1^{2}\right) - \frac{D^{-2}}{\varepsilon}\lambda_2^{-3}\lambda_1^{-2} \tag{3.21b}$$

$$E^{-} = \frac{\partial W}{\partial D^{-}} = \frac{D^{-}}{\varepsilon}\lambda_1^{-2}\lambda_2^{-2} \tag{3.21c}$$

Prestretching uniformly the DE film such that the stretch ratios in the two directions equal, let $\mu = kG$, where k is constant. Equation 3.42 becomes

$$\frac{D^{-}}{\sqrt{\varepsilon G}} = \sqrt{k(\lambda^6 - 1) + \lambda^8 - \lambda^2 - \frac{s}{G}\lambda^5} \tag{3.22a}$$

$$\frac{E^{-}}{\sqrt{G/\varepsilon}} = \sqrt{k(\lambda^{-2} - \lambda^{-8}) + 1 - \lambda^{-6} - \frac{s}{G}\lambda^{-3}} \tag{3.22b}$$

When k is assigned different constants, the stability of DE materials can be analyzed by considering the prestretch ratio as a variable. Figure 3.7 shows the relationship between $D^{-}/\sqrt{\varepsilon G}$ and $E^{-}/\sqrt{G/\varepsilon}$ when $k = 1, 2, 2.5$, and 5, respectively. In each case, s/G takes different values such as 0, 1, 2, 3, 4, and 5, and E^{-} reaches its peak values. Before E^{-} reaches its peak value, the Hessian matrix is positive-definite, while after the peak value, the Hessian matrix is negative-definite. In a word, in the peak point, det $(H) = 0$. As the value s/G increases, the nominal electric field decreases under any constant k, which implies that increasing the prestretch ratio can improve the stability of the DE actuator.

3.2.2.3 Electromechanical Stability Domain of Neo-Hookean and Mooney–Rivlin Type Dielectric Elastomer

Research was done on the stability of neo-Hookean silicone based on elastomer by R. Díaz-Calleja's group [27]. The Hessian matrix under two special loading conditions was deduced. Furthermore, they determined the stable and unstable domains of dielectric elastomers.

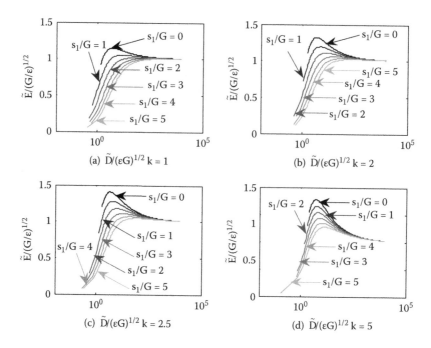

FIGURE 3.7
The nominal electric field versus the nominal electric displacement when the value of k changes. (From Liu, Y., Liu, L., Zhang, Z., Shi, L. Comment on method to analyze electromechanical stability of dielectric elastomers [*Applied Physics Letters*. 91, 061921, 2007]. *Applied Physics Letters*.2008; 93: 106101. With permission.)

The free energy function of a neo-Hookean type dielectric elastomer is

$$W(\lambda_1,\lambda_2,D^-)=\frac{\mu}{2}\left(\lambda_1^2+\lambda_2^2+\lambda_1^{-2}\lambda_2^{-2}-3\right)+\frac{D^{-2}}{2\varepsilon}\lambda_1^{-2}\lambda_2^{-2} \qquad (3.23)$$

The nominal stress and the nominal electric field are given in Equation 3.23. Let $\lambda_1=\lambda_2=\lambda$, $s>0$, and the matrix of the thermodynamic system is positive definite, so the governing equations of the dielectric elastomer's stability are

$$\lambda^6 > x^2 + 1 \qquad (3.24a)$$

$$\lambda^6 > 3x^2 - 5 \qquad (3.24b)$$

with $x = D^-/\sqrt{\varepsilon\mu}$ (see Figure 3.8).

Our group used an elastic strain energy function with two material constants to analyze the stability domain of dielectric elastomers [31].

In this case, we suppose the in-plane stretch of dielectric elastomer to be equiaxial, that is, $\lambda_1=\lambda_2=\lambda$. Due to incompressibility of the dielectric elastomer, we have $\lambda_3=\lambda^{-2}$; letting $C_2=kC_1$, $k>0$, $C_1>0$, the free energy

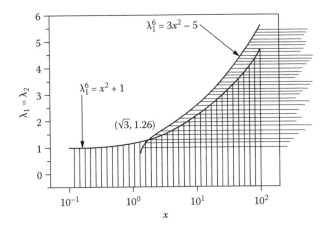

FIGURE 3.8
Plots of (λ) versus $(D^-/\sqrt{\varepsilon\mu})$ for the pull-in phenomenon and the electromechanical stabil-
ity limits in the case considered. Shaded areas represent the unstable states. (From Díaz-
Calleja et al. *Applied Physics Letters.* 2008; 93: 101902. With permission.)

function of the electromechanical coupling system can be simplified as
follows:

$$W(\lambda, D^-) = \frac{C_1}{2}(2\lambda^2 + \lambda^{-4} - 3) + \frac{kC_1}{2}(2\lambda^{-2} + \lambda^4 - 3) + \frac{D^{-2}}{2\varepsilon}\lambda^{-4} \quad (3.25)$$

The governing equations of the dielectric elastomer's stability are

$$\begin{cases} y < \sqrt{k\lambda^8 + \lambda^6 - k\lambda^2 - 1} \\ y < \sqrt{(3k\lambda^8 + \lambda^6 + 3k\lambda^4 + 5)/3} \end{cases} \quad (3.26)$$

with $y = D^-/\sqrt{\varepsilon C_1}$.
Figure 3.9 shows that for different values of k, the steady domain of dielectric elastomers is below both the curves; simultaneously, all other regions
are unstable.

3.3 Electromechanical Stability of Linear and Nonlinear Dielectric Elastomers

3.3.1 Method of Analyzing Electromechanical Stability of Dielectric Elastomers Undergoing Large Deformation

A membrane of an elastic dielectric is sandwiched between two electrodes
(Figure 3.10) [30]. In the undeformed state, the membrane has the original
dimensions L_1, L_2, and L_3. When subjected to mechanical forces F_1, F_2,

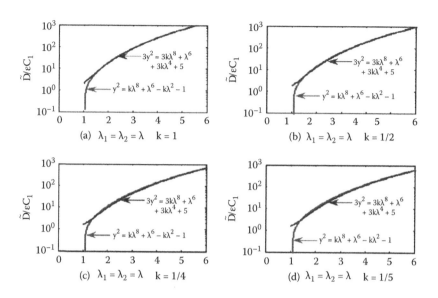

FIGURE 3.9
The steady domain of different dielectric elastomers (varying the value of k) subjected to special mechanical load are illustrated: (a) $k = 1$, (b) $k = 1/2$, (c) $k = 1/4$, (d) $k = 1/5$. (From Yanju Liu et al. Comment On electromechanical stability of dielectric elastomers [*Applied Physics Letters.* 93, 101902, 2008]. *Applied Physics Letters.* 2009; 94: 096101. With permission.)

and F_3 in the three directions, as well as an electric voltage Φ via an external circuit, the three sides deform to l_1, l_2, and l_3, and an electric charge Q flows through the external circuit from one electrode to the other. The electrodes are so compliant that they do not constrain the deformation of the dielectric.

When the sides of the membrane change by small amounts, δl_1, δl_3, and δl_3, the mechanical forces do work of $F_1\delta l_1$, $F_2\delta l_2$, and $F_3\delta l_3$. Similarly, when a small amount of charge δQ relocates from one electrode to the other through the external circuit, the electric voltage does work of $\Phi\delta Q$.

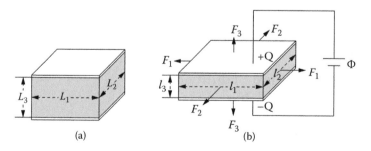

FIGURE 3.10
A membrane of an elastic dielectric is sandwiched between two compliant electrodes. (From Zhao, X. and Suo, Z. *Journal of Applied Physics.* 2008; 104: 123530. With permission.)

The dielectric is a thermodynamic system held at a constant temperature. We denote the Helmholtz free energy of the system by A. When the dielectric is in equilibrium with the applied forces and the applied voltage associated with any small change in the sides and the induced charge, any change in the Helmholtz free energy equals the work done by the applied forces and the applied voltage, namely,

$$\delta A = F_1 \delta l_1 + F_2 \delta l_2 + F_3 \delta l_3 + \Phi Q \tag{3.27}$$

If we divide Equation 3.53 by the volume of the dielectric in the reference state, L_1, L_2, and L_3, we get

$$\delta W = s_1 \delta \lambda_1 + s_2 \delta \lambda_2 + s_3 \delta \lambda_3 + \tilde{E}\tilde{D} \tag{3.28}$$

With the Helmholtz free-energy density $W = \frac{A}{L_1 L_2 L_3}$, $s_1 = \frac{F_1}{L_2 L_3}$, $s_2 = \frac{F_2}{L_1 L_3}$, and the nominal stresses $s_3 = \frac{F_3}{L_1 L_2}$, the stretches $\lambda_1 = \frac{l_1}{L_1}$, $\lambda_2 = \frac{l_2}{L_2}$, $\lambda_3 = \frac{l_3}{L_3}$, the nominal electric field $\tilde{E} = \frac{u}{L_3}$, and the nominal electric displacement $\tilde{D} = \frac{Q}{L_1 L_2}$. The nominal stresses and the nominal electric field are the partial differential coefficients, namely,

$$s_1 = \frac{\partial W(\lambda_1, \lambda_2, \lambda_3, D^{\sim})}{\partial \lambda_1} \tag{3.29a}$$

$$s_2 = \frac{\partial W(\lambda_1, \lambda_2, \lambda_3, D^{\sim})}{\partial \lambda_2} \tag{3.29b}$$

$$s_3 = \frac{\partial W(\lambda_1, \lambda_2, \lambda_3, D^{\sim})}{\partial \lambda_3} \tag{3.29c}$$

$$E^{\sim} = \frac{\partial W(\lambda_1, \lambda_2, \lambda_3, D^{\sim})}{\partial D^{\sim}} \tag{3.29d}$$

Once the free-energy function is known for a given elastic dielectric, Equation 3.29 constitutes the equations of state.

3.3.2 Electromechanical Stability of Neo-Hookean and Mooney–Rivlin Type Linear Dielectric Elastomer

Stretching membranes of a VHB elastomer by an equal amount in the two in-plane directions, Wissler and Mazza measured the permittivity as a function of the stretch [20].

$$\varepsilon(\lambda_1, \lambda_2) = \varepsilon^{\sim}[1 + c(\lambda_1 + \lambda_2 - 2)] \tag{3.30}$$

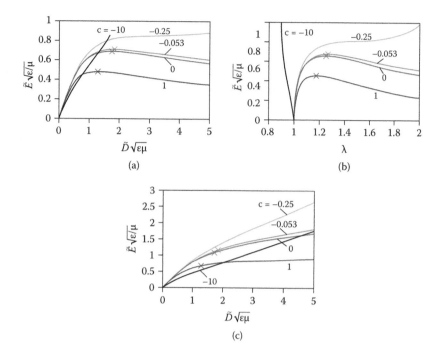

FIGURE 3.11
Behaviors of dielectric elastomers for various values of the coefficient of electrostriction. (a) Nominal electric field versus nominal electric displacement, (b) true electric field versus nominal electric displacement, and (c) nominal electric field versus actuation stretch. The critical points for electromechanical instability are marked by crosses. (From Zhao, X. and Suo, Z. *Journal of Applied Physics.* 2008; 104: 123530. With permission.)

where $c = -0.053$ and $\tilde{\varepsilon} = 4.68\varepsilon_0$, and where $\varepsilon_0 = 8.85 \times 10^{-12}$ F/m is the permittivity of vacuum.

The two in-plane stretches are equal: $\lambda_1 = \lambda_2 = \lambda$. Due to incompressibility, the free-energy function is

$$W(\lambda, D^-) = \frac{\mu}{2}(2\lambda^2 + \lambda^{-4} - 3) + \frac{\lambda^{-4}D^{-2}}{2[1 + 2c(\lambda - 1)]e^{\tilde{\ }}} \tag{3.31}$$

Let $s_1(\lambda, D^-) = 0$; further, we get $\partial W(\lambda, D^-)/\partial\lambda = 0$. Substituting Equation 3.31 into it, we have:

$$\frac{E^{\tilde{\ }}}{\sqrt{\mu/\tilde{\varepsilon}}} = \frac{\lambda^{-4}}{1 + 2c(\lambda - 1)}\frac{D^{\tilde{\ }}}{\sqrt{\mu\tilde{\varepsilon}}} \tag{3.32a}$$

$$\frac{D^-}{\sqrt{\mu\tilde{\varepsilon}}} = \frac{2(\lambda^6 - 1)[1 + 2c(\lambda - 1)]}{2 + c\lambda[1 + 2c(\lambda - 1)]^{-1}} \tag{3.32b}$$

Figure 3.11 plots Equation 3.58, using various values of the coefficient of electrostriction, c. The critical points for electromechanical instability are marked by crosses. For $c = 0$, the permittivity is independent of deformation, and the results cover those for the ideal dielectric elastomers. For $c = -0.053$, the nominal electric field reaches a peak at the stretch $\lambda \approx 1.28$. For $c = 1$, the nominal electric field reaches a peak at a smaller stretch, $\lambda \approx 1.18$.

The system free energy containing the Mooney–Rivlin elastic strain energy function with two material constants and electric energy incorporating linear permittivity has been constructed to analyze the electromechanical stability of dielectric elastomers [42].

Because of the dielectric elastomer's incompressibility, we have $\lambda_3 = \lambda^{-2}$; then the free energy function can be written as [42]

$$W(\lambda, D^-) = \frac{C_1}{2}(2\lambda^2 + \lambda^{-4} - 3) + \frac{C_2}{2}(2\lambda^{-2} + \lambda^4 - 3)$$

$$+ \frac{\lambda^{-4}D^{-2}}{2[1 + a(\lambda^{-2} - 1) + b(2\lambda + \lambda^{-2} - 3)]\varepsilon^-} \tag{3.33}$$

Introducing a dimensionless quantity k depending on the material and the activated shape, simultaneously let $C_2 = kC_1$, when C_1 is a constant, as $k = 0$, $C_2 = 0$. The system free energy function is changed to Suo's form [30]. Considering the nondimensionalized coefficient $n, b = na$, we have:

$$\frac{D^-}{\sqrt{C_1 \varepsilon^-}}$$

$$= \sqrt{\frac{2[(\lambda^6 - 1) + k(\lambda^8 - \lambda^2)]}{2 + a\lambda[n - (n+1)\lambda^{-3}]\{1 + a[2n\lambda + (n+1)\lambda^{-2} - (3n+1)]\}^{-1}} \{1 + a[2n\lambda + (n+1)\lambda^{-2} - (3n+1)]\}} \tag{3.34a}$$

$$\frac{E^-}{\sqrt{C_1/\varepsilon^-}} = \frac{\lambda^{-4}}{1 + a(\lambda^{-2} - 1) + b(2\lambda + \lambda^{-2} - 3)} \frac{D^-}{\sqrt{C_1 \varepsilon^-}} \tag{3.34b}$$

$$\frac{E^-}{\sqrt{C_1/\varepsilon^-}}$$

$$= \sqrt{\frac{2[(\lambda^{-2} - \lambda^{-8}) + k(1 - \lambda^{-6})]}{2 + a\lambda[n - (n+1)\lambda^{-3}]\{1 + a[2n\lambda + (n+1)\lambda^{-2} - (3n+1)]\}^{-1}} \{1 + a[2n\lambda + (n+1)\lambda^{-2} - (3n+1)]\}^{-1}} \tag{3.34c}$$

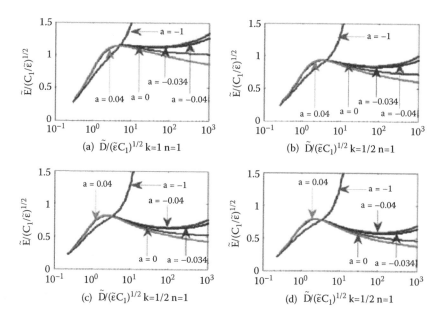

FIGURE 3.12

Relationship between the nominal electric displacement and the nominal electric field of dielectric elastomers for various values of r and k, where the stretches are equal biaxial $\lambda_1 = \lambda_2 = \lambda$ (a) $k = 1$ (b) $k = 1/2$ (c) $k = 1/4$ (d) $k = 1/5$ [42]. (From Yanju Liu et al., *Smart Materials & Structures* 2009; 18: 095040. With permission.)

Figure 3.12 illustrates the stability performances of different dielectric elastomer materials under the loading condition as $\lambda_1 = \lambda_2 = \lambda$ and $n = 1$. It shows the relationship between the nominal electric displacement and the nominal electric field of the dielectric elastomer with different values of k (1, 1/2, 1/4, and 1/5) [28,33] and different values of a (−1, −0.04, −0.034, and 0.04). Evidently, along with the decreased increase of a, the peaks of the nominal electric field decrease. However, the comparative stability performance of this kind of dielectric elastomer is even lower. When $a = 0$, neglecting the effect of deformation on the dielectric elastomer permittivity, it reduces to the analysis of the ideal dielectric elastomer [33].

Figure 3.13 shows the relationship between the stretch ratio and the nominal electric field of different dielectric elastomers under biaxial stretch with $\lambda_1 = \lambda_2 = \lambda$. Various values of k are selected, as special example namely, $k = 1, 1/2, 1/4, 1/5$. When $k = 1/2$, $a = 0$, and the nominal electric field reaches its peak, the corresponding stretch ratio (the critical value) $\lambda^c = 137$, which is consistent with the conclusion based on Reference 33.

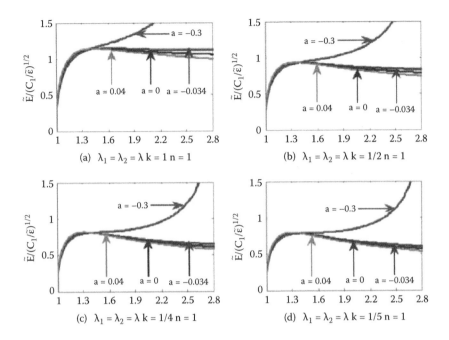

FIGURE 3.13
Relationship between the stretch ratio and the nominal electric field of different dielectric elastomers with different values of r and k, where the stretches are equal biaxial $\lambda_1 = \lambda_2 = \lambda$ (a) $k = 1$ (b) $k = 1/2$ (c) $k = 1/4$ (d) $k = 1/5$ [42]. (From Yanju Liu et al., *Smart Materials & Structures* 2009; 18: 095040. With permission.)

3.3.3 Electromechanical Stability of Nonlinear Dielectric Elastomers

The electromechanical stability of Mooney–Rivlin-type dielectric elastomers undergoing large deformation with nonlinear permittivity has been researched [34].

For $\tilde{S} > 16$, the corresponding system's free energy function is

$$W(\lambda_1, \lambda_2, \tilde{D}) = \sum_{p=1}^{N} \frac{\mu_p}{\alpha_p}\left(\lambda_1^{\alpha_p} + \lambda_2^{\alpha_p} + \lambda_1^{-\alpha_p}\lambda_2^{-\alpha_p} - 3\right) + \frac{\tilde{D}^{-2}}{2\lambda_1^2\lambda_2^2(C_1\tilde{S} + C_2)\varepsilon_0} \quad (3.35a)$$

with the material constant μ.

When $\tilde{S} > 16$, the free energy, nominal stress, and nominal electric field of the system can be expressed as follows:

$$W(\lambda_1, \lambda_2, \tilde{D}) = \sum_{p=1}^{N} \frac{\mu_p}{\alpha_p}\left(\lambda_1^{\alpha_p} + \lambda_2^{\alpha_p} + \lambda_1^{-\alpha_p}\lambda_2^{-\alpha_p} - 3\right) + \frac{\tilde{D}^{-2}}{8.96\varepsilon_0}\lambda_1^{-2}\lambda_2^{-2} \quad (3.35b)$$

We postulate that $\mu_1 = k_2\mu_2 = k_3\mu_3 = \cdots k_N\mu_N$, $k_2, k_3, \ldots k_N$ are material constants; the nominal electric field and the nominal electrical displacement can then be evaluated.

$$
\left\{
\begin{array}{l}
\left\{
\begin{array}{l}
\dfrac{D^\sim}{\sqrt{\varepsilon_0\mu_2}} = \sqrt{\dfrac{\dfrac{2k_2}{N}(\lambda^{\alpha_1-1}-\lambda^{-2\alpha_1-1}) + \dfrac{2(\lambda^{\alpha_2-1}-\lambda^{-2\alpha_2-1})}{N} + \dfrac{2k_2}{Nk_3}(\lambda^{\alpha_3-1}-\lambda^{-2\alpha_3-1})}{+\cdots+\dfrac{2k_2}{Nk_N}(\lambda^{\alpha_N-1}-\lambda^{-2\alpha_N-1}) - \dfrac{2s}{N\mu_2}}} \\[4ex]
\dfrac{E^\sim}{\sqrt{\varepsilon_0\mu_2}} = \sqrt{\dfrac{\dfrac{2k_2}{P}(\lambda^{\alpha_1-1}-\lambda^{-2\alpha_1-1}) + \dfrac{2(\lambda^{\alpha_2-1}-\lambda^{-2\alpha_2-1})}{P} + \dfrac{2k_2}{Pk_3}(\lambda^{\alpha_3-1}-\lambda^{-2\alpha_3-1})}{+\cdots+\dfrac{2k_2}{Pk_N}(\lambda^{\alpha_N-1}-\lambda^{-2\alpha_N-1}) - \dfrac{2s}{P\mu_2}}}
\end{array}
\right. \quad \lambda_1\lambda_2 \leq \\[10ex]
\left\{
\begin{array}{l}
\dfrac{D^\sim}{\sqrt{\varepsilon_0\mu_2}} = \sqrt{\dfrac{k_2c(\lambda^{\alpha_1+4}-\lambda^{-2\alpha_1+4}) + c(\lambda^{\alpha_2+4}-\lambda^{-2\alpha_2+4}) + \dfrac{ck_2}{k_3}(\lambda^{\alpha_3+4}-\lambda^{-2\alpha_3+4})}{+\cdots+\dfrac{ck_2}{k_N}(\lambda^{\alpha_N+4}-\lambda^{-2\alpha_N+4}) - \dfrac{sc}{\mu_2}\lambda^5}} \\[4ex]
\dfrac{E^\sim}{\sqrt{\varepsilon_0\mu_2}} = \sqrt{\dfrac{\dfrac{k_2}{c}(\lambda^{\alpha_1-4}-\lambda^{-2\alpha_1-4}) + \dfrac{(\lambda^{\alpha_2-4}-\lambda^{-2\alpha_2-4})}{c} + \dfrac{k_2}{ck_3}(\lambda^{\alpha_3-4}-\lambda^{-2\alpha_3-4})}{+\cdots+\dfrac{k_2}{ck_N}(\lambda^{\alpha_N-4}-\lambda^{-2\alpha_N-4}) - \dfrac{s}{c\mu_2}\lambda^{-3}}}
\end{array}
\right. \quad \lambda_1\lambda_2
\end{array}
\right.
$$

$$\text{(3.36)}$$

where

$$N = \frac{[2C_1(1+\lambda)^2 + 2C_2 + C_1(\lambda^2+\lambda)]}{[C_1(1+\lambda)^2 + C_2]^2},$$

$$P = [2C_1(1+\lambda)^2 + 2C_2 + C_1(\lambda^2+\lambda)]\lambda^3, C = 4.48.$$

Equation 3.36 describes the electromechanical stability analysis method, in which Ogden elastic strain energy is applied when the dielectric elastomer undergoes large deformation under the condition of two kinds of stretching ratios. Evidently, these functions treat the stretch ratio λ as the variable parameter. This means that the relationship between the nominal electric field and the nominal electrical displacement can be derived by changing the value of s/μ^2.

Here, it is supposed that there are only two material constants in the Ogden elastic strain energy formulation. Let $N = 2$, $\alpha_2 = -\alpha_1 = -2$, $\mu_1 = m$, and

$\mu_2 = n$. The elastic strain energy function with two material constants can be written as follows [34]:

$$W_0\left(\lambda_1, \lambda_2, \lambda_1^{-1}\lambda_2^{-1}\right) = \frac{m}{2}\left(\lambda_1^2 + \lambda_2^2 + \lambda_1^{-2}\lambda_2^{-2} - 3\right) + \frac{n}{2}\left(\lambda_1^{-2} + \lambda_2^{-2} + \lambda_1^2\lambda_2^2 - 3\right) \quad (3.37)$$

The equation of nominal stress can be rewritten in another form:

$$\frac{D^{\sim}}{\sqrt{m\varepsilon_0}} = \sqrt{\frac{2(\lambda^6 - 1)}{N} + \frac{2k(\lambda^8 - \lambda^2)}{N} - \frac{2s\lambda^5}{Nm}} \quad (3.38a)$$

where

$$N = \frac{[2C_1(1+\lambda)^2 + 2C_2 + C_1(\lambda^2 + \lambda)]}{[C_1(1+\lambda)^2 + C_2]^2}.$$

Hence, the nominal electric field is

$$\frac{E^{\sim}}{\sqrt{m/\varepsilon_0}} = \sqrt{\frac{2(\lambda - \lambda^{-5}) + 2k(\lambda^3 - \lambda^{-3})}{[2C_1(1+\lambda)^2 + 2C_2 + C_1(\lambda^2 + \lambda)]\lambda^3} - \frac{2s}{m[2C_1(1+\lambda)^2 + 2C_2 + C_1(\lambda^2 + \lambda)]\lambda^3}}$$

$$(3.38b)$$

When k is assigned different values for the variables in the stretch ratio λ, the electromechanical stability of different dielectric elastomers undergoing large deformation process can be analyzed. Figure 3.14a–d represents the relationship between $\frac{D^{\sim}}{\sqrt{m\varepsilon_0}}$ and $\frac{E^{\sim}}{\sqrt{m/\varepsilon_0}}$ when $k = 1$, $k = 1/2$, $k = 1/4$, and $k = 1/8$. In each case, s/m takes different values, so that $(0, 0.5, 1.0, 1.5, 2, 2.5)$, $E^{\sim}$ will reach peak values. The left side of curves separated by these peaks makes the Hessian matrix positive definite; conversely, the right side of curves makes the Hessian matrix negative definite. However, peaks make the Hessian matrix's det (H) = 0. With s/m increasing, the nominal electric field decreases for different values of k. This shows that prestretch can be enforced to improve the stability of dielectric elastomers.

As k increases, the critical electric field increases, indicating that larger value of k dielectric elastomer material leads to higher electrical and mechanical stability; the corresponding threshold can be achieved and the thickness of the tensile strain rate is also higher.

When $S^{\sim} > 16$, the free energy, nominal stress, and nominal electric field of the system can be expressed as follows:

$$W(\lambda_1, \lambda_2, D^{\sim}) = \frac{m}{2}\left(\lambda_1^2 + \lambda_2^2 + \lambda_1^{-2}\lambda_2^{-2} - 3\right) + \frac{n}{2}\left(\lambda_1^{-2} + \lambda_2^{-2} + \lambda_1^2\lambda_2^2 - 3\right) + \frac{D^{-2}}{8.96\varepsilon_0}\lambda_1^{-2}\lambda_2^{-2}$$

$$(3.39)$$

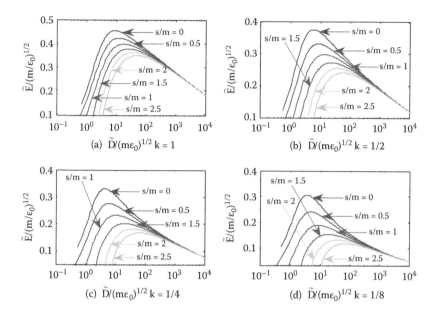

FIGURE 3.14
The nominal electric field versus the nominal electric displacement when $k = 1$, $\lambda_1 = \lambda_2 = \lambda$, $S \leq 16$. (From Yanju Liu et al. *Polymer International*. 2010; 59: 371–377. With permission.)

$$\frac{\tilde{D}}{\sqrt{m\varepsilon_0}} = \sqrt{4.48\left[(\lambda^6 - 1) + k(\lambda^8 - \lambda^2) - \frac{s}{m}\lambda^5\right]} \qquad (3.40a)$$

$$\frac{\tilde{E}}{\sqrt{m/\varepsilon_0}} = \sqrt{\frac{(\lambda^{-2} - \lambda^{-8})}{4.48} + \frac{k(1 - \lambda^{-6})}{4.48} - \frac{s}{4.48m}\lambda^{-3}} \qquad (3.40b)$$

The relationships between the nominal electric field and nominal electric displacement are shown in Figure 3.15. Similar results have been obtained previously when $\tilde{S} > 16$.

From the foregoing analysis, it is known that when $\tilde{S}$ varies from 0 to a finite value, the permittivity of DE film decreases linearly until it approaches a constant of $4.48\varepsilon_0$. For both the cases $\tilde{S} \leq 16$ and $\tilde{S} > 16$, it is clear that if the DE film is prestretched, its permittivity will decrease and the critical nominal electric field will increase (for example, $k = 1$, the critical nominal electric fields under the two kinds of stretch ratios are $0.4536\sqrt{m/\varepsilon_0}$ and $0.5424\sqrt{m/\varepsilon_0}$, respectively.) Hence, dielectric elastomers treated by prestretching show better electromechanical stability, which is consistent with the experimental result [1,5] and Suo's theoretical results [23].

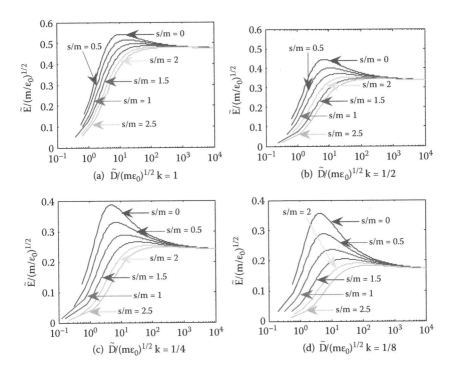

FIGURE 3.15
The nominal electric field versus the nominal electric displacement when $\lambda_1 = \lambda_2 = \lambda$, $S \le 16$. (From Yanju Liu et al. *Polymer International*. 2010; 59: 371–377. With permission.)

3.4 Reinforced Functions for Dielectric Elastomer Nanocomposites

The actuation stress and strain can be deduced as follows:

$$\sigma = \varepsilon_0 \varepsilon E^2 \tag{3.40}$$

$$\gamma = \frac{\varepsilon_0 \varepsilon E^2}{Y} \tag{3.41}$$

where σ and γ are stress and strain in the plane of the film, E is the electric field of the electrodes, Y is Young's modulus of the dielectric elastomer, ε_0 is the permittivity of vacuum, and ε is the relative permittivity (dielectric constant) of the polymer.

Dielectric elastomers have shown good electromechanical performance, including large strains (up to 380%), high elastic energy densities (up to 3.4 J/g), high efficiency, response, lifetime, and reliability. However, a considerably high activation electric field (on the order of 100 Vμm^{-1}) is required to achieve these levels, which limits the application of the dielectric elastomer. Therefore, the reduction of electric field is needed to fulfill its commercial application. From Equations 3.40 and 3.41, we can see that the actuation strain is directly proportional to the permittivity and inversely proportional to the modulus of the dielectric elastomer. Therefore, it is necessary to improve the permittivity of the dielectric elastomer actuator. At present, there are mainly three ways to improve the permittivity of the dielectric elastomer actuator: (1) ceramic particles with a high dielectric constant as filler are added to the PDMS, such as titanium dioxide (TiO$_2$), barium titanate (BaTiO$_3$), and Pb(Mg$_{1/3}$Nb$_{2/3}$)O$_3$–PbTiO$_3$ (PMN-PT); (2) conductive particles, such as carbon nanotubes, carbon black, and short fibers are added into the silicone matrix; (3) blending the elastomer with a highly polarizable conjugated polymer, such as undoped poly(3-hexylthiophene) (PHT). These three methods have both advantages and disadvantages.

3.4.1 Modification of Dielectric Elastomer Materials

3.4.1.1 Ceramic Particles as Filler

This work [43,44] focuses on the dielectric properties of a polymer–ceramic composite, consisting of a dispersion of Pb(Mg1/3Nb2/3)O3–PbTiO3 (PMN-PT) powder in a very soft silicone matrix. PMN-PT is a ceramic with one of the highest dielectric permittivities (over 28,000 at 1 kHz and room temperature) arising from its nature as a ferroelectric relaxor. A three-component poly-dimethyl-siloxane (PDMS) (TC-5005 A/B–C, BJB Enterprises Inc., U.S.) is used as the matrix. The dielectric properties of both the silicone–PMN–PT composite and the pure elastomer are studied with respect to filler content. Tensile mechanical tests are also presented that investigate the effects on the elastomer mechanical performance caused by the addition of the ceramic filler.

The (a) real and the (b) imaginary parts of the complex dielectric permittivity of the silicone composite are presented in Figure 3.16. Both the dielectric constant ε' and the dielectric loss ε'' were significantly increased by the filler; the increase depended on the frequency and composition and was particularly evident at low frequency. The real dielectric permittivity spectra were characterized by a smooth dispersive behavior, monotonically decreasing with frequency over the whole range explored. [44] The shape of the spectra of the real part of the permittivity of the composites did not substantially differ from that of the neat elastomer, whereas a slight difference was noticed in the low-frequency region of the loss factor spectra of the composites with respect to that of the neat silicone. An additional dispersive region for ε'' at low frequencies, with an increased slope, appeared on adding the PMN-PT particles, while the shape of the high-frequency region of the spectra was less influenced by the filler particles.

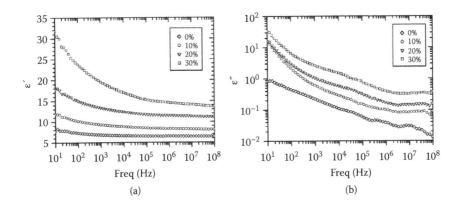

FIGURE 3.16
(a) Real and (b) imaginary parts of the complex dielectric permittivity for the silicone–PMN-PT composites at various filler volume fractions. (From Gallone G. et al. *Materials Science and Engineering C.* 2007; 110: 27.)

The stress–strain curves of the silicone composites are shown in Figure 3.17. The elastic modulus of the pure silicone is 62 kPa. [44] The stress–strain curves had the upswing at high elongations indicative of molecular chain orientation. The addition of ceramic particles increased the stiffness of the composites and decreased their stretch ability. As shown in Figure 3.18, the increase in elastic modulus was limited. The spherical shape of the PMN-PT particles helped limit the stiffening effect of the filler, which is regarded as a

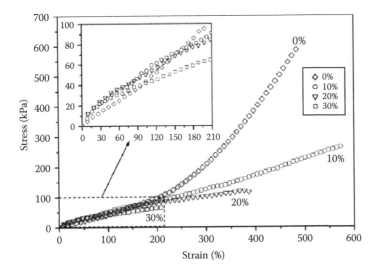

FIGURE 3.17
Nominal stress–nominal strain curves for the silicone–PMN-PT composites at various filler volume fractions. (From Gallone, G. et al. *Materials Science and Engineering C.* 2007; 110: 27.)

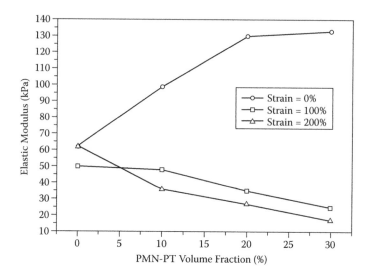

FIGURE 3.18

Elastic modulus of silicone–PMN-PT composites at three different elongations as a function of the filler volume fraction (lines are guides for the eyes). (From Gallone, G. et al. *Materials Science and Engineering C.* 2007; 110: 27.)

positive feature. That effect of filler became less evident at high loadings and is indicative of cavitation around the filler particles; the phenomenon was favored by the lack of adhesion between matrix and particles and became important at high elongations. This is why the modulus measured with high elongations decreased on increasing the filler content; this behavior is not expected to occur in compression. Cavitation was also responsible for the reduction of ultimate strains, which, however, far exceed those needed in electromechanical devices.

3.4.1.2 Conductive Particles as Filler

Under an external electric field, the dielectrics will be polarized. The macroscopic result is the orientation polarization of the material. The polar distance of polarity molecule from the permanent dipole, in the absence of an external electric field, is equal in all directions due to the thermal motion. The average polar distance of all the molecules is zero, as is shown in Figure 3.19a. When a low electric field is applied to the dielectrics, the polarity molecule will be aligned along the electric field, as shown in Figure 3.19b. When the high electric field E (Figure 3.19c) is applied to the dielectrics, there would obviously be orientation polarization of the dielectrics. Consequently, an internal electric field E', which is in the opposite direction of the external electric field, will decrease the electric intensity of the dielectrics. This requires the power

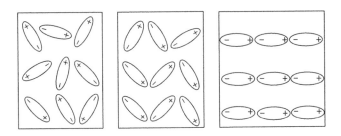

FIGURE 3.19
Polarization of the conductive particles in the silicone matrix: (a) without external electrical field, (b) with low electrical field, (c) with high electrical field.

to supply more charge to the dielectrics. The additional charge is scaled with the polarization intensity *P*. From classical electron theory, we obtain the following equations:

$$P = (\varepsilon - 1)\varepsilon_0 E \tag{3.42}$$

$$P = N\alpha \frac{\varepsilon + 2}{3} E \tag{3.43}$$

where *N* is the number of dielectric molecules per unit volume and α is the total polarizability.

Combining the preceding two equations, the Clausius–Mosotti equation is obtained as follows:

$$\frac{\varepsilon - 1}{\varepsilon + 2} = \frac{N\alpha}{3\,\epsilon_0} \tag{3.44}$$

Figure 3.20 shows the relationship between the relative permittivity ε and $N\alpha$. From the figure, we can see that the relative permittivity monotonically decreases with $3\varepsilon_0/N\alpha$. The pure silicone dielectric elastomer, as the dielectric, is under the external electric field. The relative permittivity is 4 when the parameter $N\alpha$ is $3\varepsilon_0/2$ .When the nano-sized conductive particles are blended with the silicone dielectric elastomer, both the total polarizability α and the dielectric molecules per unit volume *N* increase with the increased content of the particles. Consequently, the relative permittivity ε of the mixed dielectric also increases. In this work, two different conductive particles, carbon black and carbon nanotubes as the additive filler, are blended with the pure silicone material separately, to study the effect on the actuating behavior of the dielectric elastomer actuator.

Using a precision impedance analyzer (4294A Agilent, U.S.), the real complex permittivity $\varepsilon(\omega) = \varepsilon'(\omega) - j\varepsilon''(\omega)$ of the composites can be obtained. The real part is the relative dielectric constant, while the imaginary part is the

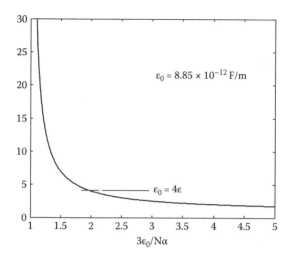

FIGURE 3.20
The relative permittivity ε versus polarizability and dielectric molecules per unit volume.

dielectric loss. For the effect of conductive filler on the decrease of dielectric breakdown strength of the composites with respect to that of the pure silicone, the following low percentages of carbon nanotube and carbon black were explored: 1, 2, 3, 4, and 5 vol%. Figure 3.21 presents the dielectric spectroscopy of the composites from 100 to 10^7 Hz. Comparing the two kinds of silicone composites (silicone–CB and silicone–CNT) to the pure silicone elastomer, it is found that both the relative dielectric constant and dielectric loss increased to varying degrees. When the content of conductive particle is below 2% in volume, the dielectric constant and dielectric loss are almost unchanged. When the content is higher than 3% in volume, obvious variation in the two factors can be observed. And when the content is 5% in volume, at the low frequency of 100 Hz, both the dielectric constant and the dielectric loss of the two kinds of silicone composites reach their peak value: ε′ is 7.87 in silicone–CNT and 7.65 in silicone–CB and ε″ is 0.97 and 0.94, respectively. From the figure, it also can be seen that both the relative permittivity and dielectric loss spectra monotonically decrease with frequency over the whole range explored. At low frequencies, due to the increase of the dielectric molecule and polarizability, the dielectric constant and dielectric loss increase. With the increase of the frequency, the molecule polarization of the composites gradually lagged behind the variation of the external electric field, which indicates the molecule polarization is no longer responsive to the shift. Consequently, ε′ and ε″ decrease with frequency. At the range of 10^2 to 10^6 Hz, the dielectric constant of the two composites shows a moderate decline. When the frequency is higher than 10^6 Hz, a drastic decline can be observed.

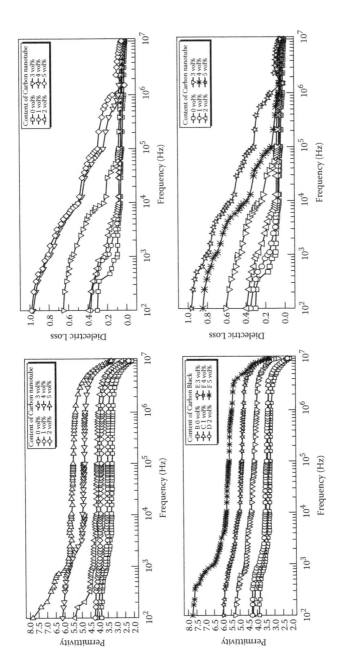

FIGURE 3.21
Dielectric spectroscopy of the composites from 100 to 10^7 Hz.

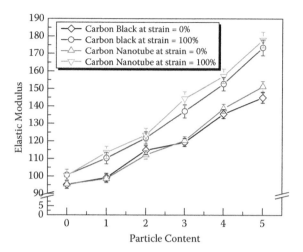

FIGURE 3.22
Elastic modulus versus concentration of conductive particle.

The results are consistent with the theoretical predictions of the dielectrics we assumed. There is no obvious difference between the two kinds of blends. The two conductive particles can improve the dielectric constant. The increased dielectric constant shows that the stress of the dielectric elastomer actuator is improved by the addition of the conductive particles. However, there is also a sharp increase in dielectric loss, which may reduce the dielectric breakdown strength of the composites.

Equation 3.2 reveals that the elastic modulus of the composite materials is a very important factor for a dielectric elastomer. How the conductive particles contribute to the mechanical properties of the composite polymers is now discussed. The tensile property of the two kinds of silicone composites is tested. Figure 3.22 presents the elastic modulus of the two composites calculated from the stress–strain curves. These curves show a nonlinear increasing trend. The elastic modulus of the composites gradually increases with elongation because of the characteristics of the crystalline polymers. The silicone rubber molecules get rearranged and crystallize with the elongation of the silicone composite samples. Consequently, the elastic modulus increases. As the dielectric elastomer, the maximum strain can reach 80% by applying a high external electric field. So we consider the elastic moduli of the silicone composites at strain to be 0% and 100%. When conductive particles are added to the raw silicone materials, the stiffness of the composites increases, which decreases their stretch ability. At strain 0%, the elastic modulus of the two kinds of silicone composites increases from 94 KPa in silicone–CB and 96 KPa in silicone–CNT to 142 and 149 KPa, respectively. At strain 100%, the elastic modulus increases from 101 KPa in silicone–CB and

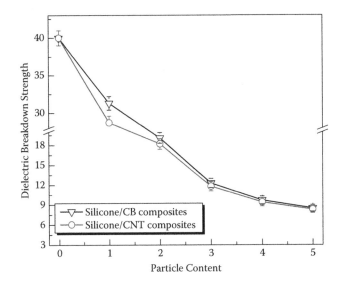

FIGURE 3.23
Dielectric breakdown strength versus concentration of the conductive particle.

99 KPa in silicone–CNT to 172 and 176 KPa, respectively. The stretch ability of silicone composites drops from 500% to 350% with mechanical loading. The less evident effect of conductive particles under high strain is indicative of cavitation around the filler particles. This phenomenon was favored by the lack of adhesion between matrix and particles and became important at high elongations.

We now discuss the effect on the dielectric breakdown strength of the composites. As shown in Figure 3.23, the dielectric breakdown strength of the two kinds of silicone composites is similar. When the content of conductive particles in the composites increases, the breakdown strength drops from 40 MV/m to a very low value. Based on previous work by Gyure et al., the following conclusions have been accepted: due to short-circuiting of some regions in the composite, causing the voltage drop to increase without conducting material, a local enhancement in the electric field occurs, allowing breakdown events to take place at a lower macroscopic electric field. This also could be explained by agglomeration effects: in the low-concentration region, particles are well separated during the casting, while at higher concentration regions, the particles are close enough to aggregate, forming large, structured clusters that short-circuit larger regions of the sample. The dielectric breakdown strength of the composites decreases with increase of carbon nanotubes. When the number of conductive particles in the composites reaches or exceeds a certain threshold, the composite material will change from an insulator to a conductor. This should be avoided.

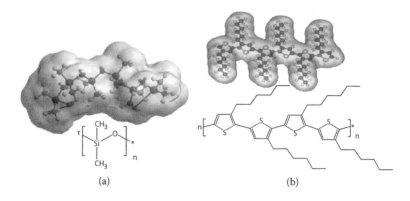

FIGURE 3.24

Chemical structures of (a) a generic poly(dimethylsiloxane) rubber and (b) poly(3-hexylthio-phene). (From Carpi, F. et al. *Advanced Functional Materials*. 2008; 235: 18. With permission.)

3.4.1.3 Blending the Elastomer with a Highly Polarizable Conjugated Polymer

Here [45], a highly polarizable conjugated polymer is blended with the silicone elastomer. The specific studied system was a poly(dimethylsiloxane) (PDMS)-based rubber (Figure 3.24a) blended with undoped poly(3-hexylthiophene) (PHT) (Figure 3.10b). Compared with previous approaches, this approach has at least two advantages: (1) it facilitates dispersion and increases homogeneity, because of the mixing of two components that are both in a liquid phase; (2) a preservation of suitable elastic modulus levels for the resulting blends can be expected as a result of the reduced modulus of the solute compound.

The frequency-dependent [45] real part (dielectric constant) and imaginary part (dielectric loss) of both the pure silicone and the composites at room temperature in the range of 10–10^8 Hz are shown in Figure 3.25. The data observed from the figure in the composites' spectra is similar to the neat poly(3-hexylthiophene). The values of the dielectric constant reduced to almost three orders of magnitude.

As expected, the dielectric constant of the PDMS–PHT composites increases dramatically by increasing the PHT content at all frequencies with respect to the pure silicone matrix [45], and a threefold increase was obtained at a low frequency, with variation from 4.5 to 14. However, this also corresponded to an analogous increase of the dielectric loss, up to a maximum value of 3 for the blend with 6 wt % of PHT. Nevertheless, the observed loss values fall in a range that typically still allows for sufficient transduction effects (see Figure 3.26).

As is evident from these results, an anomalous trend of the elastic modulus versus PHT content was obtained. In fact, a drastic reduction was achieved at 1 wt% content of PHT, while further increasing amounts of this component

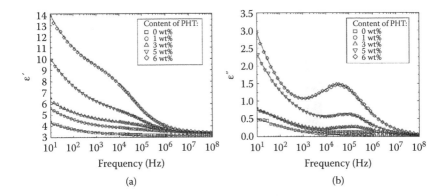

FIGURE 3.25
Frequency spectra of the relative dielectric permittivity of both pure silicone rubber and its blends with different content of poly(3-hexylthiophene): (a) real part, (b) imaginary part. (From Carpi, F. et al. *Advanced Functional Materials*. 2008; 235: 18. With permission.)

caused a progressive stiffening of the blends. A reduced cross-linking of the silicone rubber may explain the initial lowering of the elastic modulus caused by the addition of 1 wt% of poly(3-hexylthiophene). As the rubber is cured in presence of PHT, the edge-positioned S atoms of the thiophene rings in the conjugated polymer may interact with reactive sites of the silicone pre-polymer because of their lone electron pairs. This may cause a high morphological integration between the two blended polymers at a molecular scale (see Figure 3.27).

The blend of small quantities of conjugated polymer with the silicone elastomer enabled both an increase of the relative permittivity and an

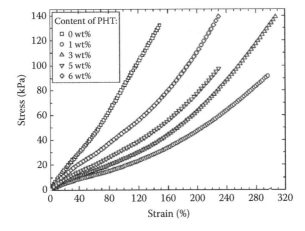

FIGURE 3.26
Stress–strain curves of both the pure silicone rubber and its blends with different content of poly(3-hexylthiophene). (From Carpi, F. et al. *Advanced Functional Materials*. 2008; 235: 18.)

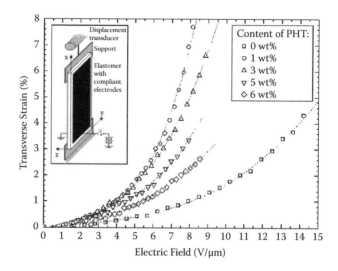

FIGURE 3.27

Electromechanical strain response exhibited by pure silicone rubber and its blends with different content of poly(3-hexylthiophene). (From Carpi, F. et al. *Advanced Functional Materials*. 2008; 235: 18. With permission.)

unexpected reduction of the tensile elastic modulus, for all the tested formulations [45]. Both these factors synergetically contributed to a remarkable increase of the electromechanical strain response. This was found to be largest at 1 wt% content of PHT, where the softening of the blend played was the most enhancing factor. Accordingly, the considered technique permitted to obtain a softened homogeneous elastomer suitable for electromechanical transduction. To the best of our knowledge, this type of result has never been reported before. [45]

3.4.2 Dielectric Elastomer Sensors

Many devices, such as electrical motors, pneumatic or hydraulic cylinders, and magnetic solenoids controlled by computers and electronics have been widely used for engineering disciplines [46]. The necessity of sensing their displacement or environment in order to quickly and accurately react to changing conditions has emerged. Sensors enable smart, self-calibrating devices, components, and structures. All sensors may be of two kinds: passive and active. There are currently many kinds of sensors, including thermocouple, piezoelectricity, magnetic, inductive, optical, capacitive, resistive, and so on. Using the sensor, the input stimulus energy can be converted to an output signal. Sensors based on the capacitive-sensing technique are strain-based. They have excellent properties such as low cost and power

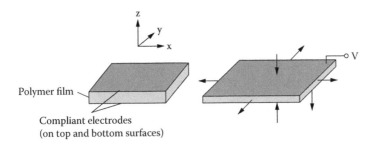

FIGURE 3.28
Operation of an EPAM device. (From Rosenthal, M. et al. *Proceedings of SPIE.* 2007; 6524: 65241F.)

usage, accuracy, good stability, resolution, speed, and temperature insensitivity. When the geometry of the electrodes or the dielectrics between the electrodes changes, the responding output signal of the capacitive sensor changes. Existing capacitive sensor configurations tend to change either the electrode gap separation or the overlap area, but typically do not change both the electrode area and gap separation simultaneously. The dielectric elastomer is a unique type of capacitor based on an elastomeric dielectric coupled with compliant electrodes that provides a capacitance change when electrically charged or when mechanical pressure is applied. When the changing capacitance of the dielectric elastomer is measured, the device can be used as a sensor [46].

The dielectric elastomer device can also act as an actuator, converting electrical energy to mechanical motion, and as a capacitive sensor by converting mechanical motion to a changing electrical signal. With proper circuit configuration, it is even possible to actuate and sense simultaneously. The dielectric elastomer sensor is unique because it simultaneously changes the dielectric thickness, d, and the area, A, as shown in Figure 3.28 [46].

The universal muscle actuator (UMA), which offers sensing capabilities, has been developed by Artificial Muscle, Inc., a high-volume manufacturer. As shown in Figures 3.29 and 3.30, the D50 UMA has a linear relationship between capacitance and both force and displacement, thus making it a simple and useful sensor [46].

In order to improve the accuracy of the dielectric elastomer sensor, a precision capacitance-measuring circuit and a linear and noise-insensitive sensor are needed. This can be achieved by designing the circuit to suppress parasitic capacitances and leakage resistance losses and by ensuring that the sensor has a low impedance output. Circuits have been designed and fabricated with these objectives in mind. In many cases, when designing the sensor circuit, it can be advantageous to integrate the capacitive sensor directly into the control circuitry in order to directly change the output signal. For example, the dielectric elastomer sensor can be used as a capacitor in an oscillator circuit; a change in the dielectric elastomer capacitance will

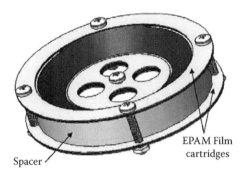

FIGURE 3.29
Universal muscle actuator D50, 50 mm diameter. (From Rosenthal, M. et al. *Proceedings of SPIE.* 2007; 6524: 65241F.)

result in a change in the output frequency of the oscillator circuit, as seen in Figure 3.31 [46].

Dielectric elastomer sensors represent a new type of capacitive sensor that offers the advantages of high strain capability, easy integration into existing components, and low cost. Dielectric elastomer actuators have been configured as capacitive sensors and have demonstrated their potential for use in capacitive pressure and displacement sensing applications. Some remaining challenges include characterizing the long-term stability of the sensors and developing appropriate measurement circuitry for the sensors. One of the fundamental advantages of dielectric elastomer technology is that it can operate as an actuator, a sensor, or a generator; by enabling actuation, sensing and generation in a single system, a new level of transducer capability can be introduced. There are also applications in which the dielectric elastomer as a

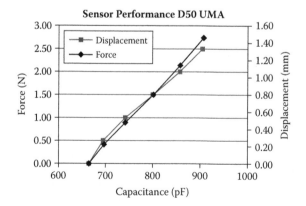

FIGURE 3.30
Measured force and displacement versus capacitance of UMA D50, one-layer device. (From Rosenthal, M. et al. *Proceedings of SPIE.* 2007; 6524: 65241F.)

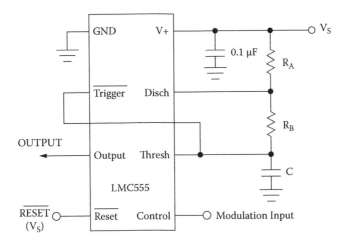

FIGURE 3.31
Pulse position modulator. (From Rosenthal, M. et al. *Proceedings of SPIE.* 2007; 6524: 65241F.)

pure capacitive sensor offers system-level advantages such as in the robotic footpad sensor application.

3.4.3 Applications of Dielectric Elastomers

Dielectric elastomer actuators, which can produce a stress–strain behavior, have shown great potential in mimicking the motion of natural muscle. These actuators are often named artificial muscles. When coupled with related technologies that may reduce the performance requirements of dielectric elastomer actuators, the commercial applications of dielectric elastomer actuators may appear sooner than anticipated in biologically inspired robots, animatronics, prosthetics, and so on [9].

The 2005 arm wrestling competition is shown in Figure 3.32 [47]. It is one of the most representative applications of dielectric elastomer actuators and other EAP technologies. The device is assembled with some DE spring roll actuators. Other artificial muscle applications have been demonstrated as well. Kornbluh et al. [15] have also reported on a mouth driven by a DE actuator. In addition, biologically inspired robots have been developed using dielectric elastomer actuators. As shown in Figure 3.33, two particularly walking robots, dubbed Flex II and MERbot, [48,49] were made of the dielectric elastomer spring roll actuators.

The fishlike airship propelled by dielectric elastomer (DE) actuators was first made at Swiss Federal Laboratories for Materials Testing and Research (EMPA). As shown in Figure 3.34, this airship is 8 m long and filled with pressurized helium. A total of four DE actuators were arranged on both sides of the airship's body and the tail. The DE actuators on the airship worked in an agonist–antagonist configuration and acted as biological muscles. When

FIGURE 3.32

Arm wrestling robot using DE spring roll actuators. (From Brochu, P. and Pei, Q.B. *Macromolecular Rapid Communications*. 2010; 31:10–36. With permission.)

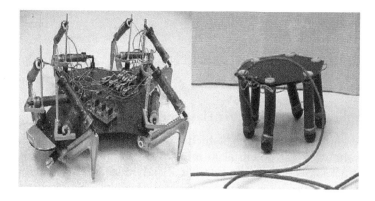

FIGURE 3.33

FLEX II and MERbot walking robots. (From Pei, Q. et al., Multiple-degrees-of-freedom electroelastomer roll actuators. *Smart Materials and Structures*. 2004; 13: N86; Pelrine et al. *Proceedings of SPIE*. 2002; 4698: 0277–786X. With permission.)

FIGURE 3.34

Model of the airship with EAP actuators in black. (From *WW-EAP Newsletter*, Vol. 11, No. 2, December 2009 [22nd issue].)

Output wheel with one-way
clutch in hub

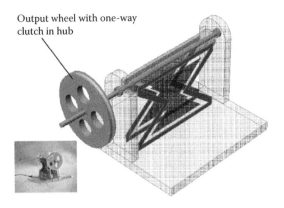

FIGURE 3.35
A simple rotary motor based on dielectric elastomers. (From Pelrine, K. et al. *Proceedings of SPIE.* 2002; 4698: 0277–786X.)

actuated by the driving voltages, the actuator on one side elongates while the actuator on the other side shrinks. Thus, the airship in the air exhibits a wavy movement and can swim freely like a fish [50].

The high energy density and fast response time of dielectric elastomer actuators make them a good choice to exploit motors. Figure 3.35 shows a prototype of a motor made of a pair of bowtie actuators, which can operate 180° out of phase to oscillate a shaft [49]. Recently, the output of the bowtie motor reached 4 W at only 100 rpm. This performance is encouraging, though it is not on par with the power capabilities of electromagnetic motors.

Braille display is one of the most important sensory functions for human perception of objects. Traditional refreshable Braille products are made from piezoelectric bimorph actuators. The complicated mechanical structure and high price of the piezoelectric Braille cells limit their further commercial application. Recently, dielectric elastomer actuators were used for refreshable Braille displays. In the 2010 Smart Structures/NDE Conference, several research groups demonstrated their Braille devices, as seen in Figures 3.36 and 3.37.

FIGURE 3.36
Braille display device using stacked dielectric elastomer actuators: (a) setup, (b) display.

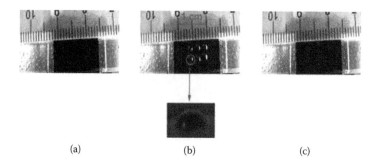

(a)   (b)   (c)

FIGURE 3.37
Braille display developed with bistable electroactive polymer. (From Yu, Z. et al., *Applied Physics Letters*. 2009; 95: 192904. With permission.)

Compared with the existing devices, these Braille devices have advantageous features such as flexibility, softness, ease of fabrication, and high power density. Leng et al. at the Harbin Institute of Technology, China, fabricated a Braille display using stacked dielectric elastomer actuators, as shown in Figure 3.36. Six stacked dielectric elastomer actuators are used as the output source. Then the output vibration is transferred to the corresponding six Braille dots according to a mechanical system. A control program is developed to demonstrate the Braille display. Another Braille product (Figure 3.37), based on a new bistable electroactive polymer (BSEP), was developed by Pei et al. [51] at the University of California, Los Angeles. A six-dot diaphragm actuator using a 30μm thick PTBA film was fabricated. The carbon nanotube was coated on both sides of the PTBA film as compliant electrode. The six dots were arrayed in a Braille-cell pattern. When a high-voltage DC or pulse current was applied, the six dots were raised to half-dome shapes. These dots can withstand up to 0.6 N force.

Except for artificial muscles, a great many other applications have been proposed to demonstrate the great potential of dielectric elastomer actuators, such as the acoustic actuator, hand-rehabilitation splints, motors, energy harvest devices, and micro-optical zoom lenses. The energy harvest device was designed and fabricated by Leng et al. (Harbin Institute of Technology), based on stacked dielectric elastomer actuators. As shown in Figure 3.38, the waves can drive the buoy up and down, which will repetitively compress the stacked dielectric elastomer actuators. Hence, the conversion of mechanical energy to electrical energy is completed again and again.

In addition, Figure 3.39 shows a heel-strike generator made of dielectric elastomer [49]. The generator, located in the heel of a shoe, could capture the energy produced in walking when the heel strikes the ground. This device was developed for the military and could produce a maximum energy of 0.8 J per cycle. In the future it is hoped to generate 1 W per foot in the course of normal walking.

FIGURE 3.38
Energy harvest device using stacked dielectric elastomer actuators.

3.5 Fabrication and Modeling of Shape Memory Polymers

SMPs are able to actively respond to an external stimulus by means of certain significantly macroscopic properties such as shape [52,53]. SMPs and their composites can recover their original shapes after large deformation when subjected to an external stimulus, such as Joule heating, light, magnetism, or moisture [54]. Among these SMPs, the thermoresponsive SMPs are the most common [55–57]. At a macroscopic level, as illustrated in Figure 3.40, the typical thermomechanical cycle of a thermoresponsive SMP consists of the following steps [58,59]: (1) fabrication of the SMPs into an *original shape*; (2) heating the SMP above the thermal transition temperature (T_{trans}) (either a glass transition temperature, T_g, or a melting temperature, T_m), and deformation of the SMP by applying an external force, cooling well below T_{trans}, removal of the constraint to obtain a temporary *predeformed shape*; (3) when needed, heating of the predeformed SMP above T_{trans}, and then recovery of the SMP towards its original shape (a *recovered shape*).

FIGURE 3.39
Heel-strike generator based on the dielectric elastomer during walking. (From Pelrine, K. et al. *Proceedings of SPIE.* 2002; 4698: 0277–786X.)

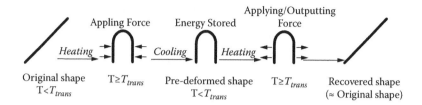

FIGURE 3.40
Schematic of shape memory effect during a typical thermomechanical cycle.

3.5.1 Synthesis of Typical Shape Memory Polymers

The basic molecular architecture of SMPs is a polymer network underlying active movement [60]. An SMP always consists of dual segments, one that is highly elastic and another that is able to significantly reduce its stiffness upon a particular stimulus. The latter can be either a molecular switch or a stimulus-sensitive domain. Upon exposure to a specific stimulus, the switching/transition is triggered and strain energy stored in the temporary shape is released, which consequently results in shape recovery [61,62].

More than 20 types of SMPs have been synthesized and widely researched in recent years. The physically cross-linked SMPs include linear polymers [63–65], branched polymers, or a polymer complex [66–68]. The shape memory effect of linear polymers is due to the phase separation and domain orientation. The typical physically cross-linked SMPs are linear block copolymers, such as polyurethanes. In polyesterurethanes, oligourethane segments are the hard elastic segments, while polyester serves as the switching segment [69,70]. Covalent netpoints can be obtained by cross-linking of linear or branched polymers as well as by (co)polymerization/poly(co)condensation of one or several monomers, whereby at least one has to be at least trifunctional. Depending on the synthesis strategy, cross-links can be created during synthesis or by postprocessing. The shape recovery can be controlled by the trans-vinylene content [71].

3.5.1.1 Thermoplastic Polyurethane Shape Memory Polymer

Since the development of the shape memory segmented polyurethane (PU) copolymer by Hayashi of Mitsubishi Heavy Industries, extensive studies on these materials have been carried out [72–75]. PU SMP consists of a two-phase structure: a hard segment and soft segment. The hard segments are formed either from a long-chain macro diol with a higher thermal transition temperature, or from diisocyanates and chain extenders. Typically, the linear polyurethane can be usually synthesized using the following process: through the reaction of difunctional, hydroxy-terminated oligoesters and -ethers with an excess of a low-molecular-weight diisocyanate, the isocyanate-terminated prepolymers can be obtained [76]. The basic structure of the diisocyanate

—R'—NHCOO—R"—OOCHN—R'—NHCOO————R—O—[—CO—(CH$_2$)$_5$—O]$_n$—H

(Hard segment, Diisocyanate + extender)　　　　　　(Soft segment, Polyol)

FIGURE 3.41
Polyurethane with microphase separation structure.

can be either aromatic or aliphatic; each type of diisocyanate has a different ability to form a semicrystalline hard segment. Then low-molecular-weight diols or diamines are incorporated as chain extenders to further couple these prepolymers. In this way, the linear, phase-segregated polyurethane or polyurethane-urea block copolymers are formed. As shown in Figure 3.41, the phase separation morphology of the molecular structure of such polyurethanes is obtained.

3.5.1.2 Thermosetting Styrene-Based Shape Memory Polymer

Styrene-based SMPs must have two phases or exhibit a cross-linked structure to exhibit the shape memory effect. The variety of methods to polymerize styrene and the wide availability of possible comonomers enable these necessary features. Styrene can be polymerized by anionic, cationic, or free controlled radical polymerization methods. In this way, a huge variety of network architectures may be designed by the proper choice of mechanism, initiator, and comonomer. Through cationic polymerization, random copolymer networks formed from renewable natural oils with a high degree of unsaturation, such as soybean oils, copolymerized with styrene and divinylbenzene, are obtained [77]. Styrene-based SMPs have been prepared through the cationic copolymerization of the following candidate materials: regular soybean oil, low saturation soybean oil (LoSatSoy), and conjugated LoSatSoy oil with various alkene comonomers, including styrene and divinylbenzene, norbornadiene, or dicyclopentadiene (comonomers used as cross-linking agents) initiated by boron trifluoride diethyl etherate or related modified initiators [78,79]. By controlling the cross-link densities and the rigidity of the polymer backbones, the SMP shows tunable T_g, mechanical properties and a good shape memory effect.

Leng et al. [80] have synthesized SMP based on styrene copolymer networks with different degrees of cross-linking. The monomers of copolymer included styrene, a vinyl compound, and a cross-linking agent (a difunctional monomer). During the synthesis, the inhibitors were removed by passing the liquid monomers through an initiator removal column. Benzoyl peroxide (BPO) was used as thermal initiator. The purified monomers (styrene and the vinyl compound) and the cross-linking agent (difunctional monomer) were mixed in varying proportions with BPO at room temperature by stirring. The mixture was then polymerized between two casting glass plates, and then placed into an oven and was kept at 75°C for 36 h. This prolonged curing time minimized shrinkage, and the heating was kept at 75°C for 48 h in order to complete the reaction. The SMP, with T_g in the range from 35°C to 55°C, exhibited good shape recovery performance. Moreover, T_g can be adjusted through the alteration of the gel content of the copolymer; T_g increased with increasing gel content. The copolymer with lower gel content showed faster recovery speed than the copolymer with higher gel content at the same recovery temperature. The largest reversible strain of the SMP reached as high as 100%, and the reversible strain of the copolymer decreased with increasing gel content.

3.5.1.3 Thermosetting Epoxy Shape Memory Polymer

Shape memory behaviors can be observed in several polymers, including polyurethane-based polymers [81,82], styrene-based polymers, epoxy polymers [83,84], and others. Out of these SMPs, epoxy SMP is a high-performance thermosetting resin possessing a unique thermomechanical property together with excellent shape memory effect. Such materials show great potential for applications in the field of smart structure materials. In addition, epoxy resin is a good candidate material for further study because its formulation can be adjusted to vary its properties to meet specific needs, which is the most significant requirement to advance the performance of a material. Controlling macroscopic properties through variation of network parameters is suggested to play an important role in the investigation of SMPs. For example, Kim [85] studied the effect of soft/hard segment phase separation and soft segment crystallization on the shape memory property in polyurethane SMP. Hu and her coworkers reported that the molecular weight of polymer network influences the shape memory effect in PU shape memory films. Studies have also been performed on the relation of network structure and shape recovery in copolymers of methyl methacyrlate and poly(ethyleneglycol) dimethacrylate [86]. However, few studies have investigated the systematic tailoring of thermomechanical properties in addition to the shape memory response for polymers.

Leng et al. [87] have synthesized epoxy SMP from an epoxy resin, a hardener, and another linear epoxy monomer. The epoxy resin was mixed with the hardener at a 1:1 weight ratio. An active linear epoxy monomer was copolymerized with the polymer matrix to tailor the network. The monomer was composed of a long linear chain of C–O bonds and two epoxy groups at the chain ends. With an increase in the linear monomer content, the glass transition temperature determined by differential scanning calorimetry ranged from 37°C to 96°C, and a decrease in the rubber modulus was obtained from dynamic mechanical analysis. The tensile test reveals that the content of the linear monomer has a profound effect on the tensile deformation behavior, which varies from a brittle to an elastomeric-like response at room temperature. These results are interpreted in terms of the soft-to-hard segment ratio and the corresponding cross-link density of the network.

3.5.1.4 *Some Novel Shape Memory Polymers*

Some types of SMPs with novel functions also attract extensive interest. Mather et al. [88] developed a two-way reversible SMP in a semicrystalline network. In this polymer, cooling-induced crystallization of cross-linked poly(cyclooctene) films under a tensile load resulted in a significant elongation, and subsequent heating to melt the network reversed this elongation (contraction). In this way, the crystallization-induced elongation on cooling and melting-induced shrinkage on heating yield a two-way shape memory effect. Peng et al. [89] demonstrated a novel SMP with two transition temperatures. For this SMP, a PMMA-PEG (poly(ethylene oxide)-poly(methyl methacrylate) (PMMA)) semi-interpenetrating network (semi-IPN) exhibits excellent shape-memory behaviors at two transition temperatures, namely, the T_m of the poly(ethylene glycol) (PEG) crystal and the T_g of the semi-IPN. In addition, Lendlein et al. [90–93] have demonstrated the triple-shape effect of polymer networks with crystallizable network segments and grafted side chains. In this SMP, the multiphased networks were synthesized by photopolymerization from PEG monomethyl ether monomethacrylate and PCL dimethacrylate as cross-linker. The triple-shape effect is a general concept that requires the application of a two-step programming process for suitable polymers (Figure 3.42a). This SMP is able to change from a (A) first shape to a (B) second shape and finally deform into a (C) third shape. The triple-shape SMP has great potential for various applications. In addition, researchers have recently proposed tunable polymer multishape memory polymers (Figure 3.42b). This type of polymer has only one broad reversible phase transition, and exhibits dual-, triple-, and quadruple-shape memory effects, all highly tunable without any change to the material composition [94].

3.5.2 Constitutive Models for Shape Memory Polymers

In many practical applications, SMPs often undergo large three-dimensional deformations. To date, an understanding of the thermomechanical behavior of SMPs is still limited to the special cases of small one-dimensional deformations [95,96]. A constitutive model that describes three-dimensional finite deformations is useful for the development of SMP components. The finite element method based on the three-dimensional constitutive model is also important for engineering analysis and design of SMP components. We introduce two thermomechanical constitutive models for SMPs based on viscoelasticity [55,56,97] and phase transition, respectively [58,98,99].

3.5.2.1 *Models Based on Viscoelasticity*

Models based on linear viscoelasticity are usually used to describe the rate-dependent behavior of polymers at a macroscopic level. In these models, materials are assumed to be combinations of elements, including springs, dashpots, or frictional elements. Some constitutive models of SMPs can be

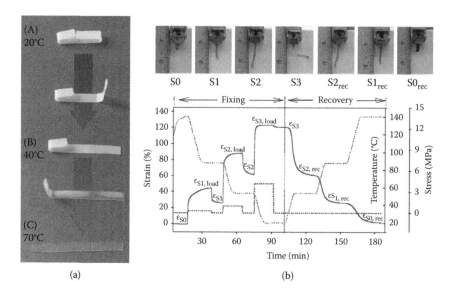

(a)   (b)

FIGURE 3.42

(a) Triple-shape (from Bellin et al. *Journal of Materials Chemistry.* 2007; 17(28):2885–91) and (b) quadruple-shape (from Xie, T. *Nature* 2010; 464(7286): 267–70) memory properties of shape memory polymers.

viewed as an extension of viscoelastic models to describe thermomechanical behaviors, such as the J. R. Chen model [100,101], F. K. Li model [102], H. Tobushi model [55,56], E. R. Abranhamson model [103], and T. D. Nguyen model [104].

With their typical constitutive model based on viscoelasticity, Tobushi et al. [55] established a linear one-dimensional constitutive equation for SMP on the basis of a standard linear viscoelastic model. In this equation, the stress–strain–temperature relationship of an SMP is expressed as

$$\begin{cases} \dot{\varepsilon} = \dfrac{\dot{\sigma}}{E} + \dfrac{\sigma}{\mu} - \dfrac{\varepsilon - \varepsilon_s}{\lambda} + \alpha\dot{T} \\ \varepsilon_s = C(\varepsilon_c - \varepsilon_l) \end{cases} \tag{3.45}$$

where σ, ε, and T denote stress, strain, and temperature, respectively. The dot denotes a time derivative. ε_s is irrecoverable strain. The temperature-related parameters E, μ, and λ represent elastic modulus, viscosity, and retardation time, respectively. The temperature-related C and ε_l are related to the process of shape fixation. They are expressed by the same function of temperature, expressed as

$$x = x_g \exp\left[a\left(\dfrac{T_g}{T} - 1\right)\right] \qquad (T_l \le T \le T_h) \tag{3.46}$$

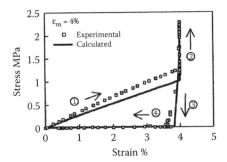

FIGURE 3.43
Relationship between stress and strain of an SMP employing Equations 3.45 and 3.46. (From Tobushi, H. et al. *Journal of Intelligent Material Systems and Structures.* 1997; 8(8):711–18.)

where x_g is the value of x at $T = T_g \cdot T_l$ and T_h are the temperatures at the starting and finishing points of the glass transition from the glassy to rubbery state in SMPs.

The shape memory thermomechanical cycle of polyurethane-based SMPs can be predicted by the constitutive equation, Equation 3.45, coupled with the material parameter function, Equation 3.46. The stress–strain curves, stress–temperature curves, and strain–temperature curves are plotted in Figures 3.43–3.45, respectively. In the first process (1), the maximum strain ε_m was applied at high temperature $T_h = T_g + 20K$. In the second process (2), maintaining ε_m constant, the specimen was cooled to $T_l = T_g - 20K$. The specimen was then unloaded at T_1 in the third process (3). Finally, in the fourth process (4), it was heated from T_1 to T_h under a no-loading condition.

In 2001, Tobushi et al. [56] developed a nonlinear constitutive equation based on the linear constitutive equation mentioned earlier. In

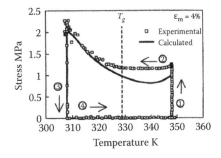

FIGURE 3.44
Relationship between stress and temperature of an SMP employing Equations 3.45 and 3.46. (From Tobushi, H. et al. *Journal of Intelligent Material Systems and Structures.* 1997; 8(8):711–18.)

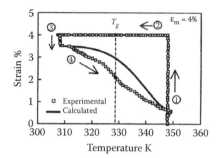

FIGURE 3.45
Relationship between strain and temperature of an SMP employing Equations 3.45 and 3.46. (From Tobushi, H. et al. *Journal of Intelligent Material Systems and Structures.* 1997; 8(8): 711–18.)

this equation, the stress–strain–temperature relationship of an SMP is expressed as

$$
\begin{cases}
\dot{\varepsilon} = \dfrac{\dot{\sigma}}{E} + m\left(\dfrac{\sigma - \sigma_y}{k}\right)^{m-1} \dfrac{\dot{\sigma}}{E} + \dfrac{\sigma}{\mu} + \dfrac{1}{b}\left(\dfrac{\sigma}{\sigma_c} - 1\right)^{n} - \dfrac{\varepsilon - \varepsilon_S}{\lambda} + \alpha\dot{T} \\[2mm]
\varepsilon_s = S(\varepsilon_c + \varepsilon_p)
\end{cases}
\tag{3.47}
$$

where σ_y and σ_c denote the elastic and viscous proportional limits of SMP. The shape memory thermomechanical cycle of an SMP was simulated by using the nonlinear constitutive equation, Equation 3.47, and the material parameter function, Equation 3.46. The stress–strain curves, stress–temperature curves, and strain–temperature curves are plotted in Figures 3.46–3.48, respectively.

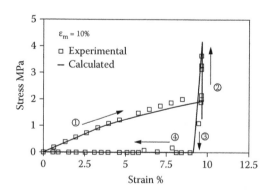

FIGURE 3.46
Relationship between stress and strain of an SMP employing Equations 3.46 and 3.47. (From Tobushi, H. et al. *Mechanics of Materials.* 2001; 33(10): 545–54.)

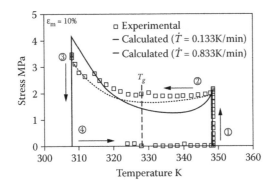

FIGURE 3.47
Relationship between stress and temperature of an SMP employing Equations 3.46 and 3.47.
(From Tobushi, H. et al. *Mechanics of Materials*. 2001; 33(10): 545–54.)

3.5.2.2 Models Based on Phase Transition

In 2006, Liu et al. [58,98] developed a constitutive model under uniaxial loading conditions of SMP. The model uses internal state variables based on experimental results and the molecular mechanism of the shape memory effect in an SMP. According to this model, there are two kinds of extreme phases, frozen and active, in an SMP at an arbitrary temperature. The fractions of frozen and active phases are defined as follows:

$$\phi_f = \frac{V_{frz}}{V}, \quad \phi_a = \frac{V_{act}}{V}, \quad \phi_f + \phi_a = 1 \tag{3.48}$$

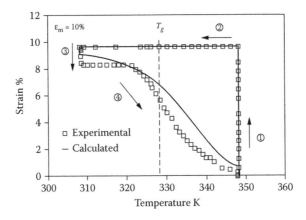

FIGURE 3.48
Relationship between strain and temperature of an SMP employing Equations 3.46 and 3.47.
(From Tobushi, H. et al. *Mechanics of Materials*. 2001; 33(10): 545–54.)

where ϕ_f and ϕ_a denote the fractions of frozen and active phases in SMP, respectively. V, V_{frz}, and V_{act} stand for the total volume, the volume of the frozen phase, and active phase, respectively.

The strain of the SMP is composed of three parts: the stored strain, mechanical elastic strain, and thermal expansion strain, expressed as

$$\varepsilon = \varepsilon_s + \varepsilon_m + \varepsilon_T \tag{3.49}$$

where ε denotes a second-order total strain tensor. ε_s, ε_m, and ε_T present second-order stored strain, mechanical elastic strain, and thermal expansion strain, respectively. The stored strain of an SMP is expressed as

$$\varepsilon_s = \int_0^{\phi_f} \varepsilon_f^e(x)\,d\phi \tag{3.50}$$

where ε_f^e denotes the entropic frozen strain. The mechanical elastic strain of an SMP is expressed as

$$\varepsilon_m = [\phi_f S_i + (1 - \phi_f)S_e] : \sigma \tag{3.51}$$

where σ denotes a second-order stress tensor. S_i is the elastic compliance fourth-order tensor corresponding to the internal energetic deformation, while S_e is the elastic compliance fourth-order tensor corresponding to the entropic deformation. The thermal expansion strain is expressed as

$$\varepsilon_T = \left\{ \int_{T_0}^{T} [\phi_f \alpha_f(\theta) + (1 - \phi_f)\alpha_a(\theta)]\,d\theta \right\} I \tag{3.52}$$

where α_f and α_a are the thermal expansion coefficients of the frozen phase and the active phase, and I is the identity tensor.

In addition, Qi and Nguyen [99] also conducted a comprehensive study of the thermomechanical behavior of the SMPs. In the 3D finite deformation constitutive model, it is assumed that the glassy phase formed during cooling has a different stress-free configuration compared to the initial glassy phase. In this sense, the deformation storage process can be captured. Therefore, the SMPs can be assumed to consist of three phases: rubbery phase, initial glassy phase, and frozen glassy phase. Recently, Chen and Lagoudas [105,106] developed a nonlinear constitutive model for SMPs to describe the thermomechanical properties under large deformation. The model is based on ideas developed in the work of Liu et al. [58]. Due to the coexisting active and frozen phases of SMP and the transition between them, they provide the underlying mechanisms for strain storage and recovery during a shape memory cycle. Their model presents excellent agreement with the experimental results of Liu et al. [58].

3.6 Multifunctional Shape Memory Polymers

As SMPs show a variety of novel properties, they have been widely researched since the 1980s. In recent years, a variety of polymers have been synthesized with shape memory effects and certain unique performance abilities have been developed. Based on the shape memory effect, some novel multifunctional SMPs or nano SMP composites have also been proposed.

The stimulus of SMPs and their corresponding composites has been carried out controllably and with programmability. They exhibit the shape memory effect and exhibit various responses to stimuli. In addition to thermal actuation, shape memory polymer composites (SMPCs) filled with functional fillers can also be actuated by other external stimuli, such as electrical resistive heating, light, or magnetic field. Moreover, it has been established that traditional thermoresponsive SMPs can also be driven by a solvent. SMP response to water or solution is due to the plasticizing effect of the solution molecule on polymer materials.

3.6.1. Electroactive *Shape Memory Polymer* Nanocomposites

3.6.1.1 Shape Memory Polymer Filled with Carbon Particles

Leng et al. [107] demonstrated that the thermosetting styrene-based SMP exhibited better mechanical properties and moisture resistance than the thermoplastic polyurethane SMP. A new system of a thermosetting styrene-based SMP filled with nanocarbon powders has been presented [108,109].

An SEM was used to observe the details of micro and nano patterns of an SMPC, particularly for the distribution of nanocarbon powders. Images of the SMPCs with various nanocarbon powder content (CB2, CB4, CB6, and CB10) are presented in Figure 3.49. Some separated nanocarbon particles, with diameters mostly from 30 to 50 nm, can be clearly observed. The distribution, aggregation, morphology, and microporosity of the nanocarbon particles greatly affect the conductivity of the resulting SMPC.

In order to determine the electrical resistivity of thin films, samples were cut into the same dimensions with two copper electrodes clamped tightly on both ends along the length direction. The volume electrical resistivity ρ was calculated as

$$\rho = \frac{RS}{l}$$

(3.53)

where R is the measured electrical resistance, S is the cross-sectional area of the sample, and l is the distance between the two copper electrodes (refer to the top insert in Figure 3.50 for the setup of the volume resistivity

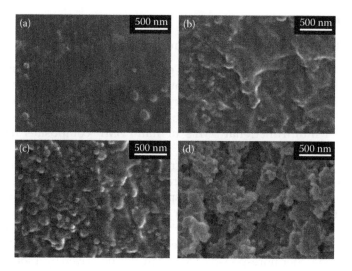

FIGURE 3.49

SEM images (SEI, 20kV, 20000.0×) of the distributions of nanocarbon powders in the SMP matrix: (a) CB2, (b) CB4, (c) CB6, and (d) CB10. (From Leng, J.S. et al. *Smart Materials. and Structures.* 2009; 18: 074003. With permission.)

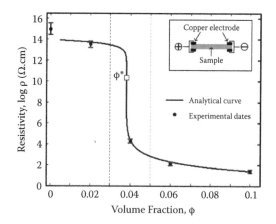

FIGURE 3.50

DC electrical resistivity of SMPC filled with nanocarbon powders. The circle symbols denote the mean values of at least three measurements for each composition from experimental results. The curves represent the calculated results according to percolation theory using the exponents of Equations 3.18 and 3.19 and a percolation threshold of $\varphi^* \approx 3.8\%$. (From Leng, J.S. et al. *Smart Materials. and Structures.* 2009; 18: 074003. With permission.)

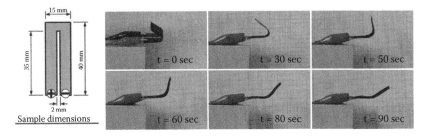

FIGURE 3.51
Sequences of the shape recovery of sample CB10 due to the passage of an electric current (30 V). (From Leng, J.S. et al. *Smart Materials. and Structures*. 2009; 18: 074003. With permission.)

measurement). The circle symbols in Figure 3.50 represent the experimental results for the electrical resistivity of the SMPCs filled with different volume fractions of nanocarbon powders. At least three samples were tested for each composition. The results indicate that the electrical resistivity of the composites with less than 3% volume fraction of nanocarbon powders is extraordinarily high (10^{14}–10^{13} Ω cm). In contrast, a sharp transition of electrical conductivity occurs between 3% and 5%, which is called the percolation threshold range. As the filler content becomes larger than 5%, the resistivity reduces to a low and stable level (10^3–10^1 Ω.cm).

It is noted that for the practical applications of an SMPCs, high and stable electrical conductivity is required. Consequently, an SMP filled with nanocarbon powders at a content of 10 vol% was employed in a shape recovery test. Figure 3.51 shows the time dependence of the shape recovery. The curved strip began to deploy at about 5 s, directly after heating by the electric current. The curved strip recovered spontaneously to about 30° at the end of 90 s (Figure 3.51; t = 90 s). The recovered shape of the SMP strip was about 75%–80% recovered compared with its original shape.

3.6.1.2 Shape Memory Polymer Filled with Hybrid Fibers

Figure 3.52 shows the typical relationship between the resistivity of composites and filler content. The data obtained for the SMP filled with microcarbon powder and the SMP filled with microcarbon powder and SCF were compared to the results found in References [110,111]. The composites containing nanoparticles have a higher conductivity than those blended with microdimension conductive filler. That is why CB was selected. As the amount of filler content increases, the volume resistivity decreases, and a synergy resulting in the CB–SCF system of low resistivity is expected. At the same time, the resistivity of the composite filled with CB is 10^4 times higher than that of the composite with CB–SCF at the same content of 7 wt%. This can be attributed to the fact that the inherent fibrillar form of SCF has a higher tendency to form a three-dimensional network in the

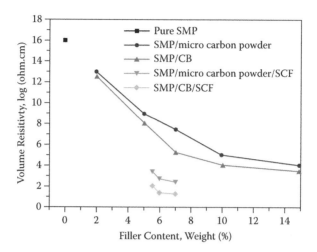

FIGURE 3.52

Resistivity of SMP matrix filled with MCP, CB, MCP/SCF, and CB/SCF systems as a function of filler content. (From Leng, J.S. et al. *Applied Physics Letters.* 2007; 91: 144105. With permission.)

composites, ensuring better electrical response than that of the particulate fillers.

The characteristic volume resistivity curves for the SMP–CB–SCF composite indicate that the SMP filled with 5 wt% CB and 0.5 wt% SCF, whose resistivity is 128.32 ohm · cm, has become a semiconductor instead of an insulator, while the resistivity of the composite containing 5 wt% CB and 2 wt% SCF is 2.32 ohm · cm, which can be defined as a conductor.

The electrically induced shape memory effect is exemplarily demonstrated for the SMP with dimensions of 112 mm × 23.2 mm × 4 mm filled with 5 wt% CB and 2 wt% SCF in Figure 3.53, where a change in shape from the

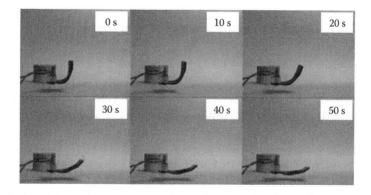

FIGURE 3.53

Electroactive SMP filled with CB–SCF induced by applying 25 V. (From Leng, J.S. et al. *Journal of Applied Physics.* 2008; 104: 104917. With permission.)

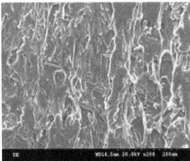

FIGURE 3.54
Morphology of (left) short carbon fibers and (right) sectional observation. (From Leng, J. et al. *Applied Physics Letters*. 2007; 91: 144105. With permission.)

temporary flexural shape to a permanent plane stripe shape occurring within 50 s is shown when a constant voltage of 24 V is applied. The rate of shape recovery was strongly dependent on the magnitude of the applied voltage and the electrical resistivity of the SMPC.

As shown in Figure 3.54, the short fibers disperse randomly. There are many interconnections between the fibers, which form the conductive networks that explain the excellent electrical conductivity of the composites filled with SCF. However, the dispersion of SCF is normally inhomogeneous within the composites; as a result, the electrical conductivity of composites filled with only SCF may not be good.

3.6.2 Solution-Driven Shape Memory Polymers

Shape memory polymers (SMPs) derive their name from their inherent ability to return to their original "memorized" shape after undergoing a shape deformation. SMPs that have been pre-deformed can be deformed to any desired shape above their transition temperature. And they must remain below, or be quenched to below, the transition temperature while maintained in the desired thermoformed shape to "lock" in the deformation. After the deformation is locked in, the polymer network cannot return to a relaxed state due to thermal barriers. The polymer will hold its deformed shape indefinitely until it is heated above its transition temperature, when the stored mechanical strain is released and the polymer returns to its pre-deformed state.

SMP is one of the most attractive shape memory materials that can change shape in response to external stimulus. However, SMPs have a number of advantages over shape memory alloys (SMAs) for particular applications. A wider range of shape recovery temperatures, from –70°C to 120°C, larger recoverable strain (up to 400%), and lower density and cost are the most important ones, among others. Unlike SMAs, in particular NiTi SMA, which can be actuated by passing an electrical current directly, SMPs are polymers that are rather difficult to heat, since they are normally nonconductive in nature. The most recent development in the actuation technique of thermo-responsive SMPs is filling functional fillers into polymer, that is, magnetic particles mixing with electrically conductive filler, by applying an alternating electric, magnetic field. However, the generation of a strong functional field requires an additional bulky system.

Recently, a new approach was identified to trigger the recovery of ther-moresponsive styrene-based SMP [112]. Instead of heating, the polystyrene SMP can recover its original shape by immersion into room temperature dimethylformamide (DMF) that is solvent-responsive for shape recovery. The recovery is due to the strong influence of solvent molecules absorbed upon immersing in DMF, which can lower the glass transition temperature (T_g) of the SMP. Hence, instead of heating the material over its T_g, the shape recovery can be initiated by the drop of T_g of the polymer. This finding also provides a simple and convenient approach to working out an SMP for use in bio-related applications, in particular, for minimally invasive surgery; for example, a laser-activated SMP microactuator, which is used for thrombus removal following ischemic stroke, can be realized by a solvent-driven SMP microactuator, when the SMP has an interaction with blood.

The DMF solvent has a strong influence on the glass transition temperature and dynamic mechanical properties of styrene-based SMPs. And the mechanism behind these phenomena is the involvement of the characteristic peak of C=O groups in interaction with polar solvent. In Figure 3.55, the shape recovery of styrene-based SMP is produced by immersing into DMF solvent for about 180 min; the SMP can recover from the predeformed "n" shape sequentially in room temperature solvent. However, the effect of electrostatic interactions of different solvents on the peak shift remains unknown. The following parts aim to identify the exact mechanism behind the effect of solubility parameter on the recovery behavior of SMPs, in particular, the interaction of different solvents with styrene-based SMP.

Fourier transform infrared (FTIR) was used to study the interaction between the SMP and its interactive solvents and qualitatively identify the effect of the solubility parameter on the shape recovery behavior. The samples used for FTIR test were thin rectangular sheets with a thickness of 1.0 mm. FTIR

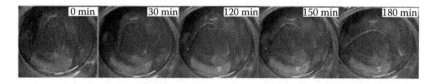

FIGURE 3.55
Shape recovery of a 2.88-mm-diameter SMP wire in DMF. The wire was bent into an n-like shape. (From Lu, H. et al. *Smart Materials and Structures.* 2009; 18: 085003. With permission.)

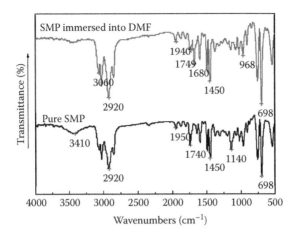

FIGURE 3.56

FTIR spectra of pure styrene-based SMP and sample (90 min of immersion). (From Lu, H. et al. *Smart Materials and Structures*. 2009; 18: 085003. With permission.)

spectra were collected by averaging 128 scans at 4 cm⁻¹ in a reflection mode from an FTIR spectrometer.

The full FTIR spectra of the SMPs (at a temperature of 20°C) after immersion for 90 min in DMF solvent were presented in Figure 3.56. The characteristic peaks that were relevant to this study had been marked and identified. As we can see, the polar bonding was evidenced in the styrene-based SMP where the infrared band of the bonded C=O stretching occurred at 1740 cm⁻¹, while the SMP immersed into DMF solvent stretched at 1749 cm⁻¹.

After immersion in interactive solvents, the infrared band of C=O stretching against immersion time and the solubility parameter are presented in Figure 3.57. Both immersion time and the solubility parameter had a remarkable effect on the C=O infrared band shift, being coincident with the infrared band shift in Figure 3.56. Furthermore, with the increase of solubility parameter, the infrared band intensity of bonded C=O stretching became more obvious as compared with that of pure polymer, which indicated that a stronger solubility parameter resulted in more C=O groups involved in polarization interaction.

Based on of experimental results shown in Figure 3.57, the mechanism behind the polar interaction is that the electrons of polar solvent molecules were absorbed by the bonded C=O of polymer, leading to the infrared band shifting to a slightly higher frequency. This reveals that the C=O bonds of solvent molecules have direct effect on the C=O bonds in the styrene-based SMP, which can be explained by the model in Figure 3.58. Where R, R', and R'' are present, alkyl, and OR' also, can be displaced by formamide. However, the absorbed solvent molecules can form two types of interactions (electronic absorption effect and conjugate effect) with C=O groups. Due to the

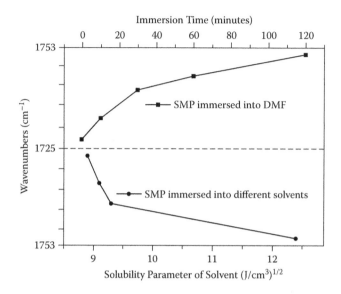

FIGURE 3.57
Infrared band of bonded C=O stretching versus immersion time and solubility parameter. (From Lu, H. et al. *Smart Materials and Structures*. 2009; 18: 085003. With permission.)

former interaction, the infrared band shifted up to a higher frequency, while the latter interaction lowers the frequency. The two interactions may work together and counteract. According to previous research, solvent in the latter interaction is more firmly bonded than that in the former interaction, largely determined by the spatial block interaction of C=O in solvent molecules. Consequently, the infrared band of C=O increases.

Utilizing the data obtained from DSC tests, we can find the relationship between decrease of T_g and weight loss of solvent. As shown in Figure 3.59, T_g decreased obviously as solvent molecule absorption increased. That may be attributed to the polar interaction and plasticizing effect between macromolecules and solvent molecules. Note that change of T_g strongly depends on the diffusion of solvent molecules. In comparison with three curves, the effect of solubility parameter on T_g was obviously quantified.

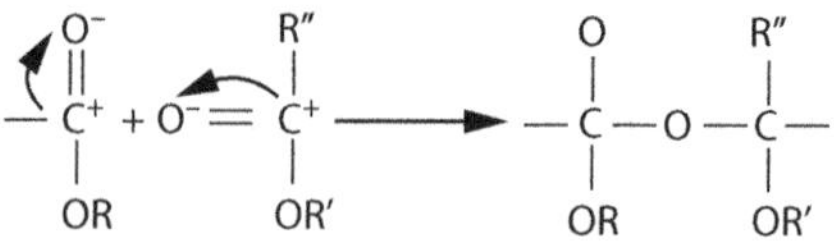

FIGURE 3.58
Effect of electron absorption on the C=O bands in polymer. (From Lu, H. et al. *Smart Materials and Structures*. 2009; 18: 085003. With permission.)

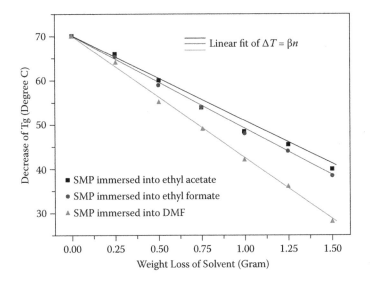

FIGURE 3.59
The effect of solvent molecules absorption on the glass transition temperature. (From Lu, H. et al. *Smart Materials and Structures.* 2009; 18: 085003. With permission.)

The experimental results revealed that the higher the solubility parameter dispersion between polymer and solvent, the more evident the decrease of T_g when polymer was immersed in an interactive solvent. That can be demonstrated by the solution theory of polymer physics.

Since T_g of the styrene-based SMP can be reduced by polar solvent, we can therefore utilize this to design solvent-driven functionally controllable SMP. The basic idea is to lower T_g of SMP by immersing it in interactive solvents. In order to verify the suggested new features, which can dramatically broaden the applications of SMPs, the effect of the solubility parameter on shape recovery behavior was experimentally validated.

The rectangular SMP samples were bent into an n-like shape at 80°C and retained this shape during cooling back to 35°C. No apparent shape recovery was found after the deformed wires were kept for one day. However, after immersing in 35°C solvents for 20 to 40 min, they started to recover gradually. With the increase of solubility parameter, the shape recovery time decreased remarkably. The dependence of the recovery behavior on the solubility parameter is due to the stronger plasticizing and polarizing effect between solvent molecules and polymeric macromolecules. After 392, 376, and 90 min immersion time, it was found that the polymer regained its original shape in the corresponding solvent. As shown in Figure 3.60, it was obvious that recovery of SMP took the shortest time in DMF solvent with the highest solubility parameter.

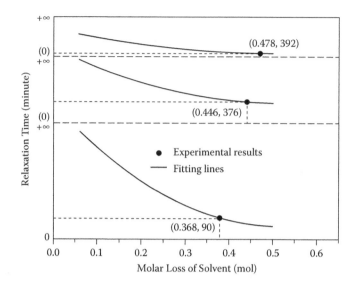

FIGURE 3.60

The relationship between relaxation time (recovery time) and molar loss of solvent. (From Lu et al. *Smart Materials and Structures.* 2009; 18: 085003. With permission.)

3.6.3 Light-Induced Shape Memory Polymers

Based on the light-thermal transition theory, the medium infrared laser light is employed to activate the SMP [113]. In order to improve the transition efficiency of light, the surface of the fiber is treated. Then the treated fiber is placed in the special location. In this case, the fiber core is contacted directly with the SMP. After that, the light emitted by a laser source is coupled into the optical fiber, and the induced heat will actuate the shape recovery of SMP.

The synthetic polymer is a thermal responsive thermoset SMP based on styrene. The absorbed laser energy can be transformed into heat and actuated by Joule heating. After the temperature approaches the glass transition temperature (T_g), the deformed SMP can recover its original shape. In this study, the T_g of SMP was analyzed by a dynamic mechanical analyzer (DMA). As shown as Figure 3.61, the temperature due to peak location is T_g. During the experiment, when the peak of the tangent delta appears, the corresponding temperature is the glass transformation temperature of SMP ($T_g = 53.7°C$).

The shape recovery process of SMP relies on thermal energy in the material, which is transformed from laser energy. An infrared light showed good performance in actuating the SMP in this study. The mode of vibration of the molecules in the material, which can change its dipole distance, is different and, therefore, various types of materials absorb infrared light at different wavelengths. Thus, an infrared analysis is proposed to select infrared light of a proper wavelength for the SMP. Figure 3.62 shows the infrared absorption

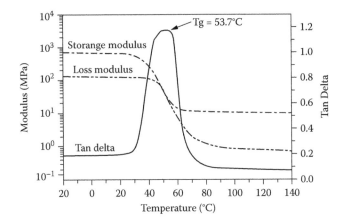

FIGURE 3.61
Storage modulus, loss modulus, and tangent (tan) delta of the shape memory polymer. (From Leng, J. et al. *Applied Physics Letters.* 2010; 96,11: 111905.)

spectrum of the SMP. A strong absorption peak of the polymer appears at 3000 cm^{-1} (λ = 3270 nm~3410 nm), which corresponds to the stretch vibration of a C–H bond in the benzene ring. So the wavelength of the infrared laser for the actuator is selected between 3 and 4 μm.

According to the former investigation in infrared spectroscopy, an infrared light with a working wave band of 2–4 μm is coupled into the fiber in SMP to study the realistic actuating effect of the infrared laser. The actuating experiment is shown in Figure 3.63, where the white belt is SMP and the thin stick is the infrared optical fiber. Clearly, the shape of the SMP is actuated gradually from a curved shape (Figure 3.63a) to relatively flat shape (Figure 3.63g).

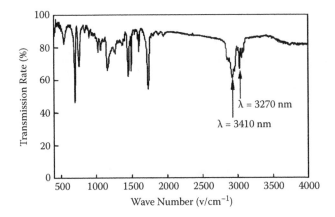

FIGURE 3.62
The IR spectrum of the thermoset shape memory polymer. (From Leng, J. et al. *Applied Physics Letters.* 2010; 96,11: 111905. With permission.)

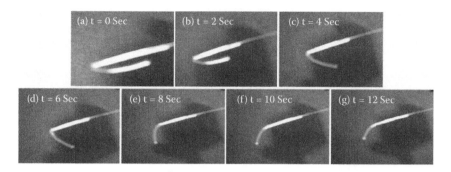

FIGURE 3.63
Shape recovery snap shots of SMP induced by infrared light. (a): Initial predeformed configuration; (b–g): deployment sequence of the actuator at $t = 2, 4, 6, 8, 10, 12$ s, respectively. (From Leng, J. et al. *Applied Physics Letters*. 2010; 96,11: 111905. With permission.)

Hence, it demonstrates that the infrared light in the selected wavelength can actuate SMP to recover its original shape.

An infrared camera was used to record the temperature distribution during the shape recovery process. The temperature of SMP around the treated optical fiber rose to 21.4°C with an increased temperature of 2.4°C compared with the ambient temperature; in contrast, the temperature of SMP at the end of the optical fiber increased to 48°C with an increased temperature of 29°C, which is above T_g, and the actuator was activated. This indicates that during the laser transmission, more light is refracted from the end of the fiber into the polymer. The SMP deploys promptly in about 12 s (see Figure 3.64).

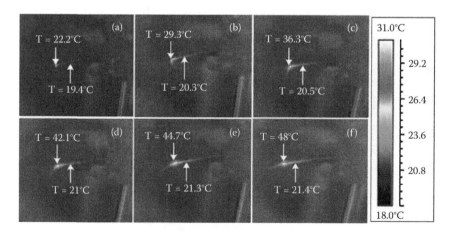

FIGURE 3.64
Temperature distributions snap shots of the SMP during the midinfrared laser-driven process. (a) $t = 2$ s, (b) $t = 4$ s, (c) $t = 6$ s, (d) $t = 8$ s, (e) $t = 10$ s, (f) $t = 12$ s. (From Leng, J. et al. *Applied Physics Letters*. 2010; 96,11: 111905. With permission.)

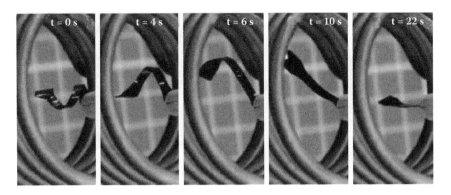

FIGURE 3.65
Series of photographs showing the macroscopic shape memory effect of SMP composite with 10 wt% content of nanoparticles. (Fe_2O_3). (From Mohr, R. et al. *Proceedings of the National Academy of Sciences USA* 2006; 103, 3540–3545.)

3.6.4 Magnetism-Induced Shape Memory Polymers

The magnetically induced shape recovery of SMPs could be realized by incorporating magnetic nanoparticles (e.g., Fe_2O_3 and Fe_3O_4) in SMPs [114–116], that is, transforming electromagnetic energy from an external high-frequency field to heat [117]. The magnetically induced SMPs can be remotely controlled to induce heating energy locally and selectively. Mohr et al. [118] investigated magnetically induced thermoplastic SMPCs filled with Fe_2O_3 nanoparticles. Iron oxide particles with average diameters at the nanoscale (20–30 nm) support a preferably homogeneous distribution within the SMP matrix. The shape recovery of SMPCs could be induced by inductive heating in an alternating magnetic field ($f = 258$ kHz; $H = 30$ kA.m^{-1}). As shown in Figure 3.65, the magnetically induced shape memory effect is well demonstrated.

In addition, Schmidt et al. [119] incorporated surface-modified superparamagnetic nanoparticles (Fe_3O_4, d ≈ 11 nm) into an SMP matrix. The thermosetting SMPC of oligo(e-caprolactone)dimethacrylate/butyl acrylate contains between 2 and 12 wt% magnetite nanoparticles serving as nano antennas for magnetic heating. The specific power loss of the particles is determined to be 30 W·g^{-1} at 300 kHz and 5.0 W. As shown in Figure 3.66, a photo series presents the electromagnetically induced shape memory effect of the sample.

3.6.5 Sensing Functions of Shape Memory Polymers

The actuation approaches play a critical role in the development of multifunctional materials that not only exhibit the shape memory effect but also perform particular functions. The triggering and subsequent actions of SMPs are sensitive to a particular stimulus, and SMPs may be used as a particular

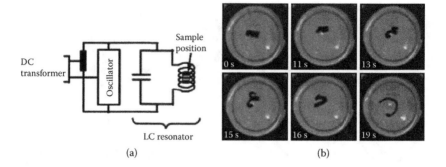

(a)   (b)

FIGURE 3.66
(a) Schematic diagram of the LC resonant circuit-based HF generator used for induction heating experiments and sample positioning; (b) photo series demonstrating the shape memory transition induced by exposure to an HF electromagnetic field. (From Schmidt, A.M. *Macromolecular Rapid Communications.* 2006; 27:1168–72. With permission.)

sensor or actuator [120–123]. For instance, the electrically conductive SMPC filled with carbon particles could also be used as a strain sensor, which is based on the relation between electrical conductivity and strain. This SMPC could monitor its own real-time deformation by testing the evolution of conductivity when it recovered its original shape. In addition, if light-induced SMPs are exposed to light of a particular wavelength, an instant reaction or deformation could result that would enable their use as both sensors and actuators. The applications of these multifunctional SMPs and SMPCs may cover a broad range of fields from biomedicine to microsensors and actuators, to MEMS/NEMS.

Electroactive SMPs have been proposed for use as strain sensors. As their fundamental research, Yang and Huang [124] investigated the effect of tensile strain on electrical resistivity. The polymer matrix used in this study is thermoplastic polyurethane [125,126]. The electrical resistivity of the SMPCs reinforced with different volume fractions of carbon powders is reported in Figure 3.67. At a volume fraction of less than 4% carbon powders, the electrical resistivity of the composite is very high. A sharp transition occurs between 4% and 7%. A further increase in the volume fraction of carbon powder reduces the gap between carbon aggregates slightly and, therefore, only a few more conductive channels are formed. Hence, the reduction in resistivity is much more gradual, and the resistivity remains almost constant.

In practical applications, conductive SMPs may be used as strain sensors based on the relationship between electrical conductivity and strain. Thus, the effect of strain on the electrical resistivity of conductive SMPs should be evaluated. As shown in Figure 3.68, electrical resistivity increases with an increase in strain. The increase of electrical resistivity is remarkable at the lower strain range below 10%. Based on the effect of strain on conductivity, the SMP can be used as a multifunctional material to fabricate strain sensors.

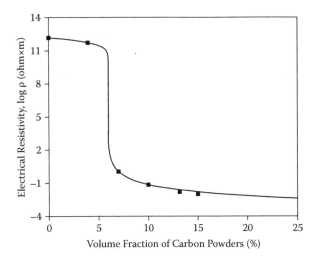

FIGURE 3.67
Electrical resistivity versus the fraction of carbon powders. (From Yang, B. Influence of moisture in polyurethane shape memory polymers and their electrical conductive composites. A dissertation for the degree of doctor of philosophy, School of Mechanical and Aerospace Engineering Nanyang Technological University, Singapore, 2005.)

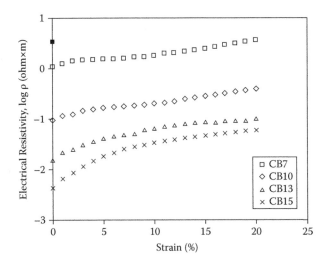

FIGURE 3.68
Effect of tensile strain on electrical resistivity. (From Yang, B. Influence of moisture in polyurethane shape memory polymers and their electrical conductive composites. A dissertation for the degree of doctor of philosophy, School of Mechanical and Aerospace Engineering Nanyang Technological University, Singapore, 2005.)

3.6.6 Shape Memory Polymer Nanocomposites Incorporated with Carbon Nanofiber Paper

To carry out shape recovery actuation of SMP by electrically resistive Joule heating, conductive carbon nanofiber (CNF) paper was introduced [127,128]. The CNF paper is manufactured with a traditional PVD process, and morphologies and structures of CNF paper at micro size are observed by scanning electron microscopy (SEM). It is found that CNF paper is a pore mass, the size of which is determined by the weight concentration of CNF.

The determination of the dynamic mechanical properties of SMP composites was performed on a dynamic mechanical analysis (DMA) instrument. The TA Instruments Q800 DMA is a thermal analytical instrument used to test the mechanical properties of many different materials. In the measurement process, the test specimen is fixed on one of several clamps, all of which were designed using finite element analysis to minimize mass and compliance. Basically, a deformation is imposed on the specimen in order to evaluate intrinsic as well as extrinsic mechanical properties of the material. In this manner, the effect of CNF paper weight concentration on the thermomechanical properties of composite was studied and analyzed at various temperatures. All experiments were performed in the dual cantilever mode at a constant heating rate of $10.0°C·min^{-1}$. The oscillation frequency was 1.0 Hz. Samples were investigated in the temperature interval from –20°C to 150°C. In the DMA measurements of composites with various weight fractions of CNF paper, the storage modulus, loss modulus, and tangent delta were recorded with respect to temperature. The storage modulus is the modulus of the elastic portion of the material while the loss modulus is the modulus of the viscous portion. The tangent delta, which is defined as the ratio of the loss modulus over the storage modulus, indicates the damping capability of a material.

As shown in Figure 3.69, the dynamic thermomechanical properties of pure SMP and an SMP composite with CNF paper concentrations of 1.2 and 2.4 g were presented. These three samples were selected to characterize the effect of different CNF weight concentrations on the evaluation change of dynamic mechanical properties of SMP samples. At a temperature of –20°C, the storage moduli of samples A, C, and E were 2756, 2121, and 1763 MPa, respectively. This result revealed that the storage modulus of the composite gradually decreases with an increase in weight concentration of CNF paper. This phenomenon can be explained by the complex interaction between polymer and CNF paper. As the SEM image shows, CNF paper is a pore mass of microscopic size, resulting in resin penetrating into or even through the paper bulk in the polymer composite fabrication. It was earlier found that the polymer content in the paper is determined by the pore size and pressure. When a thin paper (with a relative low CNF mass) is incorporated with polymer, it is easier for the polymer to penetrate through the paper, resulting in the CNF arrays and polymer being separated from each other. Thus, mechanical properties of this type of composite sometimes suffer because

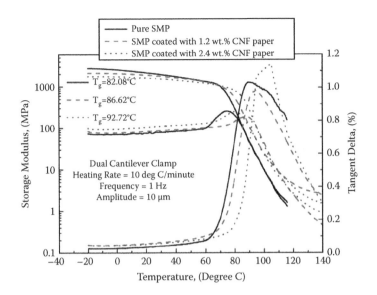

FIGURE 3.69
Storage modulus and tangent delta curves of composites as a function of temperature. (From Lu, H. et al. *International Journal of Smart and Nano Materials*. 2010, ; 1: 2–12.)

there is no effective connection in the composite. On the other hand, the CNF paper is formed by individual CNFs that are packed by the van der Waals force. This molecular force, which is always weaker than chemical bonding, causes paper to have a relatively weaker mechanical property at macroscopic size. Therefore, as the weight concentration of CNF paper increases in the composite, the mechanical properties of the composite worsen in comparison with pure polymer. At the same time, the damping property of SMP integrated with CNF paper was improved, as is evident from the tangent delta curves. The CNF paper is packed with individual CNFs, arranged as a nanomaterial, with a huge surface area. When an external force is applied, the CNF could help disperse the force and enhance damping performance of the composite.

Sometimes, DMA curves also can be used to determine the glass transition temperature (T_g) of polymeric materials. In this study, T_g is defined as the point of intersection between the storage modulus and tangent delta curves. Thus, 82.08°C, 86.62°C, and 92.72°C were marked for the corresponding pure SMP specimen, SMP composite specimen with 1.2 g CNF paper, and SMP composite specimen with 2.4 g CNF paper, respectively. This experimental result reveals that T_g reaches its critical values at higher temperature as the weight concentration of CNF paper increases. Therefore, the CNF paper plays a more positive effect in improving the damping properties of the polymer composite.

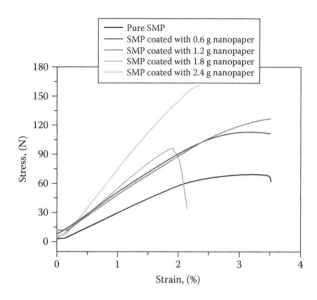

FIGURE 3.70

The flexural strength of composites versus the weight fraction of CNF nanopaper. (From Lu, H. et al. *International Journal of Smart and Nano Materials*. 2010; 1: 2–12.)

The flexural strength of the SMP composites was measured using the three-point bending test on a Zwick/Z010 static materials testing machine. The static bending test was performed at a loading speed of 2 mm·min^{-1}. The dependence of the nanopaper weight fraction on the flexural strength of the SMP composite was investigated at the testing temperature of 20°C.

Rectangular samples with dimensions of 49.5 mm × 12.1 mm × 2.7 mm were taken from the fabricated SMP composites. When the nanopaper was coated with SMP, the flexural modulus of SMP composite was higher than that of the pure SMP. As shown in Figure 3.70, the flexural moduli of the SMP specimens with various weight concentrations of nanopaper were recorded as a function of flexural strain. As the weight fraction of nanopaper increased from 0.6 to 2.4 g, the flexural moduli of SMP specimens were 70.36, 113.92, 126.08, 96.59, and 161.32 MPa, respectively. This phenomenon may be mainly attributed to the excellent mechanical properties of CNFs. Therefore, the mechanical properties of SMP composites would be dependent on the weight fraction of CNF nanopaper. However, as higher weight concentrations of CNF nanopaper are introduced to the SMP, the composite becomes more brittle and has a lower ultimate bending strain. In addition, because dispersion of CNFs in nanopaper is not uniform, the flexural strength of SMP composites will not exhibit a regular evolutionary change. In the developed SMP composites, the specimens have a relative higher flexural strength in comparison with pure SMP.

The electrical resistivity of the pure CNF papers and SMP composites was measured by the four-point probe apparatus (SIGNATONE QUADPRO

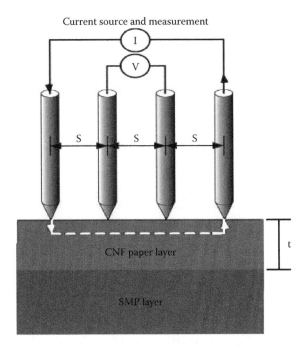

Current source and measurement

FIGURE 3.71
Schematic presentation of resistivity measurement. (From Lu, H. et al. *International Journal of Smart and Nano Materials.* 2010; 1: 2–12.)

system), which consisted of four probes in a straight line, and the constant interprobe spacing (S) was 1.56 mm. The schematic presentation of rectangle sample preparation for resistivity measurement is shown in Figure 3.71.

When a Kelvin connection is used, electric current is supplied with a pair of force connections. These generate a voltage drop across the impedance to be measured according to Ohm's law, $V = IR$. This current also generates a voltage drop across the force wires themselves. To avoid including the drop in the measurement, a pair of sense connections are made immediately adjacent to the target impedance. It is conventional to arrange the sense wires as the inside pair, and the force wires as the outside pair. The force and sense connections can theoretically be exchanged without affecting the accuracy of the technique, but this leads to uncertainty regarding exactly where within the force connection the resistance measurement begins, as the force connections are large enough to carry the necessary current, while the sense connections can be very small. The key difference between the four probe measurement method and the traditional two probe method is that the separation of current and voltage electrodes in four-point probe measurement allows the ohmmeter/impedance analyzer to eliminate the impedance contribution of the wiring and contact resistances, given that the voltage electrodes have high enough input impedance.

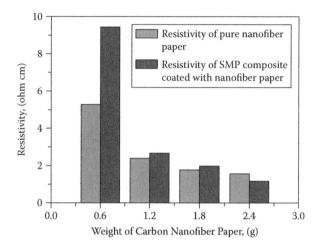

FIGURE 3.72

Comparison of electrical resistivity for pure CNF paper and its SMP composite with respect to CNF paper weight concentration. (From Lu, H. et al. *International Journal of Smart and Nano Materials*. 2010; 1: 2–12.)

The characteristic electrical resistivity as a function of weight of pure CNF papers and their composites was measured at room temperature. Based on the previous result, the electrical resistivity of sample A is considered as 10^{16} ohm · cm. The resistivity of pure CNF paper and its SMP composite is presented in Figure 3.72 with respect to CNF paper weight concentration, for comparison. The electrical conductivity of SMP coated with 1.8 g CNF paper was measured to be 1.77 ± 0.08 ohm · cm. This is lower than most previously reported conductive SMPs with comparable filler content. This low electrical resistivity allows fast activation of shape recovery by application of a constant voltage. That is because more CNFs were involved in the conductive networks, leading to the probable reduction of sites occupied by pores. This evolution also plays the same role in the electrical properties of SMP composites. However, in comparison with the two curves, the resistivity of composites with 0.6, 1.2, and 1.8 g paper is higher than that with the corresponding pure papers, while the resistivity of composites with 2.4 g paper is slightly lower than that with the corresponding papers. Perhaps the interaction between polymer and paper should be considered to account for this phenomenon. As previously mentioned, the polymer would fill into paper bulk. When the resin penetrates through the paper, the pores in the conductive paper structure will be filled with the nonconducting polymer, resulting in increase of the electrical resistivity of composites. However, when only a part of the paper is filled with polymer, the polymer provides strong bonding and acts as a bridge between conductive arrays. Meanwhile, the electrical resistivity of other part of the paper, whose conductive arrays are packed more closely, decreased. That is why the electrical resistivity of

SMP composite with high weight concentration CNF paper is lower than that with pure CNF paper.

The use of electrical resistive Joule heating to trigger the shape memory effect of SMPs is desirable for engineering applications, where it would be impossible to use heat, and is another active area of research. Some current efforts use conducting SMP composites with carbon nanotubes, short carbon fibers (SCFs), and metallic Ni powder. The shape memory effect in these types of SMPs has been shown to be dependent on the filler content, and they exhibit good energy conversion efficiency and improved mechanical properties.

Another technique being investigated involves the use of surface-modified super-paramagnetic nanoparticles. When these electromagnetic particles are introduced into the SMP matrix, remote actuation of shape transitions by electromagnetic field is possible. An example of this involves the use of oligo (e-capolactone)dimethacrylate/butyl acrylate composite with between 2% and 12% magnetite nanoparticles. Nickel and hybrid fibers have also been used with some degree of success.

Previous results on the shape memory of SMP driven by electrical resistive heating had disadvantages such as too high filler content, too high applied voltage, or drastic loss in shape memory effect. Our study employed 1.2 g conductive CNF paper, at a very low applied voltage of 12 V, and an excellent recoverability approximating to 100% of SMP composite; electrically induced shape memory effect is unequivocally demonstrated in Figure 3.73. The straight SMP (permanent shape) sheet was bent into an n-like shape at 80°C and retained this shape during cooling back to room temperature. No apparent recovery was found after the deformed sample was stored in the absence of external forces, as can be seen in Figure 3.73 (0 s). However, when a voltage of 12 V was applied for about 20 s, the composite specimen (sample C) starting conversion of the flexural shape was observed. After another 80 s, the specimen shows a

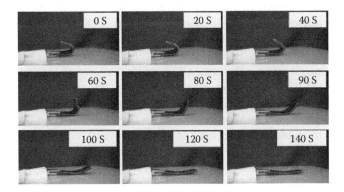

FIGURE 3.73
Demonstration of shape recovery driven by electrical resistive heating is recorded by digital camera. (From Lu, H. et al. *International Journal of Smart and Nano Materials*. 2010; 1: 2–12.)

relatively fast responsive behavior to electrical resistive heating. A change in shape from temporary to permanent was completed in 140 s. The final shape was close to the original straight shape, with some remaining flexion due to friction among the polymer chains. From this experimental outcome, it is found that the SMP composite with conductive CNF paper not only has an electroactive response, but also has a relative better shape memory effect than the polymer matrix blended with conductive filler or fillers.

3.7 Typical Applications of Shape Memory Polymers and Their Composites

As a novel kind of smart material, SMPs currently cover a broad range of application areas ranging from outer space to automobiles. Recently, they are being developed and qualified especially for deployable components and structures in aerospace. The applications include hinges, trusses, booms, antennas, optical reflectors, and morphing skins. In addition, SMPs also present additional potential in the areas of biomedicine, smart textiles, self-healing composite systems, and automobile actuators. Additionally, there are many patents connected with SMPs applications, such as grippers, intravascular delivery system, hood/seat assembly, and tunable automotive brackets.

3.7.1 Deployable Structures

During the launching of a spacecraft, the area inside the spacecraft is quite limited. Hence, the spacecraft needs lightweight, reliable, and cost-effective mechanisms for the deployment of radiators, solar arrays, or other devices. Leng et al. [59] investigated a carbon-fiber-reinforced SMP composite (SMPC) that can be used in actively deformable structures. In these structures [129], flexural deformation is the main mode of the deformation in structures with thin shells where the bending angle is almost larger than 90° but the strain is often smaller than 5%. Figure 3.74 shows the deployment process for the SMPC hinge. The entire deployment process takes about 100 s. In addition, deployment of a solar array prototype which is actuated by an SMPC hinge is demonstrated (see Figure 3.75). A voltage of 20 V is applied on the resistor heater embedded in each circular laminate. The temperature of the SMPC hinge remains at about 80°C after heating for 30 s. The deployment process takes about 80 s.

Furthermore, Composite Technology Development (CTD), Inc., has developed carbon-fiber-reinforced epoxy SMP composites (elastic memory composite, EMC). The EMC materials show very high recoverable strains. CTD has developed a deployable hinge fabricated with this EMC material.

CTD has developed an EMC boom used on a microsatellite. This extendable boom is lightweight and can support a variety of tip payloads (see Figure 3.76). It has been identified for the purpose of supporting a micro-propulsion attitude

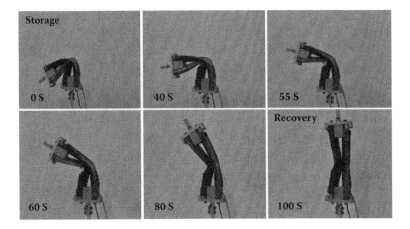

FIGURE 3.74
Shape recovery process of SMPC hinge. (From Lan, X. et al. *Smart Materials and Structures*. 2009; 18: 024002. With permission.)

control system. The EMC is the central element of the boom. To stow the boom, the longerons are in a z-like shape and, thus, the EMCs are flattened and bent in a predeformed shape [130].

Another deployable structures experiment was proposed to deploy large and lightweight solar arrays. In order to minimize the impacts to the space-craft system, the structural mass and complexity should be minimized. The structure fabricated by EMC is proposed for the deployed solar array and the actuation to deploy the solar array from its packaged configuration. As shown in Figure 3.77, the longeron booms are fabricated by the thin-film foldable tubular EMC. [130,131]

3.7.2 Morphing Structures

Flight vehicles are often proposed to be multifunctional so that they can perform more missions during a single flight, such as efficient cruising and

FIGURE 3.75
Shape recovery process of a solar array prototype actuated by an SMPC hinge. (From Lan, X. et al. *Smart Materials and Structures*. 2009; 18: 024002. With permission.)

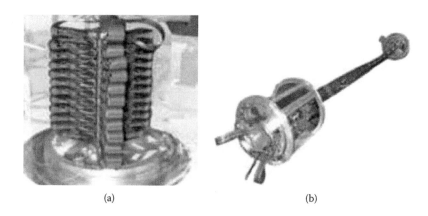

(a)   (b)

FIGURE 3.76
The proposed three-longeron EMC tubular boom in both a (a) packaged and (b) deployed configuration without the outer shroud. (From Steven, C. et al. Elastic memory composites (EMC) for deployable industrial and commercial applications, www.ctd-materials.com.)

operating in high maneuverability mode [132,133]. When the airplane moves toward other portions of the flight envelope, its performance and efficiency may deteriorate rapidly. In order to solve this problem, researchers have proposed changing the shape of the aircraft during flight [134]. By applying this kind of technology, both the efficiency and flight envelope can be improved. This is because different shapes correspond to different trade-offs between beneficial characteristics, such as speed, low energy consumption, and maneuverability. For example, the Defense Advanced Research Projects Agency (DARPA) is developing morphing technology to demonstrate such radical shape changes.

(a)   (b)

FIGURE 3.77
(a) EMC boom and (b) partially deployed model of solar arrays supported by the EMC boom. (From Taylor, R.M. et al. *48th AIAA/ASME/ASCE/AHS/ASC Structures, Structural Dynamics, and Materials Conference*, April 23–26, 2007, Honolulu, Hawaii, AIAA 2007–2269.)

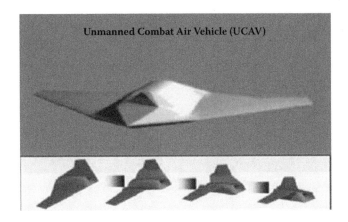

FIGURE 3.78
z-Shaped morphing wing produced by Lockheed Martin. (From Love, M.H. et al. *48thAIAA/ ASME/ASCE/AHS/ASC Structures, Structural Dynamics, and Materials Conference* 2007; *AIAA* 2007–1729: 1–12.)

As illustrated in Figure 3.78, Lockheed Martin is addressing technologies to achieve a z-shaped morphing change under the DARPA's program fund [135].

For development of morphing aircraft, finding a proper skin under certain criteria is important. Generally, a skin material for a morphing wing is necessary. In this case, SMPs show more advantages for this application. They become flexible when heated to a certain degree, and then return to a solid state when the stimulus is terminated. Since SMPs can change their elastic moduli, they could potentially be used in the aforementioned designs.

The morphing concept applied to a variable camber wing was developed by Leng et al. [136,137]. It comprises a flexible SMP skin, a metal sheet, and a honeycomb structure (Figure 3.79). A metal sheet is used to replace the

(a) Original configuration (0 deg)   (b) Morphing configuration (15 deg)

FIGURE 3.79
Photograph of the original and morphing configurations of the variable camber wing. (From Yin, W.L. et al. *SPIE International Conference on Smart Structures/NDE*, San Diego, March 8–12, 2009, USA, *Proceedings of SPIE* 7292 2009; 72921H: 1–10.)

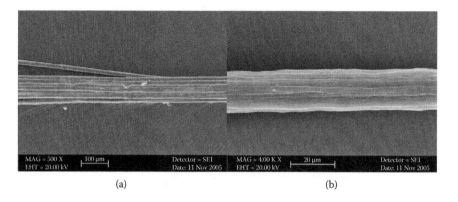

(a)   (b)

FIGURE 3.80
Scanning electron microscopy of shape memory fibers. (a) View of multifilament (500×), (b) view of an isolated single filament (4000×). (From Ji, F. et al. *Smart Materials and Structures*. 2006; 15: 1547–54. With permission.)

traditional hinges to keep the surface smooth during the camber changing. Honeycomb, which is high-strain capable in one direction without dimensional change in the perpendicular in-plane axis, provides distributed support to the flexible skin. The flexible SMP skin is covered to create the smooth aerodynamic surface. The baseline airfoil is assumed to have an NACA 0020 profile and a chord of 150 mm.

3.7.3 Shape Memory Polymer Textiles and Applications

SMP fibers can be implemented to develop smart textiles that respond to thermal stimulus [138]. They can be used for textiles: clothing in the form of shape memory fibers, shape memory yarns, and shape memory fabrics [139]. Hu et al. [72,74–76,140,141] at Hong Kong Polytechnic University have comprehensively investigated shape memory finishing chemicals and related technologies. Figure 3.80 shows morphology of a shape memory fiber. As shown in the figure, grooves can be observed, which are due to the traces left after the solution is extruded from the pinholes of the spinneret. The diameter of the single filament is measured as about 16 μm. Moreover, shape memory finishing fabrics can be produced with a coating of shape memory emulsion or combining shape memory films. When washed in hot water or dried at temperatures higher than T_g, their wrinkles will disappear, and they can recover the original flat shape. Shape memory finishing fabric can be produced by coating with shape memory emulsion or combining shape memory films. As shown in Figure 3.81, the shape-memory-finished fabric feels soft, the creases and wrinkles will disappear, and it will recover its original shape when washed in hot water or dried higher than T_g [141].

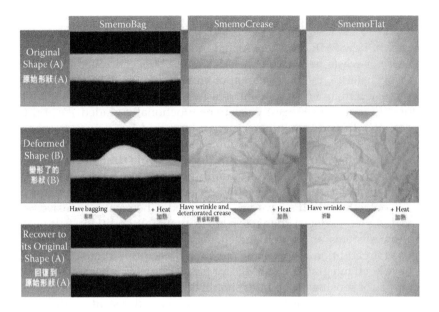

FIGURE 3.81
Shape memory recovery of cotton fabric. (From Leng, J. et al. In: Hu, J.L. *Shape Memory Polymer Textile*, CRC Press/Taylor & Francis; 2009 [In press].)

In addition, Reference [142] lists potential applications for SMPs:

1. Biomedicine and bioinspiration: SMP drug delivery system; degradable SMP materials; blood clot dissolution based on SMP actuator; self-healing composite system.

2. Deployable structures: SMP foam deployable structures, SMP reversible attachments used on automobile, access denial barriers/walls; automatic disassembly of electronic products; reusable SMP mandrel.

3. SMP packaging: packaging of thermal-sensitive products; sensors/drug/food air delivery systems.

4. Recreation/sport products: tents and camping equipment; life jacket, floating wheels; water and snow skis; surf and snow boards.

3.8 Future of Shape Memory Polymers and Electroactive Polymers

In future research, the potential directions and applications of SMPs and EAPs proposed to be developed are listed as follows.

3.8.1 Novel and Multifunctional Shape Memory Polymers

Prospective new SMP uses may include chemical energy (chemical-responsive SMP) or waste heat, two-way SMPs, and SMPs controlled in multistep ways. For instance, the deformation of an SMP may be better controlled in a step-by-step manner than by current methods, which either employ two T_g's for different transitions (two-step) or functionally graded T_g's at different locations. In addition, high actuation force and fast actuation speed in the new SMPs are necessary. In thermoresponsive SMPs, a narrow transition temperature range, down to only a few degrees, could dramatically increase the actuation speed. Alternative stimuli, for example, light, sound, chemicals, or molecular stimuli, for actuation may be developed. Wireless and remote-controllable SMPs have been proposed. Chemical-responsive SMPs can be designed to detect changes in environmental conditions. Self-healing composite systems and automatic chemical-sensing SMPs have also been proposed. In addition to the shape memory effect, built-in temperature sensors have been integrated into SMPs for temperature monitoring. SMPs with other types of sensing capabilities should be very useful.

Other prospective applications are polymer vascular stents with SMPs as the drug delivery system, smart surgical sutures, SMPs with excellent biocompatibility and/or biodegradability with adjustable degradation rates for intelligent medical devices, self-disassembling mobile phones, and shape memory toys.

3.8.2 Perspectives of Electroactive Polymers

Electroactive polymers show many interests in smart materials. Electroactive polymer actuators have great potential in the development of mechanisms (lenses with controlled configuration, mechanical lock, noise reduction), robotics (biologically inspired robots), toys and animatronics, human–machine interface (haptic interface, tactile interface, orientation indicator, artificial nose, Braille display), planetary applications (sensor cleaner, shape control gossamer structures), and so on. Though commercial applications are being pursued, there are still some difficulties to be overcome. Practical application of the dielectric elastomer actuator is a multidisciplinary task involving chemistry, materials science, electromechanics, computers, and electronics.

References

1. Pelrine, R., Kornbluh, R., Pei, Q. High-speed electrically actuated elastomers with strain greater than 100%. *Science*. 2000; 287(28): 836–39.
2. Bar-Cohen, Y. Electroactive polymers as artificial muscles-capabilities: Potentials and challenges. *Proceedings of SPIE*. 2004; 95: 148.

3. Madden, J.D., Vandesteeg, N., Madden, P.G., Takshi, A. Artificial muscle technology: physical principles and naval prospects. *IEEE Journal of Oceanic Engineering.* 2004; 29: 706.
4. Ailish O'Halloran, Fergal O'Malley, Peter McHugh. A review on dielectric elastomer actuators, technology, applications, and challenges. *Journal of Applied Physics.* 2008; 104: 071101.
5. Pelrine, R.E., Kornbluh, R.D., Joseph, J.P. Electrostriction of polymer dielectrics with compliant electrodes as a means of actuation. *Sensors and Actuators A-Physical.* 1998; 64 (1): 77–85.
6. Liwu Liu, Jiuming Fan, Zhen Zhang, Liang Shi, Yanju Liu, Jinsong Leng. Analysis of the novel strain responsive actuators of silicone dielectric elastomer. *Advanced Materials Research.* 2008; 47(50): 298–301.
7. Mirfakhrai, T., Madden, J., Baughman, R. Polymer artificial muscles. *Materials Today.* 2007; 10(4):30–8.
8. Zhang, W.Q., Lowe, C., Wissler, M., Jahne, B., Kovacs, G. Dielectric elastomers in actuator technology. *Advanced Engineering Materials.* 2005; 7: 361.
9. Brochu, P., Pei, B. Advances in dielectric elastomers for actuators and artificial muscles. *Macromolecular Rapid Communications.* 2010; 31:10.
10. Kofod, G. Dielectric elastomer actuators. Ph.D. thesis, Riso-R-1286(EN) (2001).
11. Carpi, F. et al. Dielectric elestomer cyclindrical actuators: electromechanical modeling and experiment evaluation. *Material Science and Engineering C.* 2004; 24: 555–26.
12. Carpi, F. et al. Electromechanical characterization of dielectric elastomer planar actuators: comparative evaluation of different electrode materials and different counter loads. *Sensors and Actuators A-Physical.* 2003; 107: 85–95.
13. Goulbourne, N. et al. A nonlineal model for dielectric elastomer membranes. *Journal of Applied Mechanics.* 2005; 72: 899–906.
14. Lochmatter, P. et al. Electromechanical model for static and dynamic activation of elementary dielectric elastomer actuators. *Proceedings of SPIE.* 2006; 6168: 61680F.
15. Sommer-Larsen, P. et al. Performance of dielectric elastomer actuators and materials. *Proceedings of SPIE.* 2002; 4695: 158–66.
16. Yang, E et al. Viscoelastic model of dielectric elastomer membranes. *Proceedings of SPIE.* 2005; 5759: 82–92.
17. Plante, J.S. Dielectric elastomer actuators for binary robotics and mechatronics. Ph.D. thesis, Massachusetts Institute of Technology. 2006.
18. Plante, J.S. et al. Large-scale failure modes of dielectric elastomer actuators. *International Journal of Solids and Structures.* 2006; 43: 7727–51.
19. Wissler. Modeling dielectric elastomer actuators. Ph.D. thesis, ETH, 2007, Zurich.
20. Wissler, M., Mazza, E. *Sensors and Actuators A-Physical.* 2007; A 120:185.
21. Yanju Liu, Liwu Liu, Zhen Zhang, Jinsong Leng. Dielectric elastomer film actuators: characterization, experiment and analysis. *Smart Materials and Structures.* 2009; 18: 095024.
22. Suo, Z., Zhao, X., Greene, W.H. A nonlinear field theory of deformable dielectrics. *Journal of the Mechanics and Physics of Solids.* 2008; 56: 472.
23. Zhao, X., Suo, Z. Method to analyze electromechanical stability of dielectric elastomers. *Applied Physics Letters.* 2007; 91: 061921.
24. Zhao, X., Hong, W., Suo, Z. Electromechanical hysteresis and coexistent states in dielectric elastomers. *Physical Review B.* 2007; 76: 134113.
25. Zhou, J., Hong, W., Zhao, X., Zhang, Z. Propagation of instability in dielectric elastomers. *International Journal of Solids and Structures.* 2008; 45: 3739.

26. Norrisa, A.N. Comment on method to analyze electromechanical stability of dielectric elastomers [*Applied Physics Letters*. 91, 061921, 2007]. *Applied Physics Letters*. 2007; 92: 026101.

27. Díaz-Calleja, R., Riande, E., Sanchis, M.J. On electromechanical stability of dielectric elastomers. *Applied Physics Letters*. 2008; 93: 101902.

28. Liu, Y., Liu, L., Zhang, Z., Shi, L. Comment on method to analyze electromechanical stability of dielectric elastomers [*Applied Physics Letters*. 91, 061921, 2007]. *Applied Physics Letters*. 2008; 93: 106101.

29. Kofod, G., Sommer-Larsen, P., Kronbluh, R., Pelrine, R. Actuation response of polyacrylate dielectric elastomers. *Journal of Intelligent Material Systems and Structures*. 2003; 14: 787.

30. Zhao, X., Suo, Z. Electrostriction in elastic dielectrics undergoing large deformation. *Journal of Applied Physics*. 2008; 104: 123530.

31. Yanju Liu, Liwu Liu, Liang Shi, Shouhua Sun, Jinsong Leng. Comment on electromechanical stability of dielectric elastomers [*Applied Physics Letters*. 93, 101902, 2008]. *Applied Physics Letters*. 2009; 94: 096101.

32. Jinsong Leng, Liwu Liu, Yanju Liu, Kai Yu, Shouhua Sun. Electromechanical stability of dielectric elastomer. *Applied Physics Letters*. 2009; 94: 211901.

33. Yanju Liu, Liwu Liu, Shouhua Sun, Zhen Zhang, Jinsong Leng. Stability analysis of dielectric elastomer film actuator. *Science in China Series e-Technological Sciences*. 2009; 52(9): 2715–23.

34. Yanju Liu, Liwu Liu, Shouhua Sun, Jinsong Leng. Electromechanical stability of Mooney-Rivlin-type dielectric elastomer with nonlinear variable dielectric constant. *Polymer International*. 2010; 59: 371–377.

35. Díaz-Calleja, R., Riande, E., Sanchis MJ. Effect of an electric field on the bifurcation of a biaxially stretched incompressible slab rubber. *The European Physical Journal E-Soft Matter and Biological Physics*. 2009; 30: 417–26.

36. Jian Zhu, Shengqiang Cai, Zhigang Suo. Nonlinear oscillation of a dielectric elastomer balloon. *Polymer International*. 2010; 59: 378–83.

37. Zhigang Suo, Jian Zhu, Dielectric elastomers of interpenetrating networks. *Applied Physics Letters*. 2009; 95: 232909.

38. Soo Jin Adrian Koh, Xuanhe Zhao, Zhigang Suo. Maximal energy that can be converted by a dielectric elastomer generator. *Applied Physics Letters*. 2009; 94: 262902.

39. Arruda, E.M., Boyce, M.C. A three-dimensional constitutive model for the large stretch behavior of rubber elastic materials. *Journal of the Mechanics and Physics of Solids*. 1993; 41: 389–412.

40. Díaz-Calleja, R., Riande, E., Sanchis, M.J. Effect of an electric field on the deformation of incompressible rubbers: bifurcation phenomena. *Journal of Electrostatics*. 2009; 67: 158–66.

41. Mickael Moscardo, Xuanhe Zhao, Zhigang Suo, Yuri Lapusta. On designing dielectric elastomer actuators. *Journal of Applied Physics*. 2008; 104: 093503.

42. Yanju Liu, Liwu Liu, Kai Yu, Shouhua Sun, Jinsong Leng. An investigation on electromechanical stability of dielectric elastomer undergoing large deformation. *Smart Materials and Structures*. 2009; 18: 095040.

43. Liu, Y., Shi, L., Liu, L., Zhang, Z., Leng, J. Inflated dielectric elastomer actuator for eyeball's movements: fabrication, analysis and experiments. *Proceedings of SPIE*. 2008; 6927,69271A.

44. Gallone, G., Carpi, F., Rossi D. De, Levita, G., Marchetti, A. Dielectric constant enhancement in a silicone elastomer filled with lead magnesium niobate–lead titanate. *Materials Science and Engineering C*. 2007; 110: 27.

45. Carpi, F., Gallone, G., Galantini, F., Rossi D. De. Silicone-poly(hexylthiophene) blends as elastomers with enhanced electromechanical transduction properties. *Advanced Functional Materials*. 2008; 235: 18.

46. Rosenthal, M., Bonwit, N., Duncheon, C., Heim, J. Applications of dielectric elastomer EPAM sensors. *Proceedings of SPIE*. 2007; 6524: 65241F.

47. Brochu, P., Pei, Q.B. Advances in dielectric elastomers for actuators and artificial muscles *Macromolecular Rapid Communications*. 2010; 31: 10–36.

48. Pei, Q., Rosenthal, M., Stanford, S., Prahlad, H., Pelrine, R., Multiple-degrees-of-freedom electroelastomer roll actuators. *Smart Materials and Structures*. 2004; 13: N86.

49. Pelrine, K., Pei, Q., Heydt, R., Stanford, S. Electroelastomers: Applications of dielectric elastomer transducers for actuation, generation and smart structures, *Proceedings of SPIE*. 2002; 4698: 0277–786X.

50. *WW-EAP Newsletter*, Vol. 11, No. 2, December 2009 (The 22nd issue).

51. Yu, Z., Yuan, W., Brochu, P., Chen, Z., Liu, B., Pei, Q., Large-strain, rigid-to-rigid deformation of bistable electroactive polymers, *Applied Physics Letters*. 2009; 95: 192904.

52. Liu, Y.J., Lan, X., Lu, H.B., Leng, J.S. Recent progress of smart materials. *International Journal of Modern Physics: B*. December 2010. (In press).

53. Liu, C.D., Mather, P.T. Polymers for shape memory applications. *ANTEC Proceedings*. 2003; 1962–6.

54. Tobushi, H., Hayashi, S., Kojima, S. Mechanical properties of shape memory polymer of polyurethane series: basic characteristics of stress-strain-temperature relationship, *JSME International Journal. Ser. 1, Solid Mechanics, Strength of Materials*. 1992; 35: 296–302.

55. Tobushi, H., Hashimoto, T., Hayashi, S., Yamada, E. Thermomechanical constitutive modeling in shape memory polymer of polyurethane series. *Journal of Intelligent Material Systems and Structures*. 1997; 8(8): 711–18.

56. Tobushi, H., Okumura, K., Hayashi, S., Ito, N. Thermomechanical constitutive model of shape memory polymer. *Mechanics of Materials*. 2001; 33(10): 545–54.

57. Liu, C.D., Chun, S.B., Mather, P.T., Zheng, L., Haley, E.H., Coughlin, E.B. Chemically cross-linked polycyclooctene: Synthesis, characterization, and shape memory behavior. *Macromolecules*. 2002; 35: 9868–74.

58. Liu, Y.P., Gall, K., Dunn, M.L., Greenberg, A.R., Diani, J. Thermomechanics of shape memory polymers: Uniaxial experiments and constitutive modeling. *International Journal of Plasticity*. 2006; 22: 279–313.

59. Lan, X., Wang, X.H., Liu, Y.J., Leng, J.S. Fiber reinforced shape-memory polymer composite and its application in a deployable hinge, *Smart Materials and Structures*. 2009; 18: 024002.

60. Behl, M., Lendlein, A. Actively moving polymers. *Soft Matter* 2007; 3:58–67.

61. Lendlein, A., Kelch, S. Shape-memory polymers. *Angewandte Chemie-International Edition* 2002; 41:2034–57.

62. Hayashi, S., Tobushi, H., Kojima, S. Mechanical properties of shape memory polymer of polyurethane series, *JSME International Journal Series I*. 1992; 35: 206–302.

63. Kim, B.K., Lee, S.Y., Lee, J.S., Baek, S.H., Choi, Y.J., Lee, J.O. et al. Polyurethane ionomers having shape memory effects. *Polymer*. 1998; 39: 2803–8.

64. Kim, B.K., Shin, Y.J., Cho, S.M., Jeong, H.M. Shape-memory behavior of segmented polyurethanes with an amorphous reversible phase: The effect of block length and content, *Journal of Polymer Science Part B: Polymer Physics*. 2000; 38: 2652–7.

65. Lee, B.S., Chun, B.C., Chung, Y.C., Sul, K.I., Cho, J.W. Structure and thermomechanical properties of polyurethane block copolymers with shape memory effect. *Macromolecules* 2001; 34: 6431–7.

66. Guan, Y., Cao, Y.P., Peng, Y.X., Xu, J., Chen, A.S.C. Complex of polyelectrolyte network with surfactant as novel shape memory networks. *Chemical Communications*. 2001; 17: 1694–5.

67. Liu, G.Q., Ding, X.B., Cao, Y.P., Zheng, Z.H., Peng, Y.X. Shape memory of hydrogen-bonded polymer network/poly(ethylene glycol) complexes. *Macromolecules* 2004; 37: 2228–32.

68. Cao, Y.P., Guan, Y., Du, J., Luo, J., Peng, Y.X., YIP, C.W. et al. Hydrogen-bonded polymer network—poly(ethylene glycol) complexes with shape memory effect. *Journal of Materials Chemistry*. 2002; 12; 2957–60.

69. Ma, Z.L., Zhao, W.G., Liu, Y.F., Shi, J.R. Intumescent polyurethane coatings with reduced flammability based on spirocyclic phosphate-containing polyols. *Journal of Applied Polymer Science*. 1997; 63: 1511–14.

70. Li, F.K., Hou, J.N., Zhu, W., Zhang, X., Xu, M., Luo, X.L. et al. Crystallinity and morphology of segmented polyurethanes with different soft-segment length. *Journal of Applied Polymer Science*. 1996; 62: 631–38.

71. Kunzelman, J., Chung, T., Mather, P.T., Weder, C. Shape memory polymers with built-in threshold temperature sensors, *Journal of Materials Chemistry* 2008; 18: 1082–6.

72. Hu, J.L.. Shape memory polymers and textiles. In: *Characterization of Shape Memory Properties in Polymers*, CRC Press LLC, 2007; 197–225.

73. Chen, S.J., Hu, J.L., Liu, Y.Q., Liem, H.M., Zhu, Y., Liu, Y.J. Effect of SSL and HSC on morphology and properties of PHA based SMPU synthesized by bulk polymerization method. *Journal of Polymer Science Part B: Polymer Physics*. 2007; 45: 444–54.

74. Zhu, Y., Hu, J., Yeung, K.W., Choi, K.F., Liu, Y.Q., Liem, H.M. Effect of cationic group content on shape memory effect in segmented polyurethane cationomer. *Journal of Applied Polymer Science*. 2007; 103: 545–56.

75. Zeng, Y.M., Hu, J.L., Yan, H.J. Temperature dependency of water vapor permeability of shape memory polymer. *Journal of Dong Hua University* 2002; 19: 52–7.

76. Chen, S.J., Hu, J.L., Liu, Y.Q., Liem, H.M., Zhu, Y., Liu, Y.J. Effect of SSL and HSC on morphology and properties of PHA based SMPU synthesized by bulk polymerization method. *Journal of Polymer Science Part B: Polymer Physics*. 2007; 45: 444–54.

77. Li, F., Larock, R.C. New soybean oil-styrene-divinylbenzene thermosetting copolymers. I. Synthesis and characterization. *Journal of Applied Polymer Science* 2001; 80: 658–70.

78. Li, F., Larock, R.C. New soybean oil-styrene-divinylbenzene thermosetting copolymers. III. Tensile stress-strain behavior. *Journal of Polymer Science Part B: Polymer Physics* 2001; 39: 60–77.

79. Li, F., Hanson, M.V., Larock, R.C. Soybean oil-divinylbenzene thermosetting polymers: Synthesis, structure, properties and their relationships. *Polymer*. 2001; 42: 1567–79.

80. Zhang, D.W., Liu, Y.J., Leng, J.S. Shape memory polymer networks from styrene copolymer. *SPIE International Conference on Smart Materials and Nanotechnology.* March 18–23, 2007, USA, *Proceedings of SPIE* 6423 2007:642360:1–6.

81. Tobushi, H., Hayashi, S., Hoshio, K., Miw, N. Influence of strain-holding conditions on shape recovery and secondary-shape forming in polyurethane-shape memory polymer. *Smart Materials and Structures.* 2006; 15: 1033–8.

82. Chen, S.J., Hu, J.L., Liu, Y.Q., Liem, H.M., Zhu, Y., Meng, Q.H. Effect of molecular weight on shape memory behavior in polyurethane films. *Polymer International.* 2007; 56: 1128–34.

83. Liu, G.Q., Guan, C.L., Xia, H.S., Guo, F.Q., Ding, X.B., Peng, Y.X. Novel shape-memory polymer based on hydrogen bonding. *Macromolecular Rapid Communications.* 2006; 27: 1100–4.

84. Gall, K., Kreiner, P., Turner, D., Hulse, M. Shape-memory polymers for microelectromechanical systems. *Journal of Microelectromechanical Systems.* 2004; 13: 472–83.

85. Kim, B.K., Lee, S.Y., Xu, M. Polyurethanes having shape memory effects. Polymer 1996; 7: 5781–93.

86. Yakacki, C.M., Shandas, R., Safranski, D., Ortega, A.M., Sassaman, K., Gall, K. Strong, tailored, biocompatible shape-memory polymer networks. *Advanced Functional Materials.* 2008; 18: 2428–35.

87. Leng, J.S., Wu, X.L., Liu, Y.J. Effect of linear monomer on thermomechanical properties of epoxy shape-memory polymer. *Smart Materials and Structures* 2009; 18: 095031.

88. Chung, T., Romo-Uribe, A., Mather, P.T. Two-way reversible shape memory in a semicrystalline network. *Macromolecules* 2008; 41: 184–92.

89. Liu, G., Ding, X., Cao, Y., Zheng, Z., Peng, Y. Novel shape-memory polymer with two transition temperatures. *Macromolecular Rapid Communications.* 2005; 26: 649–52.

90. Bellin, I., Kelch, S., Lendlein, A. Dual-shape properties of triple-shape polymer networks with crystallizable network segments and grafted side chains. *Journal of Materials Chemistry.* 2007; 17(28): 2885–91.

91. Bellin, I., Kelch, S., Langer, R., Lendlein, A. Polymeric triple-shape materials. *Proceedings of the National Academy of Sciences of USA.* 2006; 103: 18043–47.

92. Karp, J.M., Langer, R. Development and therapeutic applications of advanced biomaterials. *Current Opinion in Biotechnology.* 2007; 18: 454–459.

93. Behl, M., Bellin, I., Kelch, S., Wagermaier, W., Lendlein, A. One-step process for creating triple-shape capability of AB polymer networks. *Advanced Functional Materials.* 2008; 18: 1–7.

94. Xie, T. Tunable polymer multi-shape memory effect, *Nature* 2010; 464(7286): 267–70.

95. Liang, C., Rogers, C.A. One-dimensional thermomechanical constitutive relations for shape memory materials. *Journal of Intelligent Materials Systems and Structures.* 1997; 8: 285–302.

96. Morshedian, J., Khonakdar, H.A., Rasouli, S. Modeling of shape memory induction and recovery in heat-shrinkable polymers. *Macromol. Theory Simul.* 2005; 14: 428–34.

97. Bhattacharyya A, Tobushi H. Analysis of the isothermal mechanical response of a shape memory polymer rheological model. *Polymer Engineering and Science* 2000; 11: 2498–510.

98. Diani, J., Liu, Y.P., Gall, K. Finite strain 3D thermoviscoelastic constitutive model for shape memory polymers, *Polymer Engineering and Science*. 2006; 46: 486–92.

99. Qi, H.J., Nguyen, T.D., Castro, F., Yakackia, C.M., Shandas, R. Finite deformation thermo-mechanical behavior of thermally induced shape memory polymers. *Journal of the Mechanics and Physics of Solids*. 2008; 56(5): 1730–51.

100. Lin, J.R., Chen, L.W. The mechanical-viscoelastic model and WLF relationship in shape memorized linear ether-type polyurethanes. *Journal of Polymer Research-Taiwan* 1999; 6(1): 35–40.

101. Lin, J.R. Chen, L.W. Shape-memorized crosslinked ester-type polyurethane and its mechanical viscoelastic model. *Journal of Applied Polymer Science*. 1999; 73: 1305–19.

102. Li FK, Larock RC. New soybean oil-styrene-divinylbenzene thermosetting copolymers. V. Shape memory effect. *Journal of Applied Polymer Science*. 2000; 84: 1533–43.

103. Abranhamson, E.R., Lake, M.S., Munshi, N.A., Gall, K., Shape memory mechanics of an elastic memory composite resin, *Journal of Intelligent Material Systems and Structures*. 2003; 14: 623–632.

104. Nguyen, T.D., Qi, H.J., Castro, F., Long, K.N. A thermoviscoelastic model for amorphous shape memory polymers: incorporating structural and stress relaxation. *Journal of the Mechanics and Physics of Solids* 2008; 56(9): 2792–814.

105. Chen, Y.C., Lagoudas, D.C. A constitutive theory for shape memory polymers. Part I: Large deformations. *Journal of the Mechanics and Physics of Solids* 2008; 56(5): 1752–65.

106. Chen, Y.C., Lagoudas, D.C. A constitutive theory for shape memory polymers. Part II—A linearized model for small deformations. *Journal of the Mechanics and Physics of Solids*. 2008; 56(5): 1766–78.

107. Lan, X., Huang, W.M., Leng, J.S., Liu, N., Phoo, S.Y., Yuan, Q. Elektrisch leitfähige formgedächtnispolymere. *Gummi Fasern Kunststoffe* 2008; 61, 12, 784–9.

108. Leng, J.S., Lan, X., Liu, Y.J., Du, S.Y. Electroactive shape-memory polymer composite filled with nano-carbon powders. *Smart Materials and Structures* 2009; 18: 074003.

109. Lan, X., Huang, W.M., Leng, J.S., Liu, Y.J., Du, S.Y. Investigate of electrical conductivity of shape-memory polymer filled with carbon black. *Advanced Materials Research*. 2008; 47–50: 714–7.

110. Leng, J.S., Lv, H.B., Liu, Y.J., Du, S.Y. Electroactivated shape-memory polymer filled with nanocarbon particles and short carbon fibers. *Applied Physics Letters*. 2007; 91: 144105.

111. Leng, J.S., Lv, H.B., Liu, Y.J., Du, S.Y. Synergic effect of carbon black and short carbon fiber on shape memory polymer actuation by electricity. *Journal of Applied Physics*. 2008; 104: 104917.

112. Lu, H.B., Liu, Y.J., Leng, J.S., Du, S.Y. Qualitative separation of the effect of solubility parameter on the recovery behavior of shape-memory polymer. *Smart Materials and Structures*. 2009; 18: 085003.

113. Leng, J.S., Zhang, D.W., Liu, Y.J., Yu, K., Lan, X. Study on the activation of styrene-based shape memory polymer by medium-infrared laser light. *Applied Physics Letters*. 2010; 96,11: 111905.

114. Yakacki, C.M., Satarkar, N.S., Gall, K., Likos, R., Hilt, J.Z. Shape-memory polymer networks with Fe_3O_4 nanoparticles for remote activation, *Journal of Applied Polymer Science*. 2009; 112: 3166–76.
115. Hazelton, C.S., Arzberger, S.C., Lake, M.S., Munshi, N.A. RF actuation of a thermoset shape memory polymer with embedded magnetoelectroelastic particles. *Journal of Advanced Materials* 2007; 39: 35–9.
116. Vaia, R. Nanocomposites: Remote-controlled actuators. *Nature Materials*. 2005; 4: 429–30.
117. Razzaq, M.Y., Anhalt, M., Frormann, L., Weidenfeller, B. Thermal, electrical and magnetic studies of magnetite filled polyurethane shape memory polymers. *Materials Science and Engineering A-Structural Materials Properties Microstructure and Processing* 2007; 444: 227–35.
118. Mohr, R., Kratz, K., Weigel, T., Lucka-Gabor, M., Moneke, M., Lendlein, A. Initiation of shape-memory effect by inductive heating of magnetic nanoparticles in thermoplastic polymers. *Proceedings of the National Academy of Sciences USA* 2006; 103, 3540–5.
119. Schmidt, A.M. Electromagnetic activation of shape memory polymer networks containing magnetic nanoparticles. *Macromolecular Rapid Communications*. 2006; 27: 1168–72.
120. Maitland, D.J., Metzger, M.F., Schumann, D., Lee, A., Wilson, T.S. Photothermal properties of shape memory polymer micro-actuators for treating stroke. *Lasers in Surgery and Medicine* 2002; 30: 1–11.
121. Baer, G.M., Small, W., Wilson, T.S., Benett, W.J., Matthews, D.L., Hartman, J. et al. Fabrication and in vitro deployment of a laser-activated shape memory polymer vascular stent. *Biomedical Engineering Online* 2007; 6:43.
122. Maitland, D.J., Small, W., Ortega, J.M., Buckley, P.R., Rodriguez, J., Hartman, J. et al. Prototype laser-activated shape memory polymer foam device for embolic treatment of aneurysms. *Journal of Biomedical Optics* 2007; 12: 030504.
123. Monkman, G.J. Advances in shape memory polymer actuation. *Mechatronics* 2000; 10(4–5): 489–98.
124. Yang, B. Influence of moisture in polyurethane shape memory polymers and their electrical conductive composites. A dissertation for the degree of doctor of philosophy, School of Mechanical and Aerospace Engineering Nanyang Technological University, Singapore, 2005.
125. Huang, W.M., Lee, C.W., Teo, H.P. Thermomechanical behavior of a polyurethane shape memory polymer foam. *Journal of Intelligent Materials Systems and Structures*. 2006; 17: 753–60.
126. Huang, W.M., Yang, B., An, L., Li, C., Chan, Y.S. Water-driven programmable polyurethane shape memory polymer: Demonstration and mechanism. *Applied Physics Letters*. 2005; 86: 114105.
127. Lu, H.B., Liu, Y.J., Gou, J.H., Leng, J.S., Du, S.Y. Electroactive shape-memory polymer nanocomposites incorporating carbon nanofiber paper, *International Journal of Smart and Nano Materials*. 2010; 1: 2–12.
128. Lu, H.B., Liu, Y.J., Gou, J.H., Leng, J.S., Du, S.Y. Synergistic Effect of carbon nanofiber and carbon nanopaper on shape memory polymer composite, *Applied Physics Letters*. 2010; 96: 084102.

129. Lan, X., Wang, X.H., Lu, H.B., Liu, Y.J., Leng, J.S., Shape recovery performances of a deployable hinge fabricated by fiber-reinforced shape-memory polymer, *16th SPIE International Conference on Smart Structures/NDE*, San Diego, March 8–12, 2009, 1–8.

130. Steven, C. Arzbergera, Michael L. Tupper, Mark S. Lake. Elastic memory composites (EMC) for deployable industrial and commercial applications, www.ctd-materials.com.

131. Taylor, R.M., Abrahamson, E., Barrett, R., Codell, D.E., Keller, P.N. Passive deployment of an EMC boom using radiant energy in thermal vacuum, *48th AIAA/ASME/ASCE/AHS/ASC Structures, Structural Dynamics, and Materials Conference*, April 23–26, 2007, Honolulu, Hawaii, AIAA 2007–2269.

132. Keihl, M.M., Bortolin, R.S., Sanders, B., Joshi, S., Tidwell, Z. Mechanical properties of shape memory polymers for morphing aircraft applications, *SPIE Conference: Smart Structures and Materials 2005-Industrial and Commercial Applications of Smart Structures Technologies* 2005; 5762: 143–51.

133. John, S. Flanagan, Rolf C. Strutzenberg, Robert B. Myers, Jeffrey E. Rodrian Development and flight testing of a morphing aircraft, the NextGen MFX-1. *48th AIAA/ASME/ASCE/AHS/ASC Structures, Structural Dynamics, and Materials Conference*, April 23–26, 2007, Honolulu, Hawaii, AIAA 2007–1707: 1–3.

134. Browne, A.L., Johnson, N.L. Method for controlling airflow. United States Patent 7178859, 2007.

135. Love, M.H., Zink, P.S., Stroud, R.L., Bye, D.R., Rizk, S. White D. Demonstration of morphing technology through ground and wind tunnel tests. *48thAIAA/ASME/ASCE/AHS/ASC Structures, Structural Dynamics, and Materials Conference* 2007; *AIAA* 2007–1729: 1–12.

136. Yin, W.L., Fu, T., Liu, J.C., Leng, J.S. Structural shape sensing for variable camber wing using FBG sensors, *SPIE International Conference on Smart Structures/NDE*, San Diego, March 8–12, 2009, *Proceedings of SPIE* 7292 2009; 72921H:1–10.

137. Yu, K., Yin, W.L. Design and analysis of morphing wing based on SMP composite. *SPIE International Conference on Smart Structures/NDE*, San Diego, March 8–12, 2009, *Proceedings of SPIE* 7290 2009; 72900S: 1–8.

138. Ji, F., Zhu, Y., Hu, J., Liu, Y., Yeung, L., Ye, G. Smart polymer fibers with shape-memory effect. *Smart Materials and Structures*. 2006; 15: 1547–54.

139. Kobayashi, K., Hayashi, S. Shape memory fibrous sheet and method of imparting shape memory property to fibrous sheet product, United States Patent 5098776, 1992.

140. Kobayashi, K., Hayashi, S. Woven fabric made of shape memory polymer, United States Patent 5128197, 1989.

141. Leng, J.S. Du, S.Y. Shape memory polymer and multifunctional composite. In: Hu, J.L. *Shape Memory Polymer Textile*, CRC Press/Taylor & Francis; 2010, 293–313.

142. Sokolowski, W. Potential bio-medical and commercial applications of cold hibernated elastic memory self-deployable foam structures. International Symposium on Smart Materials, Nano-and Micro-Smart Systems, December 12–15, 2004, Sydney, Australia. *Proceedings of SPIE*. 2004; 5648, 397–405.

4

Magnetic Polymer Nanocomposites: Fabrication, Processing, Property Analysis, and Applications

Suying Wei, Jiahua Zhu, Pallavi Mavinakuli, and Zhanhu Guo

Lamar University

CONTENTS

4.1 Introduction

Polymers are considered excellent host matrices for composite materials. They have been extensively investigated for their potential broad applications because of their easy processability, low-cost manufacturing, good adhesion to substrates [1], and unique physicochemical properties. Polymer nanocomposites are polymer composites using nanostructured materials as the fillers. Depending on the type of filler materials, different properties can be achieved for the polymer nanocomposites. Unique physicochemical phenomena such as giant magnetoresistance (GMR) or tunneling magnetoresistance (TMR) can be created, in which the nonmagnetic conductive or insulating polymer serves as a spacer, while magnetic inorganic materials are used as the fillers [2,3]. This phenomenon is completely beyond the simple addition of the advantageous physicochemical properties of a single polymer matrix and the magnetic inorganic fillers. A typical application of magnetic polymer nanocomposites is the GMR/TMR sensor. Various combinations of polymers and magnetic nanomaterials have been explored for magnetic polymer nanocomposite fabrication, including magnetic nanoparticles and conductive polymers such as iron or polypyrrole (Ppy), magnetic nanoparticles and insulating polymers such as iron or vinyl ester and iron or Polyurethane (PU), to name a few.

Recent studies on magnetic polymer nanocomposites have shown great potential applications in high-density magnetic recording, magnetic sensors, magnetic carriers, color imaging [4–7], biomedical uses, magnetic storage, and electronic devices [6,8,9] by utilizing unique magnetic-nanosized materials. The filler-nanosized material can have a range of shapes and morphologies such as nanoparticles, nanotubes, nanoflakes, and long and short nanodiameter fibers. These are exemplified by oxide nanoparticles, carbon nanotubes, exfoliated clay particles, short carbon fibers, and long electrospun fibers, respectively. The area of the interface between the matrix and reinforcement phases is typically an order of magnitude greater than that in conventional composites. The matrix material is significantly affected by proximity to the reinforcement. This large reinforcement surface area means that the effect of a relatively small number of nanoscale fillers can have an observable effect on the macroscale properties of the composites. For example, adding carbon nanotubes improves the electrical and thermal conductivity. Other kinds of nanoparticles result in enhanced optical, electronic, and dielectric properties, or mechanical properties such as stiffness and strength. In general, the nanofillers are dispersed into the matrix during processing. The percentage (weight/mass fraction) of the nanoparticles introduced can remain very low (on the order of 0.5% to 5%) due to the low filler percolation threshold, especially for the most commonly used nonspherical, high-aspect ratio fillers (such as thin nanoplatelets, nanoflakes, or carbon nanotubes).

Magnetic nanoparticles, due to their unique magnetic and electronic properties, are particularly popular as fillers in the magnetic polymer nanocomposites. The magnetic nanoparticles include metals such as iron, cobalt, nickel and their alloys or FePt, oxides (γ-Fe_2O_3 [10–13], Fe_3O_4 [14–16], cobalt oxide [17]), iron cobalt oxide ferrite [18], manganese–zinc ferrite ($Mn_{0.68}Zn_{0.25}Fe_{2.07}O_3$) [19], barium ferrite ($BaFe_{12}O_{19}$) [20], etc.). Compared with the metallic magnetic materials, ferrite materials can have higher resistivity and still larger saturation magnetization, which enables them to be used in high-frequency applications to obtain a good frequency–permeability response [18,21]. By combining in a single material, the electrical conductivity of the conductive

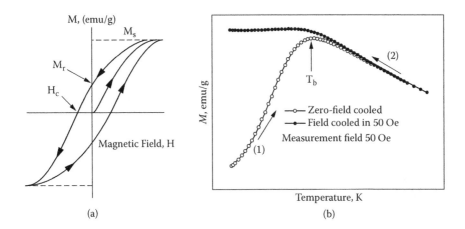

FIGURE 4.1
Schematic of the magnetic properties: (a) hysteresis loop and (b) ZFC and FC temperature-dependent magnetization curves.

polymer, and the magnetic properties of the nanoparticles, magnetopolymeric materials present many potential applications [18]. The insulating properties of the ferrites have the advantage of reducing the eddy current losses, and some of the soft ferrites have been extensively investigated for radio frequency applications such as AC/DC converters, wide-band transformers, antennas, and electromagnetic interference (EMI) shielding. Figure 4.1 illustrates the typical characterization curves for magnetic properties.

There are four important physical parameters as shown in Figure 4.1 to describe the magnetic properties of a nanomaterial—that is, saturation magnetization (M_s, emu/g), remanent magnetization (M_r, emu/g, a material's ability to retain a certain amount of residual magnetic field when the magnetizing force is removed after achieving saturation), coercivity (coercive force, H_c, Oe, the amount of reverse magnetic field that must be applied to a magnetic material to make the magnetization return to zero), and blocking temperature (T_b, K, characteristic of the transition temperature between the superparamagnetic state and the ferromagnetic state).

According to the magnetic properties, materials can be classified into soft and hard categories. Materials with coercivity (coercive force) larger than 200 Oe are called *hard materials,* and smaller than 200 Oe are called *soft materials.* In the soft materials, a unique category is called *superparamagnetic materials.* The superparamagnetic material consists of small ferromagnetic clusters (e.g., crystallites); the clusters are so small, they can randomly flip direction under thermal fluctuations. As a result, the material as a whole is not magnetized (coercivity and remanent magnetization equal to zero) except in an externally applied magnetic field (in that respect, coercivity is similar to paramagnetism).

Successful incorporation of magnetic nanoparticles into a conductive or insulating polymer matrix will definitely widen their applicability in the fields of magnetic and magnetoresistance, electronics, biomedical drug delivery, and optics. However, obtaining high-quality magnetic polymer nanocomposites is challenging. The nanoparticles tend to agglomerate at high loading, and the interaction between nanoparticles and the polymer matrix needs to be strong enough to prevent gas voids and therefore deleterious effects on the mechanical properties of the nanocomposites. Thus, appropriate surface engineering/chemistry treatment is a critical step in fabricating good nanocomposites. Another challenge facing conductive polymer matrix is the need to integrate a high fraction of nanoparticles into the polymer matrix in a strong acidic environment normally required for the conductive polymer synthesis. In this situation, the acid will etch away the nanoparticles in an aqueous system. A balance between the polymerization requirement in an acidic solution and the prevention of dissolution of reactive metal oxide nanoparticles will be a determining factor for high-quality nanocomposite fabrication.

In this chapter, fabrication and postprocessing methods for making magnetic polymer nanocomposites, including magnetic conductive and insulating polymer nanocomposites, are summarized. A variety of property analysis techniques used for the as-fabricated and processed magnetic polymer nanocomposites are discussed. The wide applications for this group of materials will also be explored. Finally, the authors offer their perspectives on the future direction of the magnetic polymer nanocomposites.

4.2 Magnetic Polymer Nanocomposite Fabrication and Processing

Various methods have been developed to fabricate and process polymer nanocomposites including ex situ methods (dispersion of the synthesized nanoparticles into a polymer solution) [22–24], in situ monomer polymerization in the presence of the nanoparticles [25–28], in situ nanoparticle formation in the presence of polymer [29], drop casting [30], dipping-coating [31], and spinning coating [32]. Here, several specific methods will be exemplified for the fabrication and process of magnetic polymer nanocomposites, including the oxidation in solution method, the electrochemical method, sequential deposition approaches, irradiation, and electrospinning methods.

4.2.1 Solution-Based Oxidation Method

Polypyrrole nanocomposites filled with γ-Fe$_2$O$_3$ nanoparticles were fabricated by the oxidation method in an aqueous solution [33]. A dispersion of γ-Fe$_2$O$_3$ nanoparticles (5–25 nm) was made by adding a desired amount of

γ-Fe$_2$O$_3$ in 20 mL deionized water under sonication. The p-toluenesulfonic acid (CH$_3$C$_6$H$_4$SO$_3$H, p-TSA, 6.0 mmol) and pyrrole (7.3 mmol, Aldrich) were added to the preceding nanoparticle-suspended solution under sonication. Ammonium persulfate (APS, oxidant, 3.6 mmol) was rapidly mixed into the preceding solution at room temperature, and the resulting solution was kept under sonication for 1 h. In addition, the effect of reaction time was investigated by using a 7 h sonication, as used previously in the study of micron-size iron oxide particles [34]. Both the mechanical stirring and the ultrasonic stirring were used to produce nanocomposites. All the products were washed thoroughly with deionized water (to remove the unreacted APS and p-TSA) and methanol (to remove oligomers). The precipitated powder was dried at 50°C. A similar solution-based oxidation method has been used to prepare polyaniline–barium ferrite nanocomposites [20].

A microemulsion polymerization method [35,36] was also reported to produce magnetic Ppy nanocomposites filled with γ-Fe$_2$O$_3$. The nanoparticles are dispersed in the oil phase. FeCl$_3$ was used as an oxidizing agent. Sodium dodecylbenzenesulfonic acid (SDBA) and butanol were used as the surfactant and cosurfactant, respectively. FeCl$_3$ (0.97 g) was dissolved in a mixture of 15 mol deionized water, SDBA (6 g), and butanol (1.6 mL). A specific amount of γ-Fe$_2$O$_3$ nanoparticles-suspended solution was added to the preceding solution for dispersion. Pyrrole was added for polymer nanocomposite fabrication in the microemulsion system. The polymerization was continued for 24 h and quenched by acetone.

4.2.2 Electropolymerization Method

Compared to an inorganic substrate, a conductive polymer has the advantages of light weight, flexibility, and easy synthesis. The conductive polymer needs to be very stable at a negative potential during electropolymerization in order to serve as an electrode. The PPy thin film doped with m-sulfobenzoic acid has high conductivity on the order of 100 S cm^{-1} in air, at relatively high temperatures, in the reductive agent, and a negative potential, which provides the opportunity for the electrodeposition application [37]. The electrochemical synthetic approach has been used to produce a multilayered structure with conductive polymer as the substrate [38]. Yan et al. used a heavily doped PPy film as a substrate for subsequent multilayered Co and Cu formations and observed magnetic-field-dependent electron transport behaviors [38].

Abe et al. [39,40] reported an Fe$_3$O$_4$ film formation by using electroplating (anodic oxidization) or electrodeless plating (air oxidation) in aqueous solution at low temperature ($T < 80°C$), and this approach promoted the formation of oxides on various substrates or in matrices. The nanoparticles used in the Ppy nanocomposite fabrication include γ-Fe$_2$O$_3$ [41] and Fe$_3$O$_4$ [42].

Jarjayes et al. have fabricated PPy–Fe$_2$O$_3$ nanocomposites in one single electrochemical step [41], rather than the other two-step deposition process [43].

The anionic ferrofluid, a stable liquid solution containing magnetic nanocrystals coated with anionic chelating agents (surfactant), rendered the needed compatibility between the nanoparticles and the polymer chains. The ferrofluid-containing PPy films were electrodeposited in a single-compartment cell on an ITO anode (2 × 3 cm) by electrolysis at 0.7 V/SCE of a nonstirred aqueous solution of 0.5 mol/L pyrrole and ferrofluid (0.1 mol/L in Fe). The deposition charge is typically 5 C/cm² for current density of around 1 mA/cm². The films were purified by washing with deionized water and acetonitrile. The films were stripped from the electrode substrate after one night free standing and dried under vacuum at 60°C for 12 h.

Yan et al. [42] have further extended the electrodeposition method to prepare multilayered PPy and magnetite (Fe_3O_4). The conductive polymer layer was electropolymerized in a one-compartment three-electrode cell with a computer-controllable EG&G potentiostat Model M273. The electrolyte was 0.1 mol/L *p*-toluenesulfonate and 0.1 mol/L pyrrole aqueous solution. Stainless steel (4 × 2 cm), Pt sheet, and Ag/AgCl (0.1 M KCl) electrodes were used as working electrode, counter electrode, and reference electrode for the PPy film formation, respectively. A galvanostatic method (consists of placing a constant current pulse on an electrode and measuring the variation of the resulting current through the solution—a way to measure the rate of an electrochemical reaction) was used for electrochemical polymerization of PPy in a current density of 1 mA/cm⁻², and the thickness of the deposited film was controlled by the passed electric charge during the electropolymerization of the thin-film growth. The working electrode covered with PPy film was washed thoroughly with distilled water and acetone, and used as working electrode for the magnetite (Fe_3O_4) film electrodeposition. The galvanostatic method was employed again in a current density of 0.5 mA cm⁻¹. The electrolyte contains 0.03 mol/L $FeSO_4$ with a pH value of 7 (adjusted with NaOH solution) at 75°C. A sandwich structure with different alternative layers was obtained by repeating the preceding process. The maximum thickness of magnetite is about 0.2 μm.

4.2.3 Two-Step Deposition Method

Forder et al. [43] have successfully prepared superparamagnetic conductive polyester textile composites (Ppy and magnetite on polyester textile fiber substrate) by a two-step deposition method, that is, magnetite deposition and PPy deposition.

Preparation of magnetite-textile fiber composites is described as follows. The cleaned and dried polymer textile substrate (average fiber diameter 10–20 μm) was immersed in the magnetite dispersion for about half an hour and air-dried for several hours. The coated textile was further dried at 60°C.

The magnetite-impregnated textile was also used for conductive polymer coating. The solution for immersion of treated textile fibers was 1,5-naphthalenedisulfonic acid (0.27 g), 5-sulfosalicylic acid (1.71 g), and $FeCl_3$ $6H_2O$

(1.82 g) in 150 mL deionized water. The magnetite-treated textile fibers (2.75 g) were immersed in the preceding solution, after which pyrrole (0.20 cm^3) was added for the polymerization. The polymerization was quenched with excess deionized water, and the product was washed and dried.

4.2.4 UV-Irradiation Technique

Poddar et al. [19] have successfully extended the UV-irradiation technique [44] for PPy formation to composite fabrication. In this technique, silver nitrate (10–15 molar%) is added to the pyrrole monomers as the electron acceptor for pyrrole photopolymerization. A photoinitiator was used to increase the polymerization rate and uniformity. A cationic photoinitiator (Cyracure UV 16992 from Dow, Inc.) was added to the pyrrole/salt solution in the amount of 0.3%. The ferrite nanoparticles in specific amount were dispersed in the preceding solution and spin coated on a glass substrate. The nanocomposite thin films were formed by overnight polymerization under UV light (365 nm) irradiation.

4.2.5 Electrospinning Method

The electrospinning method is one way to produce 1-D polymer nanocomposites. It is the most handy, low-cost, and high-speed method to make nanocomposite fibers. It is a process that involves applying a high voltage (more than 6 kV) between the tip of a needle and a collecting electrode (collector). A pendent drop of solution under surface tension will be charged, and the induced charges will be evenly distributed over the surface. The coulombic repulsive force will overcome the solution surface tension and thus force the ejection of a liquid jet, which results in the formation of a Taylor cone. The electrified jet of viscoelastic solution then undergoes a stretching process and forms thin fibers on the collector.

This electrospinning technology has attracted much attention since it was first reported in 1934 [45], and it has been used to fabricate different types of hybrid nanofibers by incorporating nanomaterials into various polymer matrices. The nanofibers produced by electrospinning have several remarkable advantages, including a small diameter (50–10 mm), high aspect ratio (the ratio of length to diameter), large specific surface area (surface area-to-volume ratio), diverse composition, unique physicochemical properties, and design flexibility for chemical/physical surface functionalization [45–48]. Together with the complex pore structure and easy fiber surface modification, electrospun nanofibers are ideal for certain applications (Figure 4.2) [45, 49–52].

The manufacturing of pure polyacrylonitrile (PAN) fibers and magnetic PAN–Fe3O4 nanocomposite fibers is explored by an electrospinning process [53]. A uniform, bead-free fiber production process is developed by optimizing electrospinning conditions: polymer concentration, applied electric voltage, feed rate, and distance from the needle tip to the collector.

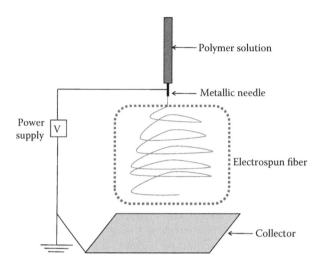

FIGURE 4.2
Electrospinning setup for preparing magnetic polymer nanocomposites.

The experiments demonstrate that slight changes in operating parameters may result in significant variations in the fiber morphology. The fiber formation mechanism for both pure PAN and the Fe_3O_4 nanoparticles suspended in PAN solutions is explained from the rheological behavior of the solution. Figure 4.3 shows that the magnetic properties of the nanoparticles in the polymer nanocomposite fibers are different from those of the dried as received nanoparticles [53].

4.3 Physicochemical Property Analysis

4.3.1 FT-IR for Functional Groups

Fourier transform infrared (FT-IR) spectrometry was widely used for characterizing functional groups in polymer nanocomposites. The changes in the spectra before and after incorporating the magnetic nanofiller into the polymer matrix indicate the formation of interactions between these two phases, which is crucial to form strong composites materials. The FTIR mode can be either transmission or attenuated total reflection (ATR), depending on the form of the composite.

Figure 4.4 shows the FT-IR spectra of the pure PPy and Fe_2O_3–PPy nanocomposites. Characteristic peaks of PPy are observed in all the samples, indicating the formation of PPy. The peaks at 1547 and 1464 cm^{-1} can be assigned to C=C and C–N stretching vibrations, respectively. The peaks at 1164 and 901 cm^{-1} are due to the C–H in-plane bending and ring deformation, respectively.

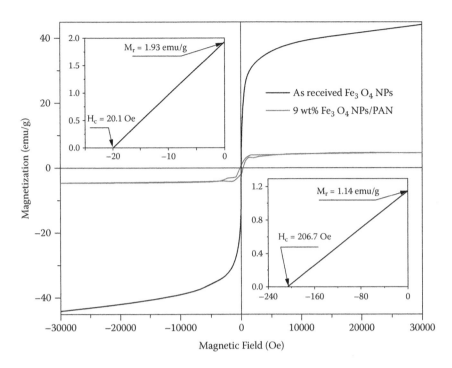

FIGURE 4.3
Hysteresis loops of (a) Fe_3O_4 nanoparticles and (b) PAN (7 wt%)/Fe_3O_4 composite nanofibers with 9 wt% particle loading at room temperature. (From D. Zhang et al. *Polymer*, 50, 4189–4198 [2009]. Reproduced with permission of Elsevier.)

Similar patterns were also observed in PPy–Fe[OH] microcomposites [34]. The obvious spectral differences between pure PPy and the composites indicate that the PPy exhibits a different chain structure, and there are physicochemical interactions working between the nanoparticles and PPy. The presence of iron oxide nanoparticles in the composite is strongly supported by the new peaks at 485 cm^{-1} and 588 cm^{-1}, as shown in Figure 4.4c,d, which are due to the stretching vibrations of iron oxide [54]. This observation indicates that a magnetic nanocomposite can be synthesized with PPy if polymerization can be achieved in a short period of time. The prolonged polymerization is characterized by the disappearance of the characteristic IR peaks of iron oxide (Figure 4.4b). The small spectrum difference between the PPy and nanocomposites with prolonged polymerization also indicates the loss of iron oxide nanoparticles.

4.3.2 Scanning Electron Microscopy (SEM) for Surface Morphology

The dispersion quality of the nanoparticles within the polymer matrix and the nanostructures of the polymer and nanocomposites were normally

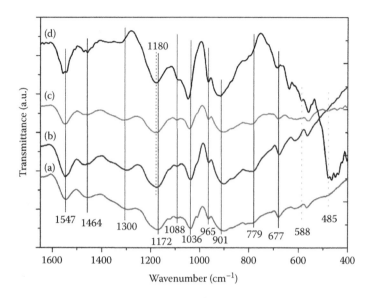

FIGURE 4.4

FT-IR spectra of (a) pure PPy without nanoparticles, (b) nanocomposite with long polymerization time, (c) nanocomposite with 20 wt% loading, and (d) nanocomposite with 50 wt% loading. (From Z. Guo et al. *J. Nanoparticle Res.* 11(6), 1441–1442 [2009]. Reproduced by kind permission from Springer Science.)

investigated by scanning electron microscopy (SEM). SEM specimens were prepared by spreading a thin layer of powder onto a double-sided carbon tape. The microstructure and crystallinity were investigated with a transmission electron microscope with an accelerating voltage of 100 keV. The samples were prepared by dispersing the powder in anhydrous ethanol, dropping some suspended solution onto a carbon-coated copper grid, and drying naturally in an ambient condition.

Figure 4.5 shows the SEM microstructures of the synthesized pure PPy and Fe$_2$O$_3$–PPy nanocomposites fabricated by the conventional method, as used for micron-size iron oxide particle-filled PPy composite fabrication [34]. The conventional method is based on mechanical stirring. In contrast with the network structure of pure PPy as shown in Figure 4.5a,b, discrete spherical nanoparticles with uniform-size distribution are observed in the nanocomposite counterparts fabricated by the conventional method, as shown in Figure 4.5c,d. However, no attraction was observed when a permanent magnet was placed nearby, and further quantitative magnetic characterization did not show any sign of magnetization in the nanocomposites. The disappearance of the magnetic nanoparticles is due to the slow dissolution over time caused by the acidic solution used in the pyrrole polymerization. This observation also suggests the formation of a porous Ppy shell rather than a solid one that can protect the magnetic nanoparticles from dissolution.

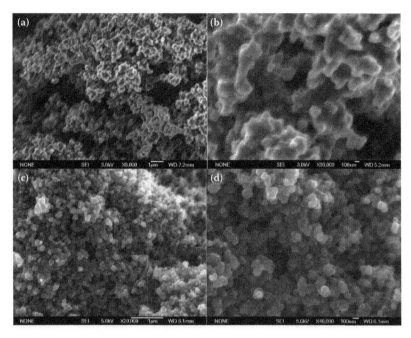

FIGURE 4.5
SEM micrographs of pure PPy formed (a) and (b) without, and (c) and (d) with iron oxide nano-particles: mechanical stirring. (From Z. Guo et al. *J. Nanoparticle Res.* 11(6), 1441–1442 [2009]. Reproduced by kind permission from Springer Science.)

For the in situ formation of the conductive magnetic nanocomposite, a short reaction time (1 h) was used to balance the PPy polymerization and the nanoparticle dissolution. Ultrasonic stirring was used rather than mechanical stirring to minimize the contamination and achieve better particle dispersion. The red particles turned black after the polymerization, indicating the formation of the PPy. Unlike the network structure as observed in the pure PPy, the SEM micrographs (shown in Figure 4.6) of the nanocomposites with different initial particle loadings show discrete nanoparticles without any obvious agglomeration. In stark contrast with the obvious loss of magnetic particles when mechanical stirring was used for 7 h polymerization, the dried nanocomposite powder in Figure 4.6 was attracted to a permanent magnet, indicating the presence of the magnetic nanoparticles.

4.3.3 Thermal Property Analysis by TGA/DSC

The thermal stability of the pure PPy and Fe_2O_3–PPy nanocomposites was investigated by TGA measurements, and results are shown in Figure 4.7. Figure 4.7a shows the thermogravimetric profiles of nanocomposites with particle loadings of 0, 20, and 50 wt%, respectively. The black nanocomposites

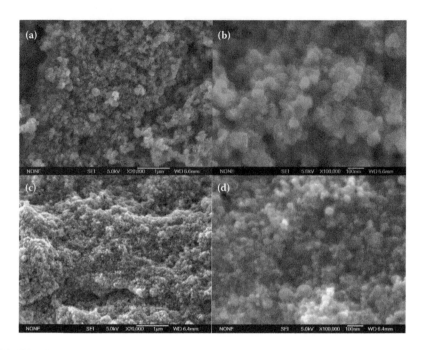

FIGURE 4.6
SEM micrographs of Fe_2O_3–PPy nanocomposite with 20 wt% (a) and (b) and 50 wt%, (c) and (d) sonication. (From Z. Guo et al. *J. Nanoparticle Res.* 11(6), 1441–1442 [2009]. Reproduced by kind permission of Springer Science.)

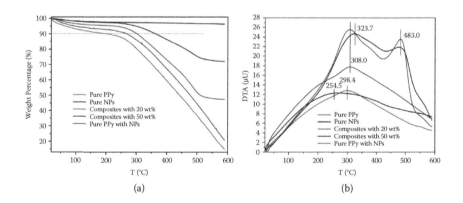

FIGURE 4.7
(a) Weight changes in TGA of pure PPy and Fe_2O_3–PPy nanocomposites with different particle loadings, and (b) corresponding DTA curves. (From Z. Guo et al. *J. Nanoparticle Res.* 11(6), 1441–1442 [2009]. Reproduced by kind permission from Springer Science.)

were observed to turn red upon test completion, characteristic of iron oxide rather than black carbon, indicating complete loss of PPy. The weight loss at temperatures lower than 120°C is due to the loss of moisture, and the major loss at temperatures higher than 240°C is due to the decomposition of PPy. The differences in the residues reflect the different amounts of iron oxide nanoparticles present. In addition, the nanocomposites show better thermal stability than that of the pure PPy. The thermal stability increases slightly with increasing nanoparticle loading, which is believed to be due to both the lower mobility of the polymer chains when the polymer chains are bounded onto the nanoparticles with strong chemical interactions [55].

The only peak at 254.5°C observed in the as-received nanoparticles, as shown in Figure 4.7b, is due to the decomposition of the iron oxide–hydroxide (goethite, FeOOH as proved by the FT-IR spectra) [56,57]. Similar to the TGA observation, a higher decomposition temperature (308.0°C) was observed in the PPy formed with the aid of nanoparticles than that (298.4°C) of the pure PPy formed without them, whereas, only one peak was observed in the pure PPy samples, and two exothermic peaks were observed in the DTA curves of the nanocomposites. These were due to the decomposition of PPy at 307°C and the possible phase transition of iron oxide at 480°C, as reported in the Fe_2O_3–PPy nanocomposites fabricated by the simultaneous gelation and polymerization (sol–gel) method [58,59], respectively. As compared with no obvious phase transition in the pure iron oxide nanoparticles, the observed phase transition was due to the intermediate product of PPy [58,59].

4.3.4 Magnetic Property Analysis

The magnetic properties of the nanocomposite were normally measured in a 9 T physical properties measurement system (PPMS) by Quantum Design. Two different measurements were normally carried out. The temperature-dependent magnetization was investigated using zero-field-cooled (ZFC) and field-cooled (FC) conditions at an applied field (e.g., 100 Oe). ZFC was done by cooling the sample first to 5 K without a field, then magnetization changes were recorded with the temperature increasing from 5 to 300 K with an applied field of 100 Oe. FC was recorded immediately after ZFC by decreasing the temperature from 300 to 5 K with a constant field of 100 Oe. Field-dependent magnetization (hysteresis loop) was tested for certain temperatures.

Figure 4.8 shows the magnetic hysteresis loops of the nanocomposites with an initial particle loading of 20 and 50 wt%, respectively. There was no hysteresis observed in the sample formed from the 7 h reaction due to the dissolution of the magnetic nanoparticles in the acidic solution. The saturation magnetization (M_s) was about 29.4 and 45.1 emu/g, based on the total weight of the nanocomposites with an initial loading of 20 and 50 wt%, respectively. The saturation magnetization of iron oxide was reported to be 74 emu/g and independent of the surface chemistry of the nanoparticles. Thus, the real particle loading of the nanoparticles in the nanocomposites with an initial

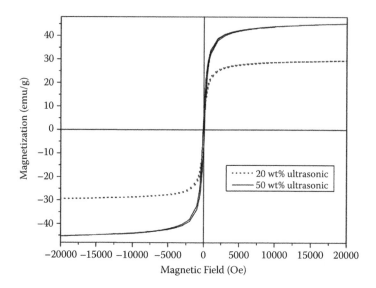

FIGURE 4.8

Hysteresis loops of nanocomposites at different loading. (From Z. Guo et al. *J. Nanoparticle Res.* 11(6), 1441–1442 [2009]. Reproduced by kind permission of Springer Science.)

particle loading of 20 and 50 wt% was calculated to be 27 and 68 wt%, respectively. The calculated particle loading based on the magnetic data is much higher than the initial particle loading, considering the partial particle loss from the dissolution during the nanocomposite fabrication. This indicates that the yield of the Ppy was lower. In other words, the pyrroles were partially polymerized into Ppy that leads to the higher particle loading. In addition, the polymer yield was observed to be lower at the higher initial particle loading than that at the lower initial particle loading.

The effect of the stirring method, that is, mechanical and ultrasonic stirring, was investigated by polymerizing 7 h as used in the mechanical stirring. The final product exhibited a strong tendency to be attracted onto a permanent magnet, indicating the presence of the iron oxide nanoparticles. Figure 4.9 shows the hysteresis loop of the as-received nanoparticles and the nanocomposite synthesized with ultrasonic stirring over 7 h and an initial particle loading of 50 wt%. The weight percentage of the iron oxide nanoparticles in the nanocomposite was estimated to be 20.2% based on the saturation magnetizations of the nanocomposite and the as-received nanoparticles. This weight percentage is much lower than the initial particle loading and the composite sample formed with a 1 h ultrasonic stirring. It is an indication that more particles were lost due to the dissolution over the long-time reaction between the nanoparticles and the protons. The coercivity was observed to be much larger

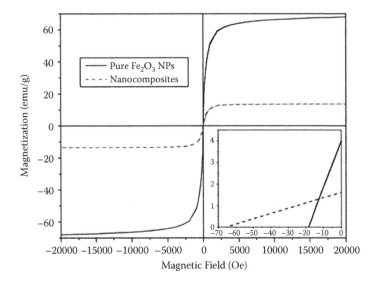

FIGURE 4.9
Hysteresis loops of pure iron oxide nanoparticles and the nanocomposite with an initial particle loading of 50 wt% and ultrasonically stirred for 7 h. (From Z. Guo et al. *J. Nanoparticle Res.* 11(6), 1441–1442 [2009]. Reproduced by kind permission of Springer Science.)

in the nanocomposite (65 Oe) than that (18 Oe) of the as-received samples, due to the dispersion of the single-domain-size nanoparticles.

4.3.5 Mechanical Property Analysis by Rheology

The rheological characterization of polymer nanocomposites is mostly oriented for better understanding the fluid dynamics of confined polymers in the nanoscale [60,61]. Researchers agree that the rheological measurements provide potentially the most sensitive methods for nanocomposites characterization, which are vital for the processing of nanocomposites than the most commonly used transmission electron microscopy (TEM) and x-ray diffraction (XRD) [62]. Dynamic rheology has been used to estimate the dispersion quality of carbon nanotubes (CNTs) in a polystyrene matrix [63]. Lozano et al. [64] observed a rheological threshold between 10 and 20 wt % for the melted CNFs–polypropylene nanocomposites. A strong correlation is found for the rheological percolation and other physical percolation properties, such as electrical conductivity in the CNTs–polystyrene nanocomposites, [65,66] mechanical properties of copper nanowire–polystyrene nanocomposites [67,68], and microwave properties in acrylic, polyurethane, and epoxy composites containing carbon nanoparticles [69]. Rheological investigation reveals a gelation (gel point) of 1.6 wt% CNTs in a polycarbonate matrix, which coincides with the percolation threshold of the electrical

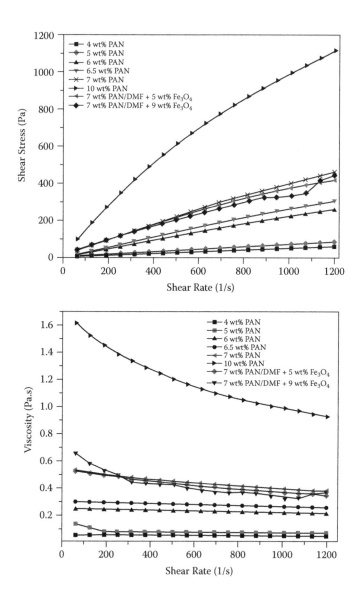

FIGURE 4.10

Rheological behavior (a) shear stress and (b) viscosity versus shear rate of pure PAN and PAN/ Fe3O4 solution system. (From D. Zhang et al. *Polymer*, 50, 4189–4198 [2009]. Reproduced with permission of Elsevier.)

conductivity and the highest strength in CNTs–polymer composites with the same loading [70].

Zhang et al. reported rheological behavior of the electrospun PAN–Fe_3O_4 nanocomposite as shown in Figure 4.10 [53]. They found that the solution viscosity (an indicator of polymer concentration) and solution surface tension play important roles in determining the morphology of the electrospun nanofibers. Under relatively low concentration conditions (4.0 wt%, 5.0 wt%, 6.0 wt%, and 6.5 wt%), the viscosity of the solutions is observed to decrease slightly. However, when the solution concentration increased to 7.0 wt% from 6.5 wt%, the viscosity decreased significantly with an increase in the shear rate. A further increase in the concentration to 10 wt% also produced another sharp drop in the viscosity. The viscosity dropped from 1.62 Pa.s to 0.92 Pa.s, when the shear rate reached 1200 s^{-1}.

Non-Newtonian properties of fluids are governed by the following power law equation:

$$\tau = K\left(\frac{\partial u}{\partial y}\right)^n, \tag{4.1}$$

where τ, is the shear stress, K is the flow consistency index, $\partial u/\partial y$ is the shear rate, and n is the flow behavior index. For Newtonian fluids, $n = 1$, and $n < 1$ for a pseudoplastic fluid. The values of n and R^2 (statistical correlation coefficient) are summarized in Table 4.1 [53].

The larger deviation of n from 1, the more non-Newtonian behavior the fluids would follow. Table 4.1 and Figure 4.10 show that solutions of lower concentration follow more Newtonian behavior at high shear rates than that of the concentrated solutions. When the concentration increased to 7.0 wt%, the PAN/DMF solution shows more pseudoplastic behavior, which exhibits the shear-thinning phenomena (the viscosity decreased nearly linearly with the increase in the shear rate). The orientation of macromolecular chains is the major cause of non-Newtonian behavior. With the increase in the shear rate, the number of the oriented polymer segments increases, which decreases the viscosity of the high concentration PAN/DMF, greatly increasing the non-Newtonian behavior [71].

TABLE 4.1

The Values of n and R^2 for PAN/DMF Solutions

Solutions	R^2	n
4 wt% PAN solution	−0.01	0.9327
5 wt% PAN solution	−0.06	0.976
6 wt% PAN solution	−0.04	0.905
6.5 wt% PAN solution	−0.05	0.857
7 wt% PAN solution	−0.12	0.921
10 wt% PAN solution	−0.18	0.947
7 wt% PAN/DMF and 5 wt% NPs solution	−0.15	0.927
7 wt% PAN/DMF and 9 wt% NPs solution	−0.23	0.96

Source: D. Zhang, A. B. Karki, D. Rutman, D. P. Young, A. Wang, D. Cocke, T. H. Ho, and Z. Guo, *Polymer,* 50, 4189–4198 (2009). Reproduced with permission of Elsevier.

4.4 Applications of Magnetic Polymer Nanocomposites

The giant magnetoresistance (GMR) effect indicates the significant change of resistance in response to an external magnetic field. Under zero external field, the magnetization directions of adjacent ferromagnetic layers in a GMR material are antiparallel due to a weak antiferromagnetic coupling between the layers. When an applied external field aligns the magnetization in adjacent layers, a lower level of resistance is observed. Since the first discovery of GMR in thin-film structures composed of alternating ferromagnetic and nonmagnetic metal layers in 1988 [72], GMR sensors have spurred many commercial applications, such as in magnetic storage (read heads in modern hard drives) [73], magnetic random access memory (MRAM), and biological detection [74,75]. Compared with the traditional metal matrix composites, polymer nanocomposites containing magnetic nanoparticles are more attractive because of their high flexibility, easy processing, and low cost.

With PPy as the substrate, Yan et al. noticed an interesting GMR behavior with a value of 4% in the cobalt and copper multilayers, as shown in Figure 4.11 [38]. The copper layer serves as a spacer for cobalt layers, similar to the conventional metal electrode system. However, there is no report regarding conductive polymer layers as ferromagnetic layer spacers. The

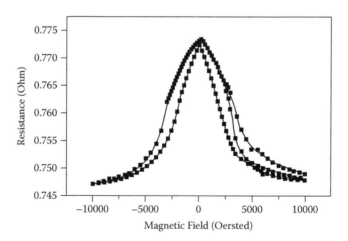

FIGURE 4.11

Resistance versus in-plane magnetic field curves at room temperature for electrodeposited (30 repeats) Co (2 nm)/Cu (3 nm) on conductive Ppy film (thickness 5 mm). (From F. Yan et al. *J. Mater. Chem.*, 12, 2606–2608 [2002]. Reproduced by permission of the Royal Society of Chemistry.)

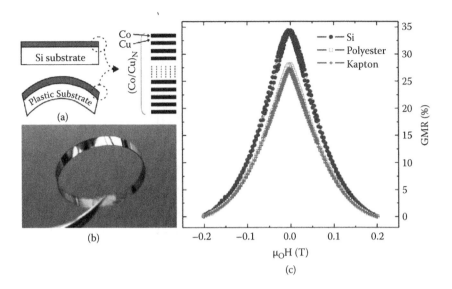

FIGURE 4.12

(a) Schematic illustration of (Co/Cu)N multilayers deposited on Si and flexible substrates, (b) a photographic image of circularly bended (Co/Cu)20 MLs deposited on polyester substrate, and (c) GMR curves of (Co/Cu)20 MLs deposited on polyester, Kapton, and thermally oxidized Si substrates. (From Y. M. Yuan-fu Chen et al. *Adv. Mater.*, 20, 3224–3228 [2008].)

reported value is lower than that of the pure metal system. Three reasons were deduced to interpret the difference, that is, the quality of the cobalt layer (with more copper in the ferromagnetic cobalt layers for PPy substrate), the roughness of the PPy thin film (rougher compared with the conventional metal), and the low conductivity of the used PPy film.

The flexibility of the plastic substrate-based GMR is shown in Figure 4.12a,b, and the observed GMR values in polyester is lower than that in the silicon and Kapton substrates as seen in Figure 4.12c [76]. By minimizing the roughness of the plastic substrate with a suitable buffer layer (2 μm photoresist), Chen et al. [76] have improved the roughness of the plastic, which is comparable to that of the conventional Si substrate, as seen in Figure 4.13a. The number of bilayers was found to have a significant effect on the GMR values (Figure 4.13b and c).

Granular polymer nanocomposites have also been investigated for possible GMR application [71,77]. The conductive iron nanoparticle-reinforced vinyl ester resin nanocomposites are observed to possess MR of only 0.9%, which increased to 1.7% after the carbonization (Figure 4.14a). The particle loading has been found to have significant effect on the GMR values (Figure 4.14b). Optimizing the particle loading has great potential to achieve further increases in GMR.

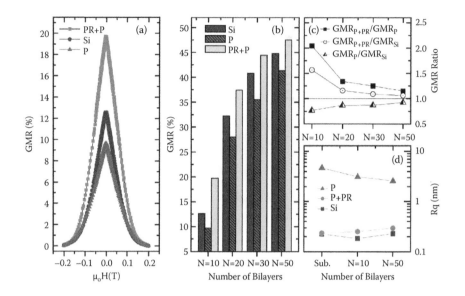

FIGURE 4.13
(a) GMR curves of (Co/Cu) 10 multilayers deposited on Si, polyester (P), photoresist-buffered polyester (PR+R) substrates. (b) GMR comparison of (Co/Cu)N MLs on various substrates with different numbers of bilayers. (c) GMR ratio of (Co/Cu)N multilayers deposited on Si, P, and PR+P substrates. (d) Root mean square roughness, Rq, of Si, P, PR+P bare substrate and Co/Cu films on corresponding substrates bare substrate and Co/Cu films on corresponding substrates. (From Y. M. Yuan-fu Chen et al. *Adv. Mater.*, 20, 3224–3228 [2008].)

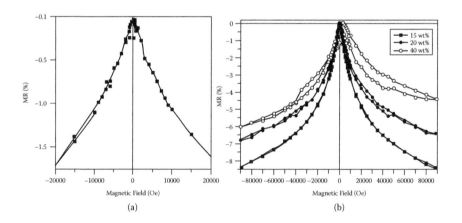

FIGURE 4.14
(a) The MR of the annealed monomer-stabilized nanoparticles and (b) room temperature MR as a function of applied field for heat-treated nanocomposites reinforced with a particle loading of 15, 20, and 40 wt %, respectively. (From Z. Guo et al. *J. Appl. Phys.*, 104, 014314 [2008]. Reprinted with permission from the American Institute of Physics.)

4.5 Summary

In this chapter, we mainly take the iron oxide-reinforced Ppy nanocomposites as an example for fabrication, processing, and physicochemical property analyses. The effect of iron oxide nanoparticles on the chemical polymerization of pyrroles in an acidic solution was investigated and found to significantly influence the morphology (size and shape) and other physicochemical properties of the Ppy. Pure discrete PPy nanoparticles with a much higher resistivity were formed over a long-time reaction in the presence of iron oxide nanoparticles. Similar to the pure PPy formed with the nanoparticles and different from the network structure of the pure PPy formed without nanoparticles, the discrete nanoparticles were observed in all the nanocomposites with an initial particle loading of 20 wt% and 50 wt%. Subsequent nanocomposites were observed to have an improved thermal stability with a higher decomposition temperature. FT-IR, and TGA/DTG analyses indicated a strong interaction between the nanoparticles and polymer matrix. PPy was observed to have a lower yield in the nanocomposite with a higher initial particle loading. The saturation magnetization in the nanocomposite with high particle loading was larger, and the conductive behavior follows the three-dimensional variable range-hopping mechanism. The presence of the nanoparticles in the nanocomposites was observed to produce a more condensed structural PPy. The decreased conductivity in the high particle loading is due to the insulating behavior of the iron oxide. Compared to the mechanical stirring, ultrasonic stirring plays a critical role in the iron oxide–PPy nanocomposite formation and provides protection of nanoparticles from proton dissolution arising from a tight PPy matrix formed around the iron oxide nanoparticles.

References

1. Z. Guo, S. E. Lee, H. Kim, S. Park, H. T. Hahn, A. B. Karki et al. *Acta Mater.*, 57, 267–77 (2009).
2. Z. Guo, H. T. Hahn, H. Lin, A. B. Karki, and D. P. Young, *J. Appl. Phys.*, 104, 014314 (2008).
3. Z. Guo, S. Park, H. T. Hahn, S. Wei, M. Moldovan, A. B. Karki et al. *Appl. Phys. Lett.*, 90, 053111 (2007).
4. X. C. Sun, N. Nava, and G. J. Reyes, *Proceed. Mater. Res. Soc.*, 704, W9.5.2W.9.5.6. (2002).
5. L. Diaz, M. Santos, C. Ballesteros, M. Marysko, and J. J. Pola, *J. Mater. Chem.* 15, 4311–4317 (2005).
6. H. Cao, G. Huang, S. Xuan, Q. Wu, F. Gu, and C. Li, *J. Alloys Compd.* 448, 272–276 (2008).
7. N. Fan, X. Ma, Z. Ju, and J. Li, *Mater. Res. Bull.*, 43, 1549–1554 (2008).

8. Z. Guo, M. Moldovan, D. P. Young, L. L. Henry, and E. J. Podlaha, *J. Electrochem. Solid-State Lett.*, 10, E31–E35 (2007).

9. X. W. Wei, G. X. Zhu, C. J. Xia, and Y. Ye, *Nanotechnology*, 17, 4307–4311 (2006).

10. S. R. Dhage, Y. B. Khollam, H. S. Potdar, S. B. Deshpande, P. P. Bakare, S. R. Sainkar, and S. K. Date, *Mater. Lett.*, 57, 457–462 (2002).

11. Z. Guo, K. Shin, A. B. Karki, D. P. Young, and H. T. Hahn, *J. Nanoparticle Res.* 11(6), 1441–1442 (2009).

12. B. Z. Tang, Y. Geng, J. W. Y. Lam, B. Li, X. Jing, X. Wang, F. Wang, A. B. Pakhomow, and X. X. Zhang, *Chem. Mater.*, 11, 1581–1589 (1999).

13. B. Z. Tang, Y. Geng, Q. Sun, X. X. Zhang, and X. Jing, *Pure Appl. Chem.*, 72, 157–162(2000).

14. J. Deng, X. Ding, W. Zhang, Y. Peng, J. Wang, X. Long, P. Li, and A. S. C. Chan, *Polymer*, 43, 2179–2184 (2002).

15. Y. B. Khollam, S. R. Dhage, H. S. Potdar, S. B. Deshpande, P. P. Bakare, S. D. Kulkarni, and S. K. Date, *Mater. Lett.*, 56, 571–577 (2002).

16. O. Y. Posudievskii and V. D. Pokhodenko, *Theor. Exp. Chem.*, 32, 213–216 (1996).

17. I. L. Radtchenko, G. B. Sukhorukov, and H. Mohwald, *Int. J. Pharm.*, 242, 219–223 (2002).

18. N. Murillo, E. Ochoteco, Y. Alesanco, J. A. Pomposo, J. Rodriguez, J. Gonzalez, J. J. del. Val, J. M. Gonzalez, M. R. Britel, F. M. Varela-Feria, and A. R. de. Arellano-Lopez, *Nanotechnology*, 15, S322–S327 (2004).

19. P. Poddar, J. L. Wilson, H. Srikanth, S. A. Morrison, and E. E. Carpenter, *Nanotechnology*, 15, S570–S574 (2004).

20. Y. Li, H. Zhang, Y. Liu, Q. Wen, and J. Li, *Nanotechnology*, 19, 105605 (2008).

21. V. R. Inturi and J. A. Barnard, *J. Appl. Phys.*, 81, 4504 (1997).

22. C. Baker, S. I. Shah, and S. K. Hasanain, *J. Magn. Magn. Mater.*, 280, 412–418 (2004).

23. F. Mammeri, E. L. Bourhis, L. Rozes, and C. J. Sanchez, *Mater. Chem.*, 15, 3787–3811 (2005).

24. P. Judeinstein and C. J. Sanchez, *J. Mater. Chem.*, 6, 511–525 (1996).

25. J. L. Wilson, P. Poddar, N. A. Frey, H. Srikanth, K. Mohomed, J. P. Harmon, S. Kotha, and J. Wachsmuth, *J. Appl. Phys.*, 95, 1439–1443 (2004).

26. E. Marutani, S. Yamamoto, T. Ninjbadgar, Y. Tsujii, T. Fukuda, and M. Takano, *Polymer*, 45, 2231–2235 (2004).

27. X. Xu, G. Friedman, K. D. Humfeld, S. A. Majetich, and S. A. Asher, *Chem. Mater.*, 14, 1249–1256 (2002).

28. J. Fang, L. D. Tung, K. L. Stokes, J. He, D. Caruntu, W. L. Zhou, and C. J. O'Connor, *J. Appl. Phys.*, 91, 8816–8818 (2002).

29. Z. Guo, L. L. Henry, V. Palshin, and E. J. Podlaha, *J. Mater. Chem.*, 16, 1772–1777 (2006).

30. M. Moniruzzaman and K. I. Winey, *Macromolecules*, 39, 5194–5205 (2006).

31. A. Sellinger, P. M. Weiss, A. Nguyen, Y. Lu, R. A. Assink, W. Gong, and C. J. Brinker, *Nature*, 394, 256–394 (1998).

32. P. Jiang and M. J. McFarland, *J. Am. Chem. Soc.*, 126, 13778–13786 (2004).

33. Z. Guo, K. Shin, A. B. Karki, D. P. Young, and H. T. Hahn, *J. Nanoparticle Res.* in press, DOI 10.1007/s11051-008-9531-8 (2009).

34. X. Li, M. Wan, Y. Wei, J. Shen, and Z. Chen, *J. Phys. Chem. B*, 110, 14623–14626 (2006).

35. X. Yang, L. Xu, S. C. Ng, and S. O. H. Chan, *Nanotechnology*, 14, 624–629 (2003).

36. K. Sunderland, P. Brundetti, L. Spinu, J. Fang, Z. Wang, and W. Lu, *Mater. Lett.*, 58, 3136–3140 (2004).
37. S. Takeoka, T. Hara, K. Yamamoto, and E. Tsuchida, *Chem. Lett.*, 25, 253 (1996).
38. F. Yan, G. Xue, and F. Wan, *J. Mater. Chem.*, 12, 2606–2608 (2002).
39. M. Abe and Y. Tamaura, *J. Appl. Phys.*, 55, 2614–2616 (1984).
40. M. Abe, T. Itoh, Y. Tamaura, Y. Gotoh, and M. Gomi, *IEEE Trans. Magn.*, 23, 3736–3738 (1987).
41. O. Jarjayes, P. H. Fries, and C. Bidan, *Synth. Met.*, 69, 343–344 (1995).
42. F. Yan, G. Xue, J. Chen, and Y. Lu, *Synth. Met.*, 123, 17–20 (2001).
43. C. Forder, S. P. Armes, A. W. Simpson, C. Maggiore, and M. Hawley, *J. Mater. Chem*, 3, 563–569 (1993).
44. O. J. Murphy, G. D. Hitchens, D. Hodko, E. T. Clarke, D. L. Miller, and D. L. Parker, Vol. United States Patent 5855755 (1999).
45. D. Li and Y. Xia, *Adv. Mater.*, 16, 1151–1170 (2004).
46. R. Chen, S. Zhao, G. Han, and J. Dong, *J. Mater Lett.*, 62, 4031–4034 (2008).
47. M. M. Demir, I. Yilgor, E. Yilgor, and B. Erman, *Polymer*, 43, 3303–3309 (2002).
48. V. Beachley and X. Wen, *Mater. Sci. Eng. C*. 29, 663–668 (2009).
49. J. H. Yu and G. C. Rutledge, *Encyclopedia of Polymer Science and Technology*, John Wiley & Sons, 2007, pp. 1–20.
50. S. Agarwal, J. H. Wendorff, and A. Greiner, *Polymer*, 49, 5603–5621 (2008).
51. T. Uyar, A. Balan, L. Toppare, and F. Besenbacher, *Polymer*, 50, 475–480 (2009).
52. Z. Sun, E. Zussman, A.L. Yarin, J. H., and Wendorff, A. Greiner, *Adv. Mater.*, 15, 1929–1932 (2003).
53. D. Zhang, A. B. Karki, D. Rutman, D. P. Young, A. Wang, D. Cocke, T. H. Ho, and Z. Guo, *Polymer*, 50, 4189–4198 (2009).
54. S. Sepulveda-Guzman, L. Lara, O. Perez-Camacho, O. Rodriguez-Fernandez, A. Olivas, and R. Escudero, *Polymer*, 48, 720–727 (2007).
55. W. Chen, X. Li, G. Xue, Z. Wang, and W. Zou, *Appl. Surf. Sci.*, 218, 215–221 (2003).
56. R. M. Cornell and U. Schwertmann, *The Iron Oxide*, 2nd ed. Wiley-VCH GmbH & Co. KGaA, Weinheim, 2003.
57. M. Pregelj, P. Umek, B. Drolc, B. Jancar, Z. Jaglicic, R. Dominko, and D. Arcon, *J. Mater. Res.*, 21, 2955–2962 (2006).
58. K. Suri, S. Annapoorni, and R. P. Tandon, *Bull. Mater. Sci.*, 24, 563–567 (2001).
59. K. Suri, A. Annapoorni, R. P. Tandon, C. Rath, and V. K. Aggrawal, *Curr. Appl. Phys.*, 3, 209–213 (2003).
60. R. Krishnamoorti, R. A. Vaia, and E. P. Giannelis, *Chem. Mater.*, 8, 1728–1734 (1996).
61. R. Krishnamoorti and E. P. Giannelis, *Macromolecules*, 30, 4097–4102 (1997).
62. L. A. Utracki, *Rapra Technology*, Shawbury, Shrewsbury, U.K., 2004.
63. Q. Zhang, F. Fang, X. Zhao, Y. Li, M. Zhu, and D. Chen, *J. Phys. Chem.*, 112, 12606–12611 (2008).
64. K. Lozano, J. Bonilla-Rios, and E. V. Barrera, *J. Appl. Polym. Sci.*, 80, 1162–1172 (2001).
65. A. K. Kota, B. H. Cipriano, M. K. Duesterberg, A. L. Gershon, D. Powell, S. R. Raghavan, and H. A. Bruck, *Macromolecules*, 40, 7400–7406 (2007).
66. F. Du, R. C. Scogna, W. Zhou, S. Brand, J. E. Fischer, and K. I. Winey, *Macromolecules*, 37, (24), 9048–9055 (2004).
67. T. K. Scott, F. D. Jack, and W. S. Francis, *J. Polym. Sci., Part B: Polym. Phys.*, 45, 1882–1897 (2007).

68. B. Lin, G. A. Gelves, J. A. Haber, and U. Sundararaj, *Ind. Eng. Chem. Res.*, 46, 2481–2487 (2007).
69. R. Kotsilkova, D. Nesheva, I. Nedkov, E. Krusteva, and S. Stavrev, *J. Appl. Polym. Sci.*, 92, 2220–2227 (2004).
70. C. Liu, J. Zhang, J. He, and G. Hu, *Polymer*, 44, 7529–7532 (2003).
71. Z. Guo, S. Park, H. T. Hahn, S. Wei, M. Moldovan, A. B. Karki, and D. P. Young, *Appl. Phys. Lett.*, 90, 053111 (2007).
72. M. N. Baibich, J. M. Broto, A. Fert, F. Nguyen Van Dau, F. Petroff, P. Eitenne, G. Creuzet, A. Friederich, and J. Chazelas, *Phys. Rev. Lett.*, 61, 2472–5 (1988).
73. A. Moser, K. Takano, D. T. Margulies, M. Albrecht, Y. Sonobe, Y. Ikeda, S. H. Sun, and E. E. Fullerton, *J. Phys. D: Appl Phys*, 35, R157–R167 (2002).
74. R. L. Edelstein, C. R. Tamanaha, P. E. Sheehan, M. M. Miller, D. R. Baselt, L. J. Whitman, and R. J. Colton, *Biosens. Bioelectron.*, 14, 805–813 (2000).
75. M. M. Miller, P. E. Sheehan, R. L. Edelstein, C. R. Tamanaha, L. Zhong, S. Bounnak, L. J. Whitman, and R. J. Colton, *J. Magn. Magn. Mater.*, 225, 138–144 (2001).
76. Y. M. Yuan-fu Chen, Rainer Kaltofen, Jens Ingolf Mönch, Joachim Schumann, Jens Freudenberger, Hans-Jörg Klauß, and Oliver G. Schmidt, *Adv. Mater.*, 20, 3224–3228 (2008).
77. Z. Guo, H. T. Hahn, H. Lin, A. B. Karki, and D. P. Young, *J. Appl. Phys.*, 104, 014314 (2008).

5

Carbon-Nanotube-Based Composites and Damage Sensing

Chunyu Li, Erik T. Thostenson, and Tsu-Wei Chou

University of Delaware

CONTENTS

5.1 Introduction

Since the discovery of carbon nanotubes was reported by S. Iijima in 1991 [1], numerous investigators have reported remarkable physical and mechanical properties [2–5]. The density of a single-walled carbon nanotube (SWCNT) is about 1.33–1.40 g/cm³, which is just half the density of aluminum. The elastic modulus of SWCNT is about 1.0 TPa and comparable to that of diamonds (1.2 TPa). The reported tensile strength (30 – 150 GPa) of SWCNT is much higher than that of high-strength steel (2 GPa). The tremendous resilience of SWCNT in sustaining bending to large angles and restraightening without damage is distinctively different from the plastic deformation of metals and brittle fracture of carbon fibers at much lower strain when subjected to the same type of deformation. The electrical current carrying capability of SWCNTs is estimated to be 1×10^9 amp/cm^2, whereas copper wires burn out at about 1×10^6 amp/cm^2. The thermal conductivity of SWCNT is predicted to be 6000 W/m·K at room temperature; this is nearly double the thermal conductivity of diamond of 3320 W/m·K. SWCNTs are stable up to 2800°C in vacuum, and 750°C in air, whereas metal wires in microchips melt at 600°C –1000°C. SWCNTs have great potential in field emission applications because they can activate phosphors at 1–3 V if electrodes are spaced 1 µm apart. Traditional Mo tips require fields of 10–100 V/µm and have very limited lifetimes. The outstanding thermal and electrical properties, combined with their high specific stiffness and strength, and very large aspect ratios, have stimulated the development of carbon nanotube-based applications for both *structural* and *functional purposes*.

The interests in applications of carbon nanotubes have encompassed nanobiotechnology, nanosystems, nanoelectronics, and nanostructured materials. The great progress in producing large numbers of nanotubes has stimulated research to create multifunctional composite materials by designing microstructures using a bottom-up approach from the nanometer scale. The expansion of length scales from nanometers (nanotube diameter), sub-micrometers (fiber–matrix interphase), micrometers (fiber diameter) to meters (finished woven composite parts) presents tremendous challenges and also opportunities for innovative approaches in the processing, characterization, and analysis/modeling of this new generation of composite materials (Figure 5.1).

To date, both single-walled and multiwalled carbon nanotubes have been utilized for reinforcing thermoset polymers (epoxy, polyimide, and phenolic), as well as thermoplastic polymers (polypropylene, polystyrene, poly methyl methacrylate [PMMA], nylon 12, poly ether ether ketone [PEEK]). The processing and characterization of these carbon nanotube-based composite systems have been detailed in other chapters in this book.

In this chapter, we mainly focus on the nanomechanics of carbon nanotubes and modeling of carbon nanotube-based composites. An atomistic modeling technique, termed *molecular structural mechanics*, is introduced first. Then, various physical properties, such as elastic modulus, shear modulus, coefficients of thermal expansion, etc., of carbon nanotubes predicted by this modeling technique are described and compared with existing experimental or theoretical results. Next, because of the importance of nanotube applications as sensors and actuators, the modeling of electromechanical coupling behaviors of carbon nanotubes is described to help explain nanotube deformations and failures under electrical charges. Then, we introduce the modeling of mechanical properties of nanotube-based composites. Two modeling techniques, constitutive modeling and multiscale modeling, are described. Further, we introduce the modeling of electrical conductivity of

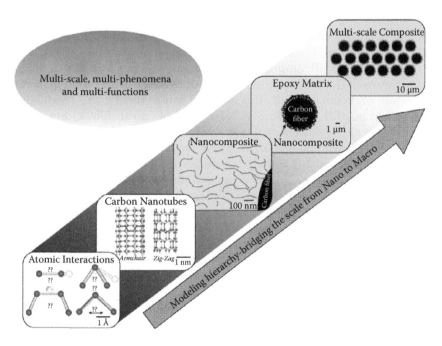

FIGURE 5.1
Multiscale nanotube-based composites.

nanotube-based composites based on percolation theory. The conductive nanotube-based composites are then used for damage self-sensing applications. Both experimental and theoretical studies are introduced. The contents of the following sections are mainly extracted from publications of the authors as indicated by the references.

5.2 Molecular Structural Mechanics Model of Nanotubes

Many researchers have pursued the analysis of carbon nanotubes by theoretical modeling. The modeling approaches can be generally classified into two categories: atomistic models and continuum models. Atomistic models include molecular dynamics (MD), tight-binding molecular dynamics (TBMD) and density functional theory (DFT), etc. Continuum models include shell models and beam models, etc. In principle, any problem associated with molecular or atomic motions can be simulated by atomistic modeling techniques. However, due to their huge computational tasks, practical applications of these atomistic modeling techniques are limited to systems containing a small number of molecules or atoms, and are usually confined to studies of relatively short-lived phenomena, from picoseconds to nanoseconds. On the other hand, continuum models neglect the detailed characteristics of nanotube chirality, and are unable to account for forces acting on the individual atoms.

Based on the demand for developing a modeling technique that analyzes the mechanical responses of nanotubes at the atomistic scale but not perplexed in time scales, Li and Chou [6] proposed a molecular structural mechanics approach. It has been successfully applied for studying nanotube properties and simulating nanotube-based composites. The idea stems from the nature of nanotubes as elongated fullerenes (named after the architect known for designing geodesic domes, R. Buckminster Fuller). By observation, they found some similarities between the molecular model of a nanotube and a space frame model of building structures. Understanding this approach needs some knowledge of molecular mechanics and structural mechanics. Let us first look at the structural characteristics of carbon nanotubes.

5.2.1 Structural Characteristics of Carbon Nanotubes

The first transmission electron microscopy evidence for the tubular nature of some nano-sized carbon filaments is believed to have appeared in 1952 in the *Journal of Physical Chemistry* in Russia. Radushkevich and Lukyanovich should be credited for the discovery of carbon nanotubes [7]. But the graphitic structure of carbon nanotubes was first observed in 1991 by the Japanese electron microscopist Sumio Iijima, who was studying the material deposited on the cathode during the arc-evaporation synthesis of fullerenes. Iijima's paper had a tremendous impact on nanotube research due to the right combination of favorable factors: a high-quality paper and a top-ranked journal, a boost received from its relation to the earlier worldwide research on fullerenes and a fully mature scientific audience ready to surf on the "nano" wave.

A single-walled carbon nanotube (SWCNT) can be viewed as a sheet of graphite that has been rolled into a tube. The atomistic structure of nanotubes can be described in terms of the tube chirality, or helicity, which is defined by the chiral vector C_h and the chiral angle θ. In Figure 5.2a, we can visualize cutting the graphite sheet along the dotted lines and rolling the tube so that the tip of the chiral vector touches its tail. The chiral vector, also known as the roll-up vector, can be described by the following equation

$$C_h = n\mathbf{a}_1 + m\mathbf{a}_2, \tag{5.1}$$

where (n, m) are the number of steps along the zigzag carbon bonds of the hexagonal lattice and $\mathbf{a}_1$ and $\mathbf{a}_2$ are unit vectors. The chiral angle

$$\theta = \arctan[\sqrt{3}m/(2n+m)], \tag{5.2}$$

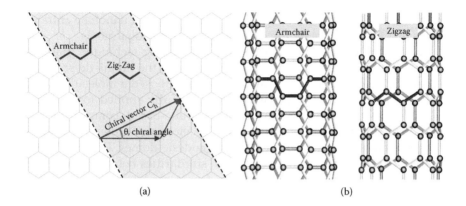

FIGURE 5.2
Illustration of (a) a graphene and (b) armchair and zigzag nanotubes. (From Thostenson et al., *Comp. Sci. Technol.* 61 [2001] 1899–1912.)

determines the amount of 'twist' in the tube. The chiral angles are 0° and 30° for the two limiting cases, which are referred to as "zigzag" and "armchair," respectively, as shown in Figure 5.2b. In terms of the roll-up vector, the zigzag nanotube is denoted by $(n, 0)$ and the armchair nanotube (n, n). The roll-up vector of the nanotube also defines the nanotube diameter:

$$d = \frac{|\mathbf{C}_h|}{\pi} = \frac{a}{\pi}\sqrt{3(n^2 + nm + m^2)}, \qquad (5.3)$$

where a is the carbon–carbon bond length.

A multiwalled carbon nanotube (MWCNT) is composed of concentric SWCNTs of different radii with closed caps at both ends. Since the nested layers are structurally independent of one another, the chirality of the layers may be different. The distance between two neighboring layers is assumed to be the same as the spacing between adjacent graphene sheets in graphite (i.e., 0.34 nm). Unlike diamonds, which assume a 3D crystal structure with each carbon atom having its four nearest neighbors arranged in a tetrahedron, within the atomic layer of a graphene, each carbon atom is covalently bonded to three neighboring carbon atoms. Three sp^2 orbitals on each carbon form σ-bonds to three other carbon atoms. One $2p$ orbital remains unhybridized on each carbon; these orbitals perpendicular to the plane of the carbon ring combine to form the π-bonding network. The atomic interactions between the neighboring layers are the van der Waals forces.

From the structural characteristics of carbon nanotubes (as well as other nanostructures), it is observed that some similarities exist between nanostructures and building structures, as shown in Figure 5.3. Thus, it is logical to anticipate that the analysis method in structure mechanics may be potentially applicable to the studies of carbon nanotubes as well as other

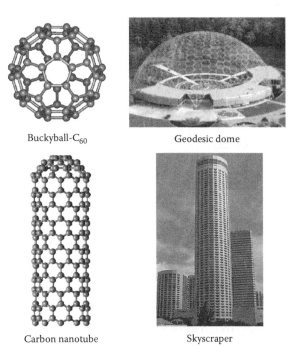

Buckyball-C_{60} Geodesic dome

Carbon nanotube   Skyscraper

FIGURE 5.3
Similarity between nanostructures and building structures.

nanostructures. The question is how to establish the linkage between a nanostructure and a building structure.

5.2.2 Molecular Mechanics of Carbon Nanotubes

To establish a linkage between a nanostructure and a building structure, we need first to have a better understanding about the molecular mechanics description of nanoscopic carbon nanotubes. From the viewpoint of molecular mechanics, a carbon nanotube can be regarded as a large molecule consisting of carbon atoms. The atomic nuclei can be regarded as material points. Their motions are regulated by a molecular force field, which is generated by electron–nucleus interactions and nucleus–nucleus interactions. Usually, the molecular force field is expressed in the form of steric potential energy. It depends solely on the relative positions of the nuclei constituting the molecule.

The general expression of the total steric potential energy, omitting the electrostatic interaction, is a sum of energies due to valence or bonded interactions and nonbonded interactions [8]:

$$U = \sum U_r + \sum U_\theta + \sum U_\phi + \sum U_\psi + \sum U_{vdw}, \qquad (5.4)$$

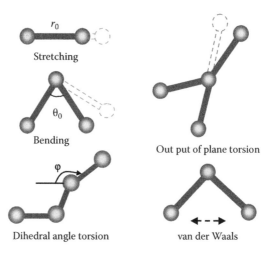

FIGURE 5.4
Interatomic interactions in molecular mechanics.

where U_r is for a bond stretch interaction, U_θ for a bond angle bending, U_φ for a dihedral angle torsion, U_ψ for an improper (out-of-plane) torsion, U_{vdw} for a nonbonded van der Waals interaction, as shown in Figure 5.4.

There has been a wealth of literature in molecular mechanics devoted to finding the reasonable functional forms of these potential energy terms [8–11]. Therefore, various functional forms may be used for these energy terms, depending on the particular material and loading conditions considered. In general, for covalent systems, the main contributions to the total steric energy come from the first four terms of Equation 5.4, which have included four-body potentials. Thus, the term of nonbounded interactions can be ignored. Under the assumption of small deformation, the harmonic approximation is adequate for describing the energy [12]. For the study of carbon nanotubes, the AMBER (Assisted Model Building with Energy Refinement) molecular force field and the Dreiding force field are commonly used because of their clear definitions and simplicity.

For example, the basic expressions of potential energies for describing the bond stretching, bond angle bending, dihedral angle torsion and inversion (out-of-plane torsion) in the Dreiding force field are

$$U_r = \tfrac{1}{2} K_r (r - r_0)^2, \tag{5.5}$$

$$U_\theta = \tfrac{1}{2} K_\theta (\theta - \theta_0)^2, \tag{5.6}$$

$$U = \tfrac{1}{2} K \{1 - \cos[2(\ - \ _0)]\}, \tag{5.7}$$

$$U_\psi = \tfrac{1}{2} K_\psi (\psi - \psi_0)^2, \tag{5.8}$$

where K_r, K_θ, K_φ, and K_ψ are force constants for the bond stretching, bond angle bending, dihedral torsion and inversion, respectively. The symbols r, θ, φ, and ψ represent the bond length, bond angle, dihedral angle, and inversion angle, respectively. The subscript 0 stands for the equilibrium position.

5.2.3 Structural Mechanic for Space Frame

Here, we consider the structural mechanics for macroscopic structures. Structural mechanics analysis enables the determination of the displacements, strains, and stresses of a structure under given loading conditions. Of the various modern structural analysis techniques, the stiffness matrix method has been by far the most generally used. The method can be readily applied to analyze structures of any geometry and can be used to solve linear elastic static problems as well as problems involving buckling, plasticity, and dynamics. In the following, let us briefly review the stiffness matrix method for linearly elastic space frame problems, which is relevant to the present studies.

For an element in a space frame as shown in Figure 5.5, the elemental equilibrium equation can be written as following [13]

$$\mathbf{Ku} = \mathbf{f} \tag{5.9}$$

where

$$\mathbf{u} = [u_{xi}, u_{yi}, u_{zi}, \theta_{xi}, \theta_{yi}, \theta_{zi}, u_{xj}, u_{yj}, u_{zj}, \theta_{xj}, \theta_{yj}, \theta_{zj}], \tag{5.10}$$

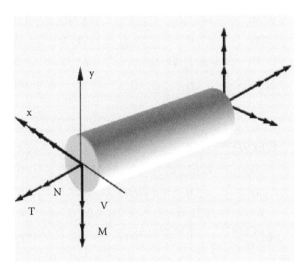

FIGURE 5.5
A beam element in a space frame. (From C. Y. Li and T. W. Chou. *Int. J. Solids Struct.* 40 [2003] 2487–2499.)

$$\mathbf{f} = [f_{xi}, f_{yi}, f_{zi}, m_{xi}, m_{yi}, m_{zi}, f_{xj}, f_{yj}, f_{zj}, m_{xj}, m_{yj}, m_{zj}], \tag{5.11}$$

are the nodal displacement vector and nodal force vector of the element, respectively, and $\mathbf{K}$ is the elemental stiffness matrix. The matrix $\mathbf{K}$ consists of the following submatrices:

$$\mathbf{K} = \begin{bmatrix} \mathbf{K}_{ii} & -\mathbf{K}_{ij} \\ -\mathbf{K}_{ij}^{T} & \mathbf{K}_{jj} \end{bmatrix}, \tag{5.12}$$

where

$$\mathbf{K}_{ii} = \begin{bmatrix} EA/L & 0 & 0 & 0 & 0 & 0 \\ 0 & 12EI_x/L^3 & 0 & 0 & 0 & 6EI_x/L^2 \\ 0 & 0 & 12EI_y/L^3 & 0 & -6EI_y/L^2 & 0 \\ 0 & 0 & 0 & GJ/L & 0 & 0 \\ 0 & 0 & -6EI_y/L^2 & 0 & 4EI_y/L & 0 \\ 0 & 6EI_x/L^2 & 0 & 0 & 0 & 4EI_x/L \end{bmatrix} \tag{5.13}$$

$$\mathbf{K}_{ij} = \begin{bmatrix} EA/L & 0 & 0 & 0 & 0 & 0 \\ 0 & 12EI_x/L^3 & 0 & 0 & 0 & -6EI_x/L^2 \\ 0 & 0 & 12EI_y/L^3 & 0 & 6EI_y/L^2 & 0 \\ 0 & 0 & 0 & GJ/L & 0 & 0 \\ 0 & 0 & -6EI_y/L^2 & 0 & -2EI_y/L & 0 \\ 0 & 6EI_x/L^2 & 0 & 0 & 0 & -2EI_x/L \end{bmatrix} \tag{5.14}$$

$$\mathbf{K}_{jj} = \begin{bmatrix} EA/L & 0 & 0 & 0 & 0 & 0 \\ 0 & 12EI_x/L^3 & 0 & 0 & 0 & -6EI_x/L^2 \\ 0 & 0 & 12EI_y/L^3 & 0 & 6EI_y/L^2 & 0 \\ 0 & 0 & 0 & GJ/L & 0 & 0 \\ 0 & 0 & 6EI_y/L^2 & 0 & 4EI_y/L & 0 \\ 0 & -6EI_x/L^2 & 0 & 0 & 0 & 4EI_x/L \end{bmatrix} \tag{5.15}$$

It is observed from the above elemental stiffness matrices that when the length, L, of the element is known, there are still four stiffness parameters, that is, tensile resistance EA, the flexural rigidities EI_x and EI_y, and the torsional stiffness GJ, need to be determined. In order to obtain the deformation of a space frame, the above elemental stiffness equations should be established for every element in the space frame, and then all these equations should be transformed from the local coordinates to a common global reference system. Finally, a system of simultaneous linear equations can be assembled according to the requirements of nodal equilibrium. By solving the system of equations and taking into account the boundary constraints, the nodal displacements can be obtained.

5.2.4 Molecular Structural Mechanics

In a carbon nanotube, the carbon atoms are bonded to each other by covalent bonds and form hexagons on the wall of the tube. These covalent bonds have their characteristic bond lengths and bond angles in a three-dimensional space. When a nanotube is subjected to external forces, the displacements of individual atoms are constrained by these bonds. The total deformation of the nanotube is the result of these bond interactions. If we consider covalent bonds as connecting elements between carbon atoms, and carbon atoms as joints of connecting elements, a nanotube could be simulated as a space frame-like structure.

For convenience, we can assume that the sections of carbon–carbon bonds are identical and uniformly round. Thus, we have sectional parameters $I_x = I_y = I$ and only three stiffness parameters, that is, EA, EI, and GJ, need to be determined for employing structural mechanics approach to conduct simulations on nanotubes. Because the deformation of a space frame results in changes of strain energy, the three stiffness parameters can be determined by considering energy equivalence. Notice that each of the energy terms in molecular mechanics (Equations 5.5–5.8) represents an individual interaction and no cross-interactions are included; we only need to consider the strain energies of structural elements under individual forces. But it should be noticed that there are four terms in potential energy and three terms in strain energy. Some modifications have to be made.

Under the assumption of small deformation, the harmonic approximation is adequate for describing the potential energies. In order to simplify the computation, a minor modification to the force field is made by merging the dihedral angle torsion and the inversion into a single term of equivalent torsion energy in the harmonic form, that is,

$$U_\tau = U_\phi + U_\psi = \tfrac{1}{2} k_\tau (- {}_0)^2. \tag{5.16}$$

Here, an approximation that the change of dihedral angle equals to the change of inversion angle is made. Because our method is based on the assumption of small deformation, this approximation is fairly reasonable. Figure 5.6 illustrates the energy variations of equivalent torsion as well as the dihedral angle

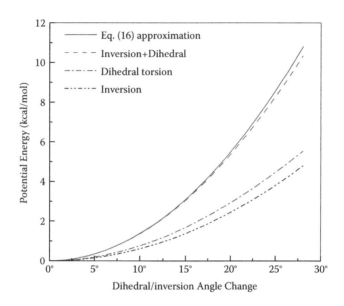

FIGURE 5.6
Combination of dihedral torsion and inversion ($k_\tau = 90$ kcal/mol·rad²). (From C. Y. Li and T. W. Chou, *J. Nanosci. Nanotechnol.* 6 [2006] 54.)

torsion and the inversion [14]. Because the equivalent force constant k_τ is adjustable, it can be seen that the approximation is very accurate for an appropriately chosen k_τ, when the angle change is small.

According to the theory of classical structural mechanics, the strain energy of a uniform beam of length L subjected to pure axial force N is

$$U_A = \frac{1}{2}\int_0^L \frac{N^2}{EA}dL = \frac{1}{2}\frac{N^2L}{EA} = \frac{1}{2}\frac{EA}{L}(\triangle L)^2, \tag{5.17}$$

where $\triangle L$ is the axial stretching deformation. The strain energy of a uniform beam under pure bending moment M is

$$U_M = \frac{1}{2}\int_0^L \frac{M^2}{EI}dL = \frac{2EI}{L}\alpha^2 = \frac{1}{2}\frac{EI}{L}(2\alpha)^2, \tag{5.18}$$

where α denotes the rotational angle at the ends of the beam. The strain energy of a uniform beam under pure torsion T is

$$U_T = \frac{1}{2}\int_0^L \frac{T^2}{GJ}dL = \frac{1}{2}\frac{T^2L}{GJ} = \frac{1}{2}\frac{GJ}{L}(\triangle \beta)^2 \tag{5.19}$$

where $\triangle \beta$ is the relative rotation between the ends of the beam.

By comparing potential energy in molecular mechanics and strain energy in structural mechanics, it is obvious that both U_r and U_A represent the stretching energy, both U_θ and U_M represent the bending energy, and both U_τ and U_T represent the torsional energy. It is reasonable to assume that the rotation angle 2α is equivalent to the total change $\Delta\theta$ of the bond angle, ΔL is equivalent to Δr, and $\Delta\beta$ is equivalent to $\Delta\phi$. Thus, a direct relationship between the structural mechanics parameters EA, EI, and GJ and the molecular mechanics parameters k_r, k_θ, and k_τ is deduced as follows:

$$\frac{EA}{L} = k_r, \quad \frac{EI}{L} = k_\theta, \quad \frac{GJ}{L} = k_\tau. \tag{5.20}$$

Equation 5.20 establishes the foundation of applying the theory of structural mechanics to the modeling of single-walled carbon nanotubes. As long as the force constants k_r, k_θ, and k_τ are known, the sectional stiffness parameters EA, EI, and GJ can be readily obtained.

For multiwalled carbon nanotubes, van der Waals forces between coaxial nanotube layers must be considered. The van der Waals force is a nonbonded interaction, and it can be an attraction force or a repulsion force. The attraction occurs when a pair of atoms approach each other within a certain distance. The repulsion occurs when the distance between the interacting atoms becomes less than the sum of their contact radii. These interactions are often modeled using the general Lennard–Jones 6–12 potential, which provides for a smooth transition between the attraction and repulsion regions.

The general Lennard–Jones 6–12 (LJ) potential is commonly expressed as

$$U(r) = 4\varepsilon\left[\left(\frac{\sigma}{r}\right)^{12} - \left(\frac{\sigma}{r}\right)^{6}\right], \tag{5.21}$$

where, r is the distance between interacting atoms, ε and σ are the Lennard–Jones parameters. For carbon atoms the Lennard–Jones parameters are $\varepsilon = 0.0556$ kcal/mol and $\sigma = 0.34$ nm. The potential $U(r)$ is usually truncated at an interatomic distance of 2.5σ without significant loss of accuracy, that is, no interactions are evaluated beyond this distance. Based on the Lennard–Jones potential, the van der Waals force between interacting atoms can then be written as

$$F(r) = -\frac{dU(r)}{dr} = 24\frac{\varepsilon}{\sigma}\left[2\left(\frac{\sigma}{r}\right)^{13} - \left(\frac{\sigma}{r}\right)^{7}\right]. \tag{5.22}$$

The van der Waals force acting along the connecting line between two interacting atoms can be simulated by a truss rod that connects the two interacting atoms with rotatable end joints. The truss rod thus transmits only tensile or compressive forces as given by Equation 5.22. In an MWCNT, an atom situated in one of these layers may form interacting pairs with several atoms in other layers as long as the distance between the pair of atoms is less than 2.5σ. But for convenience in computation, the van der Waals interactions may only be assumed between the nearest neighboring layers.

The truss rod connecting the atoms is a nonlinear element with its load-displacement relationship characterized by the van der Waals force. The load-displacement curve consists of two distinct stages in loading and unloading. This characteristic of load-displacement curve brings about some numerical difficulties in the simulation of the nonlinear behavior of the truss rod. Thus, the selection of a suitable nonlinear analysis method is needed. The available methods include arc-length method [15], work control method [16], and a generalized displacement control method [17]. The generalized displacement control method has been demonstrated to be more accurate and numerically stable at the critical and snap-back points.

Some researchers conducted further derivations based on Equation 5.20 and obtained bond parameters such as diameter, Young's modulus, shear modulus, and even Poisson's ratio. These parameters were then used in commercial software of finite element method, such as Abaqus, Ansys, etc. But that practice should be performed cautiously because the deduced parameters may have no clear physical meaning.

Also, it should be pointed out that the molecular structural mechanics introduced above is a versatile method for nanostructured materials, although it is introduced using carbon nanotubes as an example. Compared with other refined atomistic simulation methods, such as classical molecular dynamics or tight-binding molecular dynamics, the present approach possesses two advantages. One is that it can be used for both static and dynamic simulations. Another is that its computational efficiency for obtaining mechanical properties of nanomaterials is much higher than the efficiencies of other methods.

5.2.5 Static Properties of Carbon Nanotubes

The physical properties of carbon nanotubes are sensitive to their diameter and chirality. Graphite is considered to be a semimetal, but it has been shown that carbon nanotubes can be either metallic or semiconducting, depending on tube chirality. The influence of chirality on the mechanical properties of carbon nanotubes has also been reported [18,19].

Due to the difficulties in experimental characterization of nanotubes, computer simulation has been regarded as a powerful tool for predicting material properties of carbon nanotubes. Among the available modeling techniques, molecular dynamics simulation has been used most extensively. However, molecular dynamics simulations need to consider the thermal vibration of atoms, and thus the time step is usually in the order of femtoseconds. It is not an efficient method for simulating long time or static problems such as stiffness and strength properties. On the other hand, the molecular structural mechanics approach introduced above is very effective for studying such problems because it does not involve the time scale.

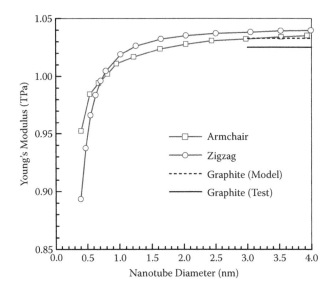

FIGURE 5.7
Young's moduli of single-walled carbon nanotubes. (From C. Y. Li and T. W. Chou. *Int. J. Solids Struct.* 40 [2003] 2487–2499.)

5.2.5.1 Young's Modulus of Carbon Nanotubes

As stated earlier, a carbon nanotube can be viewed as a sheet of graphite that has been rolled into a tube. To verify the reliability and efficiency of the molecular structural mechanics approach to the modeling of carbon nanotubes and obtain useful information concerning the selection of molecular force field constants, the Young's modulus of graphite was first calculated by Li and Chou [6]. In their computation, the force constants, $k_r/2 = 469$ kcal/mol·$Å^2$ and $k_\theta/2 = 63$ kcal/mol·rad^2, were adopted, and the initial carbon–carbon bond length was taken as 1.421 Å. The graphene thickness required for calculating the tensile stress was taken as the interlayer spacing of graphite, that is, 0.34 nm. The computed Young's moduli for multilayered graphite and a graphene sheet are about 1.03 TPa, which is in excellent agreement with the experimental data (1.025 TPa) [20].

Li and Chou conducted simulations on armchair and zigzag carbon nanotubes. The molecular force constants are chosen as $k_r/2 = 469$ kcal/mol·$Å^2$ and $k_\theta/2 = 63$ kcal/mol·rad^2, and $k_\tau/2 = 20$ kcal/mol·rad^2. Figure 5.7 displays the variations of the Young's modulus with nanotube diameter. The general tendency is that the Young's modulus increases with increasing tube diameter. For smaller tubes ($d < 1.0$ nm), the Young's modulus exhibits a strong dependence on the tube diameter. The lower Young's modulus at smaller diameter may be attributed to the higher curvature, which results in a more significant distortion of C–C bonds. As the nanotube diameter increases, the effect of curvature diminishes gradually, and the Young's

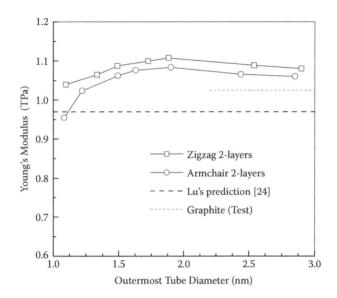

FIGURE 5.8
Young's moduli of multiwalled carbon nanotubes. (From C. Y. Li and T. W. Chou, *Comp. Sci. Technol.* 63 [2003] 1517–1524.)

modulus approaches that of graphite as predicted by the present method and reported in the literature. The effect of nanotube chirality seems to be not significant.

Li and Chou also conducted simulations on MWCNTs. As shown in Figure 5.8, there is a small variation in Young's modulus for MWCNTS. The difference between maximum and minimum Young's modulus values is about 0.15 TPa. The average Young's modulus is 1.05 ± 0.05 TPa and 1.08 ± 0.02 TPa for armchair and zigzag MWCNTs, respectively. A general conclusion that can be drawn is that the Young's moduli of MWCNTs are generally higher (~7% in maximum) than those of SWCNTs. That means the van der Waals force has some effect on the tensile modulus of MWCNTs. Regarding the experimental results of Young's moduli of MWCNTs, Wong et al. [21] and Salvetat et al. [22] reported values of 1.28 ± 0.59 TPa and 0.81 ± 0.41 TPa, respectively, using AFM-based experiments. These data are comparable to Li and Chou's predictions [23].

5.2.5.2 Shear Modulus of Carbon Nanotubes

Due to the difficulty in experimental techniques, there is still no report on the measured values of shear modulus of carbon nanotubes. Theoretical predictions on the shear modulus of carbon nanotubes are also very few. Lu [24] predicted the shear modulus for carbon SWCNTs by using an empirical lattice dynamics model, and concluded that the shear modulus (~0.5 TPa) is comparable to that of diamond and is insensitive to tube diameter and tube chirality. Popov et al. [18] also used the lattice dynamics model and derived an analytical expression for the shear modulus. Their results indicated that the shear moduli of carbon SWCNTs are about equal to that of graphite for large radii but are less than that of graphite at small radii, and the tube chirality has some effect on the shear modulus for small tube radii.

Using the molecular structural mechanics approach, Li and Chou [6] calculated the shear modulus of carbon nanotubes by considering nanotubes subjected to a torsional moment at one end (the torsion is only acting on the outermost layer) and constrained at the other end. The shear modulus is obtained based on the theory of elasticity at the macroscopic scale,

$$S = \frac{TL_0}{\theta J_0},$$ (5.23)

where T stands for the torque acting at the end of a nanotube, L_0 is the length of the tube, θ is the torsional angle, and J_0 is the cross-sectional polar inertia of the nanotube. In the calculation of the polar inertia J_0, the wall thickness of tube layer was taken as of 0.34 nm. In addition, the force field constants are $k_r/2 = 469$ kcal/mol Å^2, $k_\theta/2 = 63$ kcal/mol rad^2, $k_\tau/2 = 20$ kcal/mol rad^2, and Lennard–Jones parameters: $\sigma = 3.4$ Å, $\varepsilon = 0.0556$ kcal/mol.

Figure 5.9 shows the computational results of the shear moduli of SWCNTs and two-layered armchair and zigzag MWCNTs. On average, the shear modulus of an MWCNT is about 0.4 TPa and is lower than that of an SWCNT,

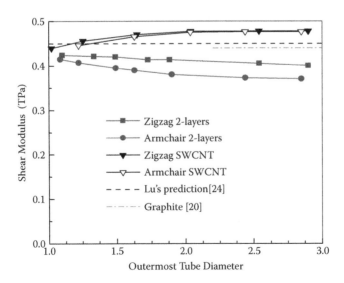

FIGURE 5.9
Shear modulus of nanotubes. (From C. Y. Li and T. W. Chou, *Comp. Sci. Technol.* 63 [2003] 1517–1524.)

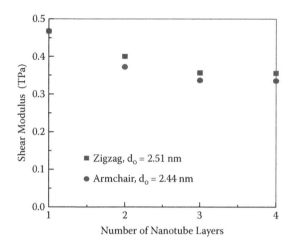

FIGURE 5.10
The effect of tube layers. (From C. Y. Li and T. W. Chou, *Comp. Sci. Technol.* 63 [2003] 1517–1524.)

which is about 0.45 TPa. It is also observed that the trends of variation of shear moduli with tube diameter are different for MWCNTs and SWCNTs. The shear moduli of SWCNTs first increase with the increase of tube diameter and then gradually become insensitive to the tube diameter. For MWCNTS, however, the shear moduli decrease with the increase of tube diameter. The effect of tube chirality is not significant.

Figure 5.10 summarizes the effect of number of tube layers on shear modulus. For both zigzag and armchair MWCNTs, the shear moduli are reduced when the number of tube layers increases. This is due to the fact that in the present simulation, it is assumed that only the outermost tube layer is subjected to applied torque. Because of the weak van der Waals forces between the neighboring tube layers, the inner tube layers do not contribute as effectively as the outermost tube layer in resisting the applied torque. However, all the layers have been taken into account in the calculation of cross-sectional polar inertia, which increases with the increase of the number of tube layers.

5.2.5.3 Transverse Elastic Modulus of Carbon Nanotubes

Knowledge of transverse elastic properties of nanotubes is essential in the study of the interfacial residual stresses and the failure mechanism of nanocomposites, as well as the mechanical performance of nanowire templates, hydrogen containers, and nanogears based on carbon nanotubes. The radial deformability of multiwalled carbon nanotubes was first examined by Ruoff et al. [25]. Their experiments showed that van der Waals forces between adjacent nanotubes can deform them substantially, destroying the cylindrical symmetry. Hertel et al. [26] also observed a shape change of nanotube cross sections, from circular to elliptical, at the point of overlap between two contacting nanotubes. Lordi and Yao [27] studied the response of carbon nanotubes to asymmetrical radial compressive forces by experiments and computer simulations. They concluded that the elasticity and resilience of nanotubes depend on the tube radius and the number of tube layers under compression. Shen et al. [28] characterized the mechanical properties of multiwalled nanotubes in radial direction using nanoindentation tests with a scanning probe microscope. They reported that the radial modulus increased from 9.7 to 80.0 GPa with increasing compressive stress, and the compressive strength was estimated to be higher than 5.3 GPa. Yu et al. [29] also conducted nanoindentation tests on multiwalled carbon nanotubes using a tapping-mode atomic force microscope, and reported the effective radial elastic modulus range of ~0.3–4.0 GPa. Single-walled carbon nanotubes were reported to be more deformable in the radial direction than multiwalled carbon nanotubes [30]. However, there is still a lack of quantitative measurement of the radial modulus of single-walled carbon nanotubes. Tang et al. [31] reported the deformation of single-walled carbon nanotubes under hydrostatic pressure using a diamond anvil cell and in situ x-ray diffraction. It was shown that single-walled carbon nanotubes maintain linear elasticity under hydrostatic pressure up to 1.5 GPa.

Li and Chou [32] simulated elastic deformation of SWCNTs in transverse directions using the molecular structural mechanics approach. A hydrostatic pressure loading condition was considered because the radial and circumferential deformations occurred simultaneously, and the stress and strain fields are uniform under such loading conditions. The radial (r) and circumferential (θ) stresses under external pressure P were approximated using the concept of continuum mechanics:

$$\sigma_r = P, \quad \sigma_\theta = PR/t, \tag{5.24}$$

respectively, where R is the tube radius and t, the wall thickness. If the effect of Poisson's ratio is ignored, the radial (E_r) and circumferential (E_θ) elastic moduli can be expressed as

$$E_r = \sigma_r/\varepsilon_r, \quad E_\theta = \sigma_\theta/\varepsilon_\theta, \tag{5.25}$$

where ε_r and ε_θ are the strains in the radial and circumferential directions, respectively, and are calculated by $\varepsilon_r = \varepsilon_\theta = u_r/R$ with u_r standing for the radial displacement. The radial displacement was found to be a linear function of hydrostatic pressure for both armchair and zigzag nanotubes.

Figure 5.11 shows the variation of transverse Young's modulus of single-walled carbon nanotubes with tube diameter. It can be seen that E_θ is almost independent of the nanotube diameter and the tube chirality has a very minor effect. The magnitude of E_θ is about 1.0 TPa, which is the commonly

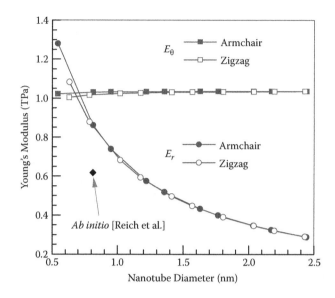

FIGURE 5.11
Elastic modulus of SWCNTs in transverse directions. (From C. Y. Li and T. W. Chou, *Comp. Sci. Technol.* 63 [2003] 1517–1524.)

accepted value for carbon nanotube axial Young's modulus. This result is consistent with the in-plane isotropic elastic nature of a graphene sheet. Figure 5.11 also shows that the radial Young's modulus is highly dependent on the tube diameter, and it deceases rather rapidly with increasing tube diameter. This is due to the increase of radial strain with increasing tube diameter at a given external pressure. Similar to the case of circumferential Young's modulus, the tube chirality also has no effect on the radial Young's modulus when the tube diameter is greater than 1.2 nm.

For comparison, Figure 5.11 also shows the ab initio result reported by Reich et al. [33]. Due to the formidable computational task involved in the ab initio calculation, only the radial modulus of a single-walled carbon nanotube with a diameter of 0.8 nm was given, and its value is 650 GPa.

5.2.6 Dynamic Properties of Carbon Nanotubes

Dynamic properties of carbon nanotubes have been paid much attention. The thermal vibration of carbon nanotubes was used for estimating the elastic properties of carbon nanotubes. Treacy et al. [34] first estimated Young's modulus of multiwalled nanotubes by measuring their thermal vibrations using transmission electron microscopy. Krishnan et al. [35] applied the same technique for predicting Young's modulus of single-walled carbon nanotubes. The basic assumption in these experiments is that the nanotube, whether single-walled or multiwalled, is a single-layer continuum shell.

Also, continuum shell models were used for studying the vibration behavior of carbon nanotubes. Sohlberg et al. [36] theoretically studied the vibration of carbon nanotubes by simply assuming nanotubes as solid slender rods. Yoon et al. [37] investigated the intertube vibration of multiwalled carbon nanotubes by a multiple-elastic beam model. Snow et al. [38] considered the vibration responses of single-walled carbon nanotubes as atomic force microscope probes, and also assumed the nanotube as a continuum shell with Young's modulus of 1.0 TPa.

But one of the most important potential applications of carbon nanotubes is as nanoresonators in nanoelectromechanical systems. Resonators are key components in signal processing systems. Reduction in the size of a resonator enhances its resonant frequency and reduces its energy consumption. So far, the highest frequency nanomechanical resonator (1.029 GHz) was fabricated from SiC using optical and electron-beam lithography. The potential of using nanotubes in practical nanomechanical resonators provides the impetus for the simulation of their dynamic behavior. Li and Chou [39,40] examined the feasibility of using carbon nanotubes as nanoresonators. It was predicted that the fundamental frequencies of cantilevered or bridged single-walled carbon nanotubes as nanomechanical resonators could reach the level of 10 GHz–1.5 THz depending on the nanotube diameter and length. Thus, the high potential of using carbon nanotubes for nanomechanical resonators is unmistakable.

For the simulation of nanotube dynamic behaviors, the above mentioned continuum models, while useful in some instances, are not adequate for a basic understanding of the effect of nanotube size and structures. The molecular structural mechanics method provides a much needed tool for examining the vibrational behavior of nanotubes. In the following text, we introduce this method using a multiwalled carbon nanotube as an example.

For determining the fundamental frequencies and vibrational modes of a multiwalled carbon nanotube, the nested tube layers are simulated by equivalent space frame-like structures and intertube van der Waals interactions by truss rods. According to the theory of structural dynamics, the equation of motion for the free vibration of an undamped structure is

$$[M]\{\ddot{y}\} + [K]\{y\} = \{0\}, \tag{5.26}$$

where $[M]$ and $[K]$ are, respectively, the global mass and stiffness matrices, and $\{y\}$ and $\{\ddot{y}\}$ are, respectively, the nodal displacement vector and acceleration vector.

The global mass matrix $[M]$ can be assembled from the elemental mass matrix. By considering the atomistic feature of a carbon nanotube, the masses of electrons are neglected and the masses of carbon nuclei ($m_c = 1.9943 \times 10^{-26}$ kg) are assumed to be concentrated at the centers of atoms, that is, the joints of beam members. Due to the extremely small radius ($r_c = 2.75 \times 10^{-5}$ Å) of the carbon atomic nucleus, the coefficients in the mass matrix corresponding to flexural rotation and torsional rotation, $\frac{2}{3} m_c r_c^2$, are assumed to be zero. Only the coefficients corresponding to translatory displacements are kept. Thus, the elemental mass matrix $[M]^e$ is given by

$$[M]^e = \mathrm{diag}\left[\frac{m_c}{3} \quad \frac{m_c}{3} \quad \frac{m_c}{3} \quad 0 \quad 0 \quad 0\right], \qquad (5.27)$$

The factor $1/3$ in the elements of the elemental mass matrix is introduced because the three bonds of a carbon atom connect with the three nearest neighboring atoms, which ensures that the nodal mass has the value of a single atom in the assembled global mass matrix $[M]$.

The global stiffness matrix $[K]$ is assembled from the elemental stiffness matrix $[K]^e$, that is, $[K] = \sum_{e=1}^n [K]^e$, where n is the number of elements. The assembling procedure follows the node-related technique in the finite element method. There are two kinds of elemental stiffness matrix $[K]^e$. One is for the beam element in tube layers, that is,

$$[K]_b^e = \begin{bmatrix} [k_{ii}] & [k_{ij}] \\ [k_{ji}] & [k_{jj}] \end{bmatrix}, \qquad (5.28)$$

where the submatrices $[k_{ii}]$, $[k_{ij}]$, $[k_{ji}]$, and $[k_{jj}]$ are stiffness coefficients related to the cross-sectional parameters of the beam element $i - j$ and can be found in Reference 6. Another elemental stiffness matrix is for the truss rod representing van der Waals interaction, that is,

$$[K]_r^e = \begin{bmatrix} [k_{ii}] & [k_{ij}] \\ [k_{ji}] & [k_{jj}] \end{bmatrix}. \qquad (5.29)$$

The submatrices in Equation 5.29 are

$$[k_{ii}] = [k_{jj}] = \begin{bmatrix} [A] & [O] \\ [O] & [O] \end{bmatrix}, \quad [k_{ij}] = -[k_{ii}] \qquad (5.30)$$

where

$$[A] = \begin{bmatrix} \alpha & 0 & 0 \\ 0 & 0 & 0 \\ 0 & 0 & 0 \end{bmatrix}, \quad \alpha = 24\frac{\varepsilon}{\sigma^2}\left[26\left(\frac{\sigma}{r}\right)^{14} - 7\left(\frac{\sigma}{r}\right)^8\right], \quad [O] = [0]_{3\times3}. \qquad (5.31)$$

For minimizing the computational work, the orders of the global stiffness matrix and mass matrix are reduced by using the static condensation method.

The fundamental frequencies f and mode shapes are then obtained from the solution of the eigenproblem

$$([K]_s - \omega^2 [M]_s)\{y_p\} = 0, \tag{5.32}$$

where $[K]_s$ and $[M]_s$ are the condensed stiffness matrix and condensed mass matrix, respectively, $\{y_p\}$ is the displacement vector corresponding to the primary coordinates, that is, the translatory displacements of carbon atoms, and $\omega = 2\pi f$ is the angular frequency.

The vibration of a nanotube is usually studied under certain end constraints, depending on the physical application of the nanotube. Figure 5.12 displays the computational results of the fundamental frequencies of single-walled CNTs at a fixed nanotube diameter. It is observed that the fundamental frequencies of finite length CNTs with the aspect ratio (length/diameter) studied can reach more than 10 GHz, and bridged CNTs have much higher fundamental frequencies than cantilevered CNTs. For both end boundary conditions, the fundamental frequency of CNTs increases with decreasing nanotube aspect ratio. Figure 5.12 also shows that the fundamental frequency depends on nanotube diameter even if the nanotube aspect ratio is kept unchanged. The general tendency is that the fundamental frequency decreases with an increasing tube diameter. Thus, for nanotubes of the same aspect ratio, it is desirable to select the ones with smaller diameter for obtaining higher vibrational frequencies.

Another issue relevant to nanomechanical resonators is the vibrational mode associated with the fundamental frequency. For bridged nanotubes, it is desirable that the vibration mode is in the form of a half sine wave for the convenience of experimental detection. Figures 5.13 and 5.14 display some vibrational modes of single-walled and double-walled carbon nanotubes, respectively. It is noted that the first, and even the second, vibration modes are in the shape of a half sine wave. However, this particular shape of vibration mode occurs only when the nanotube length is not very small. For shorter nanotubes, the third and fourth vibrational modes would interchange with the first and second modes. The minimum nanotube length for maintaining a half sine wave vibration mode depends on the nanotube diameter and the number of nanotube layers. Li and Chou's simulations indicated that the limiting aspect ratios are 4.0 and 3.0, respectively, for single-walled and double-walled nanotubes.

For the vibration of double-walled carbon nanotubes, noncoaxial vibration modes are clearly observed, as shown in Figure 5.14. This phenomenon was first reported by Yoon et al. [37] based on an elastic continuum multishell model. Li and Chou's MSM simulation results agree with Reference 37 in that the vibration modes associated with the fundamental frequencies are almost coaxial, and noncoaxial vibrations are excited at higher frequencies. The MSM simulations identify the noncoaxial vibration initiating at the third resonant frequency, which is usually much higher than the first two resonant frequencies. Thus, the noncoaxial vibration does not significantly diminish the value of DWCNTs as effective high-frequency nanoresonators.

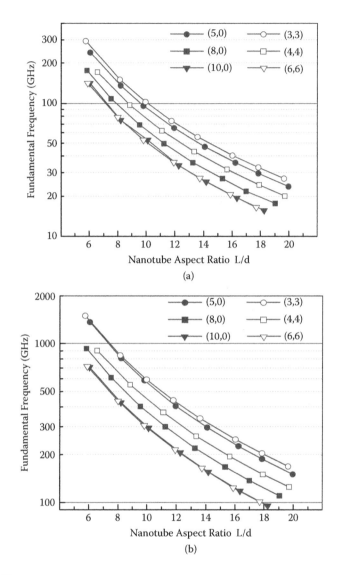

FIGURE 5.12
Natural frequency of single-walled nanotubes: (a) cantilevered nanotubes and (b) bridged nanotubes. (From C. Y. Li and T. W. Chou, *Phys. Rev. B* 68 [2003] 073405.)

5.2.7 Thermal Properties of Carbon Nanotubes

Since many applications of carbon nanotubes significantly depend on their thermodynamic parameters, increasing attention has been paid to the thermal properties of carbon nanotubes. Here, we only focus on specific heat and thermal expansion, which are two fundamental thermal characteristics.

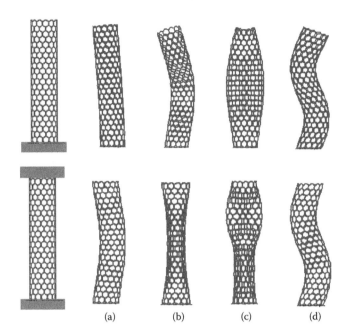

FIGURE 5.13
Vibrational modes of single-walled nanotubes. (From C. Y. Li and T. W. Chou, *Phys. Rev. B* 68 [2003] 073405.)

5.2.7.1 Specific Heat

Several theoretical and experimental studies have been devoted to the determination of the specific heat of carbon nanotubes. Benedict et al. [41] first theoretically predicted a linear relationship between the specific heat of carbon nanotubes and the temperature under the condition that both the nanotube diameter and temperature are sufficiently small. But experimental measurements by Yi et al. [42] demonstrated a strikingly linear temperature dependence of the specific heat over a broad range of temperatures (10–300 K) and tube diameters (20–40 nm) of multiwalled carbon nanotubes (MWCNTs). The specific heat measured by Mizel et al. [43] on MWCNTs and on ropes of single-walled carbon nanotubes (SWCNTs) shows a linear temperature dependence of the specific heat in the temperature range of 100–200 K, but a roughly quadratic dependence at low temperature (<50 K). The experiments of Hone et al. [44] on SWCNT ropes in the temperature range of 2–300 K also reached similar conclusions as in Reference 43. Lasjaunias et al. [45] extended the measurements on SWCNT ropes down to much lower temperature and found that the specific heat follows triplicate temperature dependence at temperatures in the range of 4.5 K to 0.3 K.

Numerical studies have also been conducted for the specific heat of carbon nanotubes. Popov [46] calculated the low temperature specific heat of

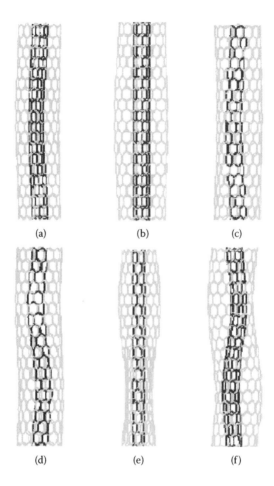

FIGURE 5.14
Vibrational modes of double-walled nanotubes: (a) half-wave coaxial, (b) half-wave breathing, (c) half-wave noncoaxial, (d) full-wave coaxial, (e) full-wave breathing, and (f) full-wave noncoaxial.

MWCNTs and isolated or bundled SWCNTs with force-constant dynamic models and found that the temperature dependence of the low-temperature specific heat changes from square-root form to linear form with the increase of SWCNTs in a bundle or tube layers in an MWCNT. Cao et al. [47] calculated the phonon spectrum and specific heat of SWCNTs based on the model of lattice dynamics with force-constant matrix. They concluded that the specific heat increases with the increase of nanotube diameter, and it is generally larger for armchair tubes than for zigzag tubes. Dobardzic et al. [48] applied a symmetry-based force constant approach to calculate the phonon dispersion and the specific heat of isolated SWCNTs, which shows variable diameter dependence at different temperatures. A continuum shell model has been adopted by Zhang et al. [49], and it was concluded that the specific heat is mainly contributed by vibrations along the length of an SWCNT.

Up to now, almost all experimental measurements on specific heat were conducted on SWCNT bundles or MWCNTs at low temperature. There are only a few theoretical studies on the specific heat of isolated SWCNTs. Li and Chou [50] studied the specific heat of finite-length isolated SWCNTs and MWCNTs by using the molecular structural mechanics method. The dependence of the specific heat on temperature, tube diameter, and tube chirality was systematically investigated.

The studying of thermal properties has to begin with the normal coordinates of atomic vibration and the basic concept of quantum mechanics. For the atomic vibration of a single-walled carbon nanotube with n carbon atoms, the mass of electrons is negligible compared with the mass of a carbon nucleus ($m_c = 1.9943 \times 10^{-23}$ g). The mass of carbon nuclei can be assumed to be concentrated at the centers of the atoms. Thus, the nanotube is a lumped system. Also, due to the extremely small radius ($r_c = 2.75 \times 10^{-5}$ Å) of the carbon atomic nucleus, the kinetic energy corresponding to the flexural rotation and the torsional rotation can be omitted. Hence, the kinetic energy of the nanotube is given by

$$T = \frac{1}{2}\dot{\mathbf{q}}^{\mathrm{T}}\mathbf{M}\dot{\mathbf{q}}, \tag{5.33}$$

where $\dot{\mathbf{q}} = \{\dot{q}_1, \dot{q}_2, \ldots, \dot{q}_{3n-p}\}$ is the vibrational velocity vector corresponding to translatory displacements, with p denoting the number of constraints; the superscript T stands for the transpose of a vector, and $\mathbf{M} = diag\{m_c, m_c, \ldots, m_c\}$ is the global mass matrix. The potential energy of the nanotube is a function of vibrational displacements of carbon atoms. For small vibrational displacements, the potential energy can be expanded in a Taylor series. By neglecting terms higher than the second order and considering that the potential energy is a minimum when all atoms are in their equilibrium positions, the potential energy reduces to

$$V = \frac{1}{2}\mathbf{q}^{\mathrm{T}}\mathbf{K}\mathbf{q}, \tag{5.34}$$

where $\mathbf{q} = \{q_1, q_2, \ldots, q_{3n-p}\}^T$ is the vibrational translatory displacement vector, and $\mathbf{K}$ is the global stiffness matrix, which is assembled from the beam stiffness matrix $\mathbf{K}^e$. The general form of the beam stiffness matrix has been detailed before, and its stiffness coefficients are related to the cross-sectional parameters EA, EI, and GJ in Equation 5.20.

The substitution of the kinetic energy T and potential energy V given, respectively, by Equations 33 and 34 into the Lagrange equation of motion for conservative systems,

$$\frac{d}{dt}\left(\frac{\partial T}{\partial \dot{\mathbf{q}}}\right) + \frac{\partial V}{\partial \mathbf{q}} = 0, \tag{5.35}$$

yields a system of equations of motion, that is,

$$\mathbf{M}\ddot{\mathbf{q}} + \mathbf{K}\mathbf{q} = 0, \tag{5.36}$$

where $\ddot{\mathbf{q}} = \{\ddot{q}_1, \ddot{q}_2, \dots, \ddot{q}_{3n-p}\}^T$ is the acceleration vector of atomic vibrations. The general solution of Equation 5.36 is of the form

$$\mathbf{q} = \mathbf{A}\exp[-i(\omega t - \alpha)] \tag{5.37}$$

where $\mathbf{A}$ is a vector of the vibrational amplitude, ω is the angular frequency, and α is the phase angle. The natural angular frequencies of carbon atoms in the nanotube, which are our main concerns, can then be obtained from the solution of the following eigenproblem:

$$\det |\mathbf{K} - \omega^2 \mathbf{M}| = 0. \tag{5.38}$$

Corresponding to each natural frequency $\omega_i (i = 3n - p)$, there is a natural mode A_i. These natural modes of vibrations possess orthogonality properties about the mass matrix $\mathbf{M}$ and the stiffness matrix $\mathbf{K}$ and can be normalized. By using a matrix containing these normalized modes, $\mathbf{R}$, the system of equations of motion, Equation 5.36, can be uncoupled into a set of independent equations,

$$\ddot{\mathbf{Q}} + \omega\mathbf{Q} = 0, \tag{5.39}$$

where $\omega = diag[\omega_1^2, \omega_2^2, \cdots \omega_{3n-p}^2]$, and $\mathbf{Q} = \{Q_1, Q_2, \dots, Q_{3n-p}\}^T = \mathbf{R}^{-1}\mathbf{q}$ is termed the *normal coordinate vector*.

In terms of the normal coordinates, the Hamiltonian function of the nanotube becomes

$$H = T + V = \frac{1}{2}\sum_{j=1}^{3n-p}[\dot{Q}_j^2 + \omega_j^2 Q_j^2]. \tag{5.40}$$

Therefore, each term of the Hamiltonian function represents a simple harmonic oscillator. Based on the theory of quantum mechanics, these oscillators can be considered as quantum mechanical oscillators or phonons. The quantization can be achieved by introducing the momentum operator corresponding to $\dot{Q}_j$, that is,

$$\dot{Q}_j = -i\hbar \frac{\partial}{\partial Q_j} \tag{5.41}$$

where $\hbar$ is Planck's constant. The Schrödinger equation of the oscillator j is then

$$\left(-\frac{\hbar^2}{2} \frac{\partial^2}{\partial Q_j^2} + \frac{1}{2} \omega_j^2 Q_j^2 \right) \psi_{v_j}(Q_j) = E_{v_j} \psi_{v_j}(Q_j), \tag{5.42}$$

where E_{v_j} is the eigenvalue and ψ_{v_j} the eigenfunction of the oscillator in the state with quantum number v_j. The eigenvalues are given by

$$E_{v_j} = (v_j + \frac{1}{2})\hbar\omega_j. \tag{5.43}$$

E_{v_j} in Equation 5.43 stands for the discrete harmonic oscillator energy spectrum. It means that an oscillator may be excited only to one of an equally spaced set of energy levels.

According to statistic thermodynamics, the vibrational partition function of these phonons can be written as

$$Z = \prod_{j=1}^{3n-p} \frac{\exp(-\hbar\omega_j/2k_BT)}{1-\exp(-\hbar\omega_j/k_BT)}, \tag{5.44}$$

where k_B is the Boltzmann's constant, and T stands for temperature. The internal energy of the nanotube is then

$$E = k_BT^2 \left(\frac{\partial \ln Z}{\partial T} \right) = \sum_{j=1}^{3n-p} \frac{\hbar\omega_j[\exp(\hbar\omega_j/k_BT)+1]}{2[\exp(\hbar\omega_j/k_BT)-1]}. \tag{5.45}$$

Hence, the specific heat is

$$C_V = \frac{\partial E}{\partial T} = k_B \sum_{j=1}^{3n-p} \frac{(\hbar\omega_j/k_BT)^2 \exp(\hbar\omega_j/k_BT)}{[\exp(\hbar\omega_j/k_BT)-1]^2}. \tag{5.46}$$

From the above procedure for the derivation of the specific heat, it can be seen that there are some similarities between our approach originated from the molecular structural mechanics method and lattice dynamics, especially in the quantization of harmonic oscillators. The advantages of the present

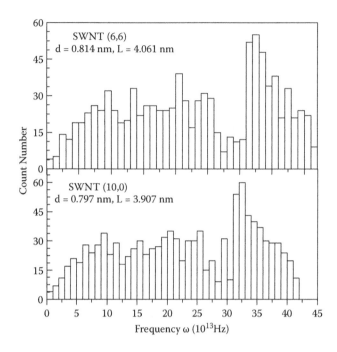

FIGURE 5.15
Vibrational frequencies of armchair and zigzag SWCNTs. (From C. Y. Li and T. W. Chou, *Phys. Rev. B*, 71 [2005] 075409.)

method are found in the following aspects. First, it directly uses the force field constants from computational chemistry to account for the interatomic actions. These force field constants have been extensively used, and they are kept unchanged in our studies of both static and dynamic behavior of carbon nanotubes. In contrast, the force constants in lattice dynamics usually need to be adjusted by fitting the experimental data. Second, the present approach considers multiple atomic interactions, including bond stretching, bond bending, and bond torsion. In contrast, lattice dynamics only considers central forces and angular forces, corresponding to bond stretching and bond bending, respectively. Finally, the present method treats nanotubes as having finite length, and the effect of the nanotube length should be considered. But in lattice dynamics, the nanotubes are always assumed to be infinitely long.

Using the modeling technique developed above, the vibrational frequencies of isolated SWCNTs can be calculated. Figure 5.15 displays the distributions of vibrational frequencies of an armchair SWCNT and a zigzag SWCNT. These SWCNTs are of almost the same diameter (~0.8 nm) and the same length (~4.0 nm). It can be seen that there are some differences in the distribution and span of the frequency spectrum between the two types of SWCNTs.

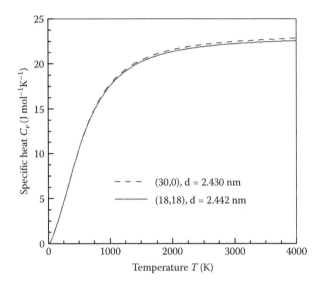

FIGURE 5.16
The variation of specific heat of SWCNTs with temperature. (From C. Y. Li and T. W. Chou, *Phys. Rev.* B, 71 [2005] 075409.)

The specific heat can be calculated using Equation 5.46. Figure 5.16 illustrates the temperature dependence of the specific heat for a large range of temperature (0–4000 K). The extremely high temperature is used only to demonstrate that the specific heat approaches the value $3k_B$ (24.9 J/mol·K) as expected from the law of Dulong and Petit. It also can be seen that the effect of tube chirality on the temperature dependence of specific heat appears only at high temperature and is not very significant.

Figure 5.17 displays the variation of specific heat at low temperatures. It is seen that the specific heat curves exhibit three regimes. At very low temperatures, the specific heat of carbon nanotubes shows an exact T^3 behavior (Figure 5.18). This means that the Einstein–Debye phonon model, which successfully predicted the low-temperature cubic dependence of specific heats of solids, is still valid for carbon nanotubes. The second is the quadratic part (roughly T < 150 K), and another is the linear part (roughly 150 K < T < 600 K). This result agrees to some extent with existing theoretical predictions and experimental measurements. Benedict et al. [41] predicted a linear relationship between the specific heat and the temperature when the nanotube diameter and temperature are small, but no specific diameter and temperature ranges were given. Yi et al. [42] observed that a linear temperature dependence of the specific heat exists over a large range of temperature (10–300 K) for MWCNTs with diameters of 20–40 nm. The measurements of Mizel et al. [43] also indicated that the specific heat of both MWCNTs and SWCNT ropes shows a linear temperature dependence in the range of

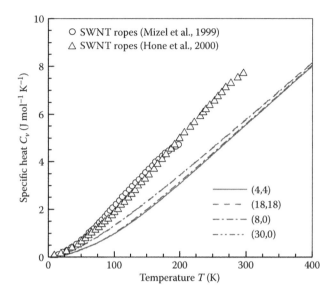

FIGURE 5.17
Comparison of specific heat for single-walled carbon nanotubes at low temperatures with experimental results on nanotube ropes. (From C. Y. Li and T. W. Chou, *Phys. Rev. B*, 71(2005) 075409.

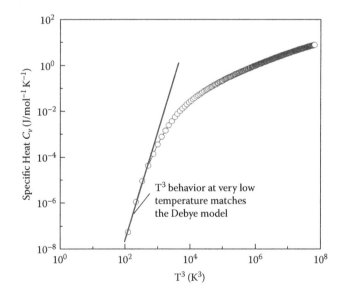

FIGURE 5.18
Nanotube specific heat at very low temperatures. (From C. Y. Li and T. W. Chou, *Materials Science and Engineering* A 409 (2005) 140–144.)

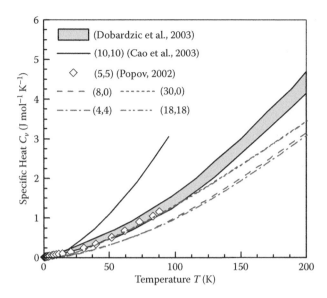

FIGURE 5.19
Comparison of specific heat of SWCNTs with other theoretical results. (From C. Y. Li and T. W. Chou, *Phys. Rev.* B, 71 (2005) 075409.)

100–200 K, and a roughly quadratic dependence at temperatures below 50 K. Our computational results indicate that the upper limit of the linear temperature dependence of SWCNT specific heat is around 600 K. For comparison, some experimental results of SWCNT ropes are displayed in Figure 5.17. The present computational results are in good agreement with the experimental results [43,44] for temperatures below 50 K. The discrepancies increase with the increase of temperature, perhaps due to the fundamental difference between SWCNT ropes and individual SWCNTs.

Figure 5.19 presents comparisons between our modeling results and some existing theoretical results for individual SWCNTs. The present results agree well with the computational results of Popov [46] and Dobardzic et al. [48]. It should be noted that the data points representing the results of Popov [46] in Figure 5.19 were taken from the original smooth curve. The results of Dobardzic et al. [48] are represented by the band in Figure 5.19. which covers the range of nanotube diameters from 2.8Å to 20.5Å in their computations.

The present method can readily deal with the issue of specific heat of finite length SWCNTs. Figure 5.20a displays the temperature dependence of specific heat for armchair SWCNTs of the same diameter but different lengths. It can be seen that the effect of nanotube length increases with increasing temperature. Comparing with Figure 5.16, it can be concluded that, for relatively short tubes, the effect of nanotube length is even greater than that of nanotube chirality. However, the length effect at a given temperature reduces gradually with the

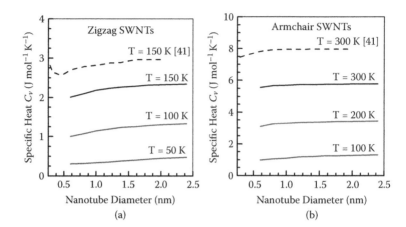

FIGURE 5.20
The effect of tube diameter on (a) zigzag (b) armchair SWCNT specific heat. (From C. Y. Li and T. W. Chou, *Phys. Rev. B*, 71 [2005] 075409.)

increase of nanotube length. This tendency can be seen in Figure 5.20b. Also, at a higher temperature, the effect of nanotube length diminishes at a longer length. Our results further indicate that zigzag SWCNTs have the same effects of tube length on specific heat as armchair nanotubes. We also conclude that, for the aspect ratios of armchair and zigzag SWCNTs examined in this research, the effect of nanotube length on specific heat becomes negligible if the nanotube length/diameter ratios are higher than 20.

Li and Chou [51] also studied the heat capacity of multiwalled carbon nanotubes by using the molecular structural mechanics modeling technique. The heat capacity is derived from the vibrational partition function, which is directly expressed in terms of the vibrational frequencies of carbon atoms. The modeling results indicate that the heat capacity of carbon nanotubes is mainly dependent on temperature, and slightly dependent on tube chirality and tube length at higher temperatures. The effects of tube diameter and the number of tube layers are negligible. Our results in measuring the heat capacity of carbon nanotubes coincide with the predictions of the Einstein–Debye phonon model at low temperature, as well as the law of Dulong and Petit at extremely high temperatures.

5.2.7.2 Thermal Expansion

The coefficient of thermal expansion (CTE) of carbon nanotubes is a key property relevant to their physical and mechanical behavior. Due to the challenge in nanoscale experiments and modeling, studies on the CTE of carbon nanotubes are fairly limited. Most of the experiments focused on the CTE of MWCNTs. Ruoff and Lorents [52] reported that the experimentally measured radial CTE of MWCNTs is essentially identical to the axial CTE and suggested that the CTE is isotropic in defect-free CNTs. Using x-ray diffraction, Bandow [53] measured the MWCNT thermal expansion in the radial direction in the temperature range of 10 to 320 K and found that the radial CTE is almost the same as the CTE of graphite in the thickness direction. Maniwa et al. [54] also used x-ray diffraction technique and found that the radial CTE of MWCNTs is in the range of about 1.6×10^{-5} to $2. \times 10^{-5}$/K. Meanwhile, the x-ray diffraction studies of Yosida [55] and Maniwa et al. [56] on SWCNT bundles indicated that the CTE in the direction normal to the bundle axis is negative in a large temperature range. There is still no experimental report on the CTE of individual single-walled carbon nanotubes.

A few theoretical studies have been performed on the CTE of SWCNTs. Raravikar et al. [57] investigated the CTE of (5,5) and (10,10) armchair SWCNTs by molecular dynamics simulations. The CTE in the radial direction of the SWCNT was found to be less than that in the axial direction, and tubes with smaller diameter show larger radial and axial CTEs. Schelling and Keblinski [58] studied the thermal expansion of carbon structures, including graphite, diamonds, and SWCNTs, using both MD simulations and lattice dynamics calculations. Their results indicated that the axial CTE of SWCNTs is negative at room temperature and positive at high temperatures. Kwon et al. [59] also performed MD simulations on SWCNTs and reported that SWCNTs contract axially up to ~900 K, and the minimum value of the axial CTE is -1.2×10^{-5}/K. Differing from these theoretical studies, Jiang et al. [60] developed an analytical method to determine the CTE of SWCNTs based on the interatomic potential and the local harmonic model. Their results showed that both the axial and radial CTEs of SWCNTs are negative at low temperature but positive at high temperature.

Basically, the research on the CTE of CNTs is still in the initial stages. The available results are limited and not in agreement with each other. In the following text, we introduce the studies of Li and Chou [61] on the CTE of individual SWCNTs by using the molecular structural mechanics method.

The theoretical basis of thermal expansion, which is one of the directly observable properties of a solid, can be interpreted from the standpoint of statistical thermodynamics. In expressing thermal expansion, the basic thermodynamic relationship is

$$\left(\frac{\partial V}{\partial T} \right)_P = -\left(\frac{\partial S}{\partial P} \right)_T, \tag{5.47}$$

where V is the volume of the material, S the entropy, T the absolute temperature, and P the pressure. Then the coefficient of volume thermal expansion α can be obtained as follows:

$$\alpha = \frac{1}{V} \left(\frac{\partial V}{\partial T} \right)_P = -\frac{1}{V} \left(\frac{\partial S}{\partial P} \right)_T. \tag{5.48}$$

Analogous expressions hold for the coefficients of linear thermal expansion. For SWCNTs, the axial CTE α_a and radial CTE α_r can be defined as

$$\alpha_a = \frac{1}{L}\left(\frac{\partial L}{\partial T}\right)_P, \quad \alpha_r = \frac{1}{D}\left(\frac{\partial D}{\partial T}\right)_P, \tag{5.49}$$

where L and D stand for, respectively, the length and diameter of a nanotube. It is convenient to consider the axial CTE and the radial CTE separately. In calculating the axial CTE, it is assumed that only pressure P_a is applied along the axial direction on the cross-section of an SWCNT. The change of the cross-sectional area of the SWCNT can be omitted provided that the axial strain is small enough. Likewise, for calculating the radial CTE, a pressure P_r is assumed to act on the cylindrical surface of an SWCNT. The change of the nanotube length in this case can also be omitted, provided that the radial strain is sufficiently small. Therefore, the axial and radial CTEs can be obtained as

$$\alpha_a = \frac{1}{L}\left(\frac{\partial L}{\partial T}\right)_P = \frac{1}{\pi Dt \cdot L}\left(\frac{\pi Dt \cdot \partial L}{\partial T}\right)_P = \frac{1}{V}\left(\frac{\partial V}{\partial T}\right)_P = -\frac{1}{V}\left(\frac{\partial S}{\partial P_a}\right)_T \tag{5.50}$$

$$\alpha_r = \frac{1}{D}\left(\frac{\partial D}{\partial T}\right)_P = \frac{1}{\pi Lt \cdot D}\left(\frac{\pi Lt \cdot \partial D}{\partial T}\right)_P = \frac{1}{V}\left(\frac{\partial V}{\partial T}\right)_P = -\frac{1}{V}\left(\frac{\partial S}{\partial P_r}\right)_T. \tag{5.51}$$

Here, the wall thickness t of an SWCNT is taken as 0.34 nm.

For the convenience of calculations, Equations 5.50 and 5.51 can be expressed approximately in forms of first-order difference,

$$\alpha_a = -\frac{1}{V}\left(\frac{S}{P_a}\right)_T, \tag{5.52}$$

$$\alpha_r = -\frac{1}{V}\left(\frac{S}{P_r}\right)_T. \tag{5.53}$$

Thus, if we apply two different values of the same kind of pressure (axial tension or radial pressure) on an SWCNT and then calculate the change of entropy, the CTEs of the SWCNT can be readily obtained from Equations 5.52 and 5.53.

The entropy of an SWCNT is calculated by using the quantized molecular structural mechanics modeling technique developed by Li and Chou [50]. According to statistical thermodynamics, the vibrational partition function of phonons can be written as

$$Z = \prod_{j=1}^{3n-p} \frac{\exp(-\hbar\omega_j/2k_B T)}{1 - \exp(-\hbar\omega_j/k_B T)}, \tag{5.54}$$

where k_B is Boltzmann's constant, and T stands for temperature. The entropy of the nanotube is then obtained from

$$S = k_B \ln Z + k_B T \left(\frac{\partial \ln Z}{\partial T} \right). \tag{5.55}$$

In order to obtain the change of entropy, we consider a two-step approach. In the first step, the nanotube is free of strain ($P_1 = 0$). In the second step, the nanotube is under certain strain resulted from the applied pressure ($P_2 \neq 0$). For the first step, the frequencies of atomic vibration can be directly obtained from Equation 5.38. Then the entropy S_1 for the first step can be calculated from Equation 5.55. However, for the second step, two computational substeps need to be performed. The first substep is to calculate the atomic displacements when the nanotube is subjected to a uniform end stress (for the axial CTE) or radial pressure (for the radial CTE). After obtaining displacements of all atoms in the nanotube, the new position of each atom is given by adding its displacements to the original coordinates of the atom. In the second substep, the new coordinates are used in Equation 5.38 for calculating the frequencies of atomic vibration of the nanotube under strain. The entropy S_2 for the second step can then be obtained by Equation 5.55. Having obtained the difference of the entropies ($\Delta S = S_2 - S_1$) under two different pressure conditions ($\Delta P = P_2 - P_1$), the axial CTE and the radial CTE are finally obtained from Equations 5.52 and 5.53, respectively.

Figure 5.21 displays the variation of the axial CTE of SWCNTs with temperature, where both armchair and zigzag SWCNTs show the same trend of temperature dependence. The smaller-diameter nanotubes have slightly higher axial CTEs, especially at higher temperatures. The CTE data at high temperature are less than those of Raravikar et al. [57] from MD simulations, which are 3.9×10^{-6}/K and 2.4×10^{-6}/K for (5,5) and (10,10) SWCNT, respectively. The results of Raravikar et al. [57] are in agreement with those of Schelling and Keblinski [58] from lattice dynamics calculations, which are 3.84×10^{-6}/K and 2.52×10^{-6}/K for (5,5) and (10,10) SWCNT, respectively. But neither Raravikar et al. nor Schelling and Keblinski reported any temperature dependence of the axial CTE. Figure 5.21 also gives the recent results of Jiang et al. [60]. It can be seen that the temperature dependence of Li and Chou's results generally follows the same trend as that of Jiang et al. It has been reported that graphite has negative in-plane CTE at low temperature and positive axial CTE at high temperature. But Li and Chou's results on SWCNTs do not show a negative CTE region below room temperature.

Figure 5.22 illustrates Li and Chou's results of the radial CTE about SWCNTs. Similar to the case of axial CTE, the trends of temperature dependence of the radial CTEs for both armchair and zigzag SWCNTs are also similar. But the tube diameter in this case has a more significant effect on the radial CTE. For an SWCNT of small diameter, the radial CTE is less than the

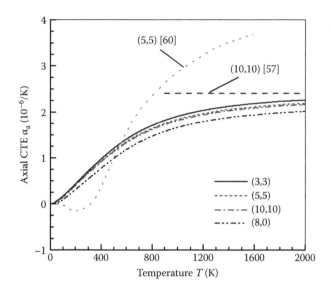

FIGURE 5.21
Variation of the axial CTE of SWCNTs (tube length: 2 nm). (From C. Y. Li and T. W. Chou, *Phys. Rev.* B 71 [2005] 235414.)

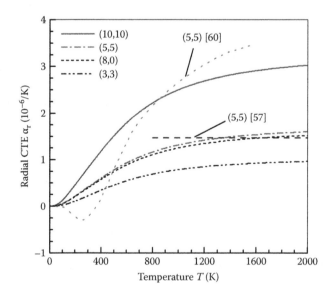

FIGURE 5.22
Variation of the radial CTE of SWCNTs (tube length: 2 nm). (From C. Y. Li and T. W. Chou, *Phys. Rev.* B 71 (2005) 235414.)

axial CTE. This trend is reversed in the case of an SWCNT of large diameter. Raravikar et al. [57] reported that the radial CTEs for (5,5) and (10,10) SWCNT using MD simulations are 3.5 10^{-6}/K and 0.8×10^{-6}/K, respectively. By lattice dynamics calculations, Schelling and Keblinski [58] predicted the radial CTE values of 1.47×10^{-6}/K for (5,5) SWCNT and 0.48×10^{-6}/K for (10,10) SWCNT. Again, those results are independent of temperature. Furthermore, Li and Chou's results indicate that the radial CTE is positive for the entire temperature range. This prediction agrees with the existing results for the CTE of graphite in the direction perpendicular to the basal plane (c-axis). Jiang et al. [60], on the other hand, reported that radial CTE of SWCNTs is negative below room temperature.

5.3. Electromechanical Behaviors of Carbon Nanotubes

Carbon nanotubes have unique electrical properties. Metallic tubes have conductivities and current densities that meet or exceed the best metals, and semiconducting tubes have mobilities and transconductances that meet or exceed the best semiconductors. Because of the huge economic incentives to shrink microelectronic devices further to nanoscale, carbon nanotubes may play the same role one day as silicon does in electronic circuits, because at a molecular scale, silicon and other standard semiconductors cease to work. Although there still exist challenges in fabrications, experiments over the past several years have given researchers hope that wires and functional devices tens of nanometers or smaller in size could be made from nanotubes and incorporated into electronic circuits that work far faster and on much less power than those existing today.

So far, CNTs have been explored to be used as nanoresonators, nanosensors, nanooscillators, nanotweezers, nanoactuators, nanorelays, nonvolatile random access memories, etc. Electromechanical coupling is an important characteristic of carbon nanotubes in these applications. In the coming section, we will introduce the modeling of deformation and failure resulting from electrical charge injection.

5.3.1 Charge Distribution On Carbon Nanotubes

When carbon nanotubes are connected at both ends via conducting electrodes, they have the capacity of carrying high density current. If a nanotube is connected to a conducting electrode at only one end, charge accumulates on the nanotube under the applied potential. An effective and precise analysis of charge distribution on the nanotube is critical to understanding their electromechanical characteristics. Therefore, researchers have employed various methods attempting to determine the charge distribution of nanotubes under externally applied electric fields.

Lou et al. [62] investigated the electronic properties of nanotubes in strong electric fields by using a density functional cluster method. Their calculations indicated electrons on the outer side of a carbon nanotube are redistributed in the presence of an external electric field while the electrons on the inner side are relatively unaffected. Rotkin et al. [63] conducted a full electrostatics calculation of single-walled nanotubes by using the boundary element method. The calculated charge densities were compared with results obtained from quantum mechanical computations. It was shown that results from the different approaches are almost identical; slight differences are due to pure quantum effects, like quantum beating at the ends of a finite length nanotube. Luo et al. [64] performed density functional theory (DFT) calculations of the electronic structures of charged single-walled nanotubes. They paid particular attention to the electrostatic potential distribution and the energy variation with respect to the number of extra electrons added to the nanotube. Their results indicated that, when extra electrons are added, the potential between the tube ends is lifted, and its value at infinity remains zero; hence, potential barriers are formed at the nanotube ends. Keblinski et al. [65] reported a combinational analysis of DFT calculations and classical electrostatics for studying voltage-induced electrostatic charging on finite length nanotubes. It was concluded that the charged nanotube exhibits essentially classical charge distribution with a significant charge concentration at the tube ends. Zheng and coworkers [66] investigated the charge distribution and electrostatic potential along a single-walled carbon nanotube under field emission experimental conditions by carrying out a quantum mechanical simulation. They found that a single layer of carbon atoms is sufficient to shield most of the electric field except for penetration at the tip. Their results showed that induced charges in the segment between tube ends are only a few percentage of the total number of electrons accumulated at the tube ends.

More recently, Li and Chou [67] developed a simple and accurate atomistic moment method (AMM) for analyzing charge distribution on nanotubes in an external electric field. This method is essentially based on the moment method in classical electrostatics. Its basic concept is to solve integral equations in electromagnetics by dividing the integral domain into subsections with equal areas and expanding the unknown function in a set of basis functions. The basis functions are known and hence can be integrated. The coefficients of the basis function can be obtained by solving algebra equations. For the conductive CNT in an external electric field, the induced charges are distributed on its outer surface. Because of the strong attraction of nucleus, the charge distributed in the vicinity of a carbon atom can be considered to be concentrated at the atom. Thus, the nanotube becomes a charge carrier with point charges at the positions of carbon atoms. Here, we introduce the atomistic moment method using a freestanding armchair single-walled

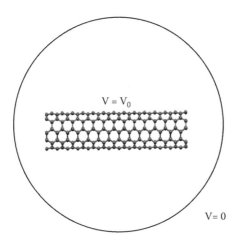

FIGURE 5.23
Freestanding nanotubes in an electric field.

nanotube as an example. The configuration is shown in Figure 5.23. The voltage on the nanotube is assumed to be V_0.

For a freestanding nanotube in an infinite electric field, the only boundary condition is that the whole outer surface of the nanotube is kept to the potential V_0. If the nanotube has n atoms and the point charges on the nanotube are denoted as q_j, ($j = 1, 2, \ldots n$), the potential at an arbitrary atomic position is given by

$$V(\mathbf{r}_i) = \sum_{j=1}^{n} \frac{q_j}{4\pi\varepsilon_0 |\mathbf{r}_i - \mathbf{r}_j|},$$ (5.56)

where V is the electric potential, $\mathbf{r}_i$ denotes the position of atom of interest, $\mathbf{r}_j$ represents the location of the charged atom, and ε_0 is the permittivity of vacuum ($\varepsilon_0 = 8.854 \ C^2/N \cdot m$).

For each atom in the nanotube, an equation can be established based on Equation 5.56. By assuming that the whole outer surface of the nanotube is at an equipotential V_0, the n equations can be written in matrix form, that is,

$$[A]\{q\} = \{V_0\},$$ (5.57)

where $\{q\}$, $\{V_0\}$ are the charge vector and the potential vector, respectively, and $[A]$ is the $n \times n$ order matrix, in which the elements are expressed as:

$$a_{ij} = \frac{1}{4\pi\varepsilon_0 |\mathbf{r}_i - \mathbf{r}_j|} = \frac{1}{4\pi\varepsilon_0 \sqrt{(x_i - x_j)^2 + (y_i - y_j)^2 + (z_i - z_j)^2}}$$ (5.58)

where x, y, and z are the Cartesian coordinates of carbon atoms.

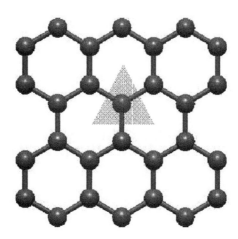

FIGURE 5.24
Electric charge distributed on a triangular area surrounding a carbon atom.

Note that if the atom of interest and the charged atom are interchanged, Equation 5.58 remains the same. Thus, the matrix element a_{ij} only depends on the distance between atom i and atom j, and there is no need to distinguish between the atom of interest and the charged atom. Therefore, [A] is a symmetric matrix.

However, when the atom of interest coincides with the charged atom, that is, $i = j$, the diagonal terms in matrix [A] becomes infinite if Equation 5.58 is used. An alternate equation must be established for obtaining the diagonal terms a_{ii}. For this purpose, we assume that the charge is distributed over the triangular surface area around the carbon atom, as shown in Figure 5.24. Here, the actual three-dimensional surface is approximated as a flat triangle because of the small area involved. The charge density is $\sigma = q_i/s$, where $s = 3\sqrt{3}b^2/4$ denotes the triangular area and $b = 0.142$ nm is the carbon bond length. The voltage at atom i resulted from this charge distribution σ can be expressed as

$$V(\mathbf{r}_i)_i = \iint_s \frac{q_i/s}{4\pi\varepsilon_0\sqrt{\xi^2 + \eta^2}}\, d\xi\, d\eta. \tag{5.59}$$

The diagonal terms can be extracted from Equation 5.59 as

$$a_{ii} = \frac{1}{4\pi\varepsilon_0 s} \iint_s \frac{1}{\sqrt{\xi^2 + \eta^2}}\, d\xi\, d\eta. \tag{5.60}$$

A closed form expression of the above equation can be obtained:

$$a_{ii} = \frac{b\sqrt{3}\ln(2 + \sqrt{3})}{4\pi\varepsilon_0 s}. \tag{5.61}$$

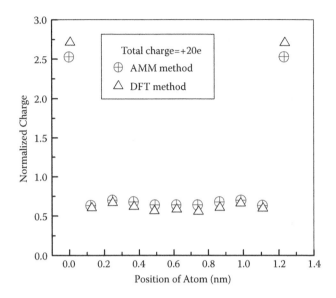

FIGURE 5.25
Normalized charge profile. (From C. Y. Li and T. W. Chou, *Appl. Phys. Lett.* 89 [2006] 063103.)

Having given the nondiagonal and diagonal elements of the square matrices, the unknown charges at each atom can be obtained by solving the linear system of equations (Equation 5.57). The total charge Q on the nanotube under a voltage V_0 can then be obtained by:

$$Q = \sum_{i=1}^{n} q_i.$$

(5.62)

Li and Chou's calculations showed that the charges on atoms located at tube ends are much higher than those on atoms in the central portion of the tube. Figure 5.25 compares the computational results of Li and Chou with the DFT calculations of Keblinski et al. [65]. It indicates that classical electrostatics is applicable to nanostructures. Li and Chou also showed that both nanotube diameter and length have significant effect on the charge distribution (Figure 5.26). If the nanotube is in a semi-infinite electric field, the position of nanotubes will also affect the charge distribution.

Some researchers have also considered the charge distribution on multiwalled nanotubes. Ke and Espinosa [68] reported their studies of electrostatic charge of multiwalled nanotubes based on classical electrostatics and using the boundary element method. Their results showed that when the nanotube length is beyond a "characteristic length," the charge distribution in the central part gradually approaches the uniform distribution of an infinitely long cylinder, and the charge distribution at the ends is

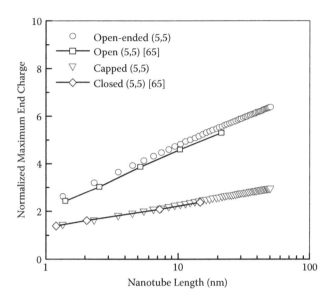

FIGURE 5.26
Tube end charge normalized by average charge.

independent of the nanotube length. They suggested the charge distribution of a finite length nanotube can be modeled with the uniform charge distribution corresponding to an infinite nanotube plus concentrated charges at the ends.

Compared with theoretical modeling, the experimental studies of charge distribution on carbon nanotubes are very few. Chen et al. [69] studied the radial charge distribution of double-walled nanotubes by chemically doping them. Their resonant Raman scattering experiments indicated an electrostatic favoritism of doping the outer layer with 90% of the charge residing on the outer tube. Dorozhkin and Dong [70] reported the use of a low-energy electron point source microscope for real-time quantitative mapping of the linear charge density distribution along nanotube ropes. More recently, Paillet et al. [71] investigated the static charge distribution in individual single-walled nanotubes by using electrostatic force microscopy. It was concluded that charges are distributed uniformly along the nanotubes, and the charge density is in the range of 1.0 electron/nm for a 2.5 nm nanotube under a voltage of 3.5 V. This seems to be in reasonable range comparing to theoretical results, except they did not mention the charge concentration.

Based on these studies surveyed, it can be concluded that classical electrostatics is still applicable to structures at the nanoscale. Although the amount of electric charges accumulated on a nanotube depends on factors such as electric field and nanotube diameter, the profile of charge distribution is mainly affected by nanotube length. The shorter the length, the higher the charge concentration at tube ends. For nanotubes that are sufficiently long, the portion of the tube with high charge concentration relative to the entire tube length becomes rather small, and the approximation used by Ke and Espinosa [68] can be justified.

5.3.2 Electromechanical Actuation of Carbon Nanotubes

There also have been some theoretical efforts to understand the mechanical response of charged carbon nanotubes. Sun et al. [72] used the DFT calculations to predict the charge-induced strain for isolated single-walled carbon nanotubes. They found that the nanotube charge–strain relationship is similar to that of graphene if the nanotube diameter is sufficiently large. For nanotubes of small radii, the charge–strain curves are influenced by nanotube chirality. Gartstein et al. [73] studied the charge-induced strain of nanotubes by using a simplified electron lattice model. They considered the gap energy modulation by external strains, dimensional and torsional deformation caused by charge injection, and stretch-induced torsion. Only low charge injection levels were considered and the coulombic effects were ignored. It was predicted that nanotube deformations are anisotropic and strongly dependent on nanotube structure. Hartman et al. [74] conducted an investigation of electromechanical coupling in carbon nanotubes by focusing on phonon frequency shifts as a result of charge injection. Their computational results for single-walled carbon nanotubes using a tight-binding model and experimental results for multiwalled carbon nanotubes using Raman spectroscopic measurements concluded that the phonon frequency shifts depend only on the amount of injected charge per atom, without significant difference due to the positive or negative sign of the charge. It means both positive and negative charges result in the same kind of deformation. Furthermore, the ab initio and analytical tight-binding calculations of Verissimo-Alves et al. [75] concluded that semiconducting nanotubes expand regardless of the sign of injected charge and metallic nanotubes initially contract when the positive charge is at a low level but eventually expand when the positive charge reaches a higher level.

Guo and Guo [76] demonstrated an exceptionally large electrostrictive deformation in single-walled nanotubes by using Hartree-Fock (H-F) and density functional quantum mechanics simulations. The charge-induced axial strains for both armchair and zigzag nanotubes were predicted to be greater than 10% for electric field strength within 10 V/nm. This large axial deformation of nanotubes is attributed to two aspects of geometrical changes: the elongation of the C–C bonds and the distortion of the carbon hexagonal rings. Their detailed investigation showed that with increasing electric field strength there was no systematical change in bond angles for contribution to axial extension, but the C–C bond lengths were found to be sensitive to the electric field. Comparing the axial deformation ratio, they concluded that the C–C bond elongation is the only significant contribution to the electrostrictive deformation.

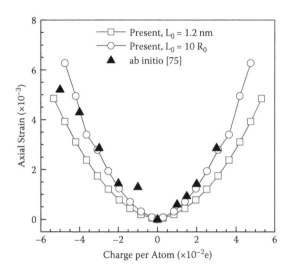

FIGURE 5.27
Charge-induced axial strain. (From C. Y. Li and T. W. Chou, *Nanotechnology* 17 [2006] 4624–8.)

More recently, Li and Chou [77,78] theoretically investigated the electro-mechanical coupling effect in nanotubes. The electrostatic interactions between charged carbon atoms were calculated based on the coulombic law. The charge-induced deformations in both axial and radial directions were obtained by using the molecular structural mechanics method and considering the electrostatic interactions as external loads acting on carbon atoms. Figure 5.27 shows the variation of the axial strain of an SWCNT with the level of injected charge. For comparison with existing results, we selected a zigzag SWCNT (11, 0) and an armchair SWCNT (5, 5). The results of our simulation indicate that the axial strain increases with increasing charge level. The axial strain is an even function of the extra charge per atom. Thus, the nanotube is elongated when extra electrons or holes are injected. Verissimo-Alves et al. [75] have studied the charge-induced strain of SWCNTs with unit cell lengths by using ab initio and analytical tight-binding calculations. Their results for the (11, 0) and (5, 5) SWCNTs are also shown in Figure 5.27 for comparison. It can be seen that our prediction of charge-induced strain is in very good agreement with the ab initio results. Figure 5.28 shows the variation of charge-induced axial strain with the nanotube aspect ratio. The axial strain increases with the nanotube aspect ratio. The effect is especially obvious when the aspect ratio is small but diminishes at larger aspect ratios. We conclude that, for the diameters of nanotubes studied here, the charge-induced strain becomes insensitive to nanotube aspect ratios when they are greater than 10. This result seems to be desirable because in practical applications SWCNTs usually have an aspect ratio of 100 or greater.

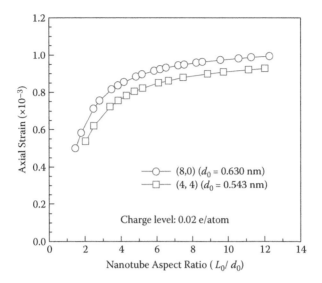

FIGURE 5.28
The effect of tube length on charge-induced axial strain. (From C. Y. Li and T. W. Chou, *Nanotechnology* 17 [2006] 4624–8.)

Li and Chou's results also show that the radial deformation was not uniform for nanotubes and a trumpet shape of open-ended nanotube can be seen, just like those shown by Keblinski et al. [65] and Luo and Wu [79]. For capped nanotubes, the nonuniformity of the radial deformation was not as obvious as that of open-ended nanotubes. Li and Chou [78] gave a quantitative demonstration of nanotube deformations (Figure 5.29). It was shown that the large radial strain was confined to a small region near the nanotube ends for an open-ended nanotube. The majority of the tube length between the ends had an almost uniform radial strain, and the gradient of the radial strain at the vicinity of tube ends was very significant. However, the radial strain of the capped nanotube was much smaller than that of open-ended nanotube at roughly the same charge level. The distribution of radial strain was also very different from that of the open nanotube, and the transition was more gradual, from the smaller strain at the middle to the larger strain at the ends. For both open-ended and capped nanotubes, the radial strain increases with increasing charge level.

5.3.3 Charge-Induced Failure of Carbon Nanotubes

An issue closely tied to the electromechanical coupling of carbon nanotubes is nanotube stability. If the current in a nanotube from an applied voltage exceeds the maximum allowable value, an electrical breakdown of the nanotube would occur. Although metallic nanotubes are efficient one-dimensional conductors that can sustain current densities up to $10^9 \sim 10^{10}$ A/cm^2, they

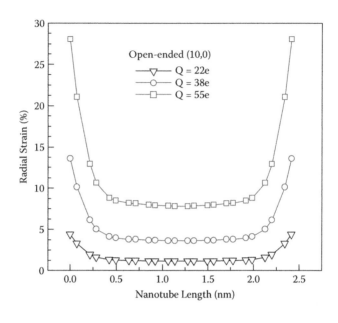

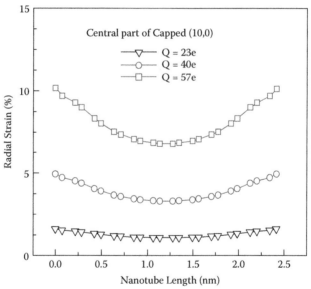

FIGURE 5.29
The variation of radial strain along the nanotube length. (From C. Y. Li and T. W. Chou, *Carbon* 45 [2007] 922–30.)

ultimately fail at high enough currents. Thus, the study of the current-carrying capability of nanotubes is essential for assessing the lifetime of nanotube-based nanodevices and, in some cases, for constructively utilizing the electrical breakdown.

Theoretical studies on the stability of carbon clusters can be traced back to the 1980s. Raghavachari and Binkley [80] conducted ab initio calculations for understanding the nature of the structures, stabilities, and the fragmentation behavior of small carbon clusters ($C_2 - C_{10}$). The study on the stability of nanotubes began about a decade later after nanotubes were considered to be ultimate field electron emitters with potential applications in flat panel displays. Lee et al. [81] investigated the mechanism of disintegration of nanotubes in strong electric fields using the ab initio density function formalism. Their simulations showed that nanotubes could withstand electric fields up to 20 V/nm before the unraveling of carbon chains at the exposed edge. Keblinski et al. [65] and Luo and Wu [79] examined the effect of charge on the stability of single-walled nanotubes by using DFT calculations. Their studies indicated that nanotubes deform into a trumpet shape because of the significant charge concentration at the tube ends when the extra charge injected on the nanotube is large. Nanotubes were found to become gradually unstable beginning from the tube ends with the increase of extra charge. For a 100-atom armchair (5, 5) nanotube, Keblinski et al. [65] found that the nanotube became unstable when the total extra charge reached +30e and the positively charged carbon atoms at the tube ends were ejected. Luo and Wu [79] reached a similar conclusion for a 96-atom zigzag (6, 0) nanotube that no stable structure was obtained when the extra charge was 29e.

Recently, Li and Chou [78] simulated the charge-induced failure of carbon nanotubes by gradually increasing the applied electric voltage. Figure 5.30 shows the change of carbon bond length after the nanotubes are charged. The increase in bond length is not uniform along the length of the nanotube. For the open-ended nanotube, the largest increment occurs at the few bonds located at the end of the tube. For the capped nanotube, the closer the bonds are to the ends, the longer are their bond lengths. But there is no abrupt jump in the bond length.

The steric energy of a carbon–carbon bond becomes large when the bond is stretched. It can be expected that the carbon bond would be broken when the extra charge reaches a certain level. The nanotube would become unstable at that critical charge level, and the electrical failure would occur. To determine the critical charge level, the criterion for carbon bond breaking needs to be first established.

The interatomic bonds in carbon nanotubes are hybridized sp^2 bonds. Each carbon atom uses the sp^2 hybrids to form σ bonds with two other carbon atoms. The π bond created by the overlap of p_z atomic orbitals is much weaker than the σ bond. The bond dissociation energy for a σ bond is much greater than that for a π bond. Also, the bending potential of bond angle does not contribute to the stretching energy and has little effect on bond breaking,

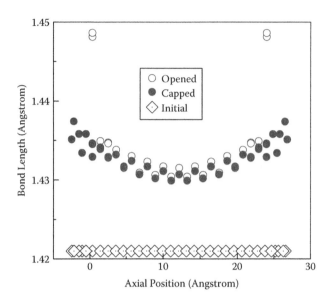

FIGURE 5.30
Bond length changes when a nanotube is charged ($V_0 = 30$ V). (From C. Y. Li and T. W. Chou, *Carbon* 45 [2007] 922–30.)

although it plays an essential role in establishing an equilibrium configuration of the nanotube. Thus, the criterion for bond breaking in carbon nanotubes can be simplified as:

$$E_r = k_r(r - r_0)^2 \geq E_{dissociation},$$ (5.63)

where k_r is the force field constant for bond stretching, $E_{dissociation}$ stands for the bond dissociation energy, and r_0 and r are the bond length before and after stretching, respectively. In Li and Chou's simulations, the k_r and $E_{dissociation}$ are taken as 469 kcal/mol·Å² and 5.2 eV/bond, respectively.

The bonding status of an armchair nanotube at the critical charge level is shown in Figure 5.31. The highlighted atoms show the dissociation of bonds, and the breaking of carbon bonds first appeared at the ends for open-ended nanotubes. Half of the bonds inclined to the axial direction at the very ends break simultaneously. The critical levels of total extra charge for the 96 atoms (6, 0) and 110 atoms (5, 5) nanotubes were 39e and 34e, respectively. Their findings are comparable to the existing results based on DFT calculations [65,79]. The study of Li and Chou [78] indicated that the occurrence of bond breaking in nanotubes with end caps is somewhat different from that of open-ended tubes. For a capped (10, 0) nanotube at their critical charge level, the bond breaking first appeared at the end caps near the pentagon at the very ends, but the pentagon itself was not broken (Figure 5.32). Their results revealed that

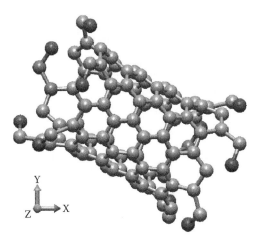

FIGURE 5.31
An armchair (5, 5) open-ended nanotube at the critical charge level. (From C. Y. Li and T. W. Chou, *Carbon* 45 [2007] 922–30.)

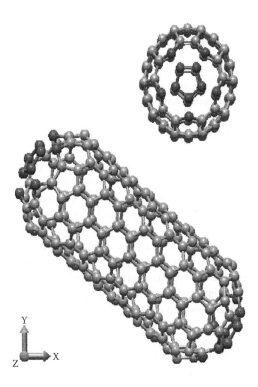

FIGURE 5.32
A capped zigzag (10, 0) nanotube at the critical charge level. (From C. Y. Li and T. W. Chou, *Carbon* 45 [2007] 922–30.)

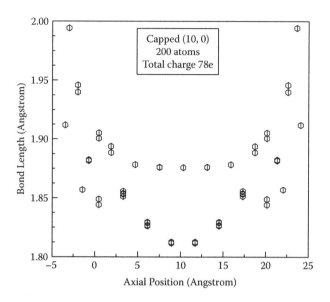

FIGURE 5.33

Distribution of bond lengths of a capped nanotube at the critical charge level. (From C. Y. Li and T. W. Chou, *Carbon* 45 [2007] 922–30.)

the end-caps can enhance the stability of the nanotubes. Li and Chou [78] also examined the distribution of bond lengths at the critical bond-braking status. Figure 5.33 shows that the length of carbon bonds at the critical status is in the range of about 1.8–2.0 Å, and the second largest bond length also exists in the cap region, just beside the broken bonds.

These bonds would be broken subsequently if the charge process continued. Li and Chou's simulation results are consistent with the experimental findings reported by Wang et al. [82], who observed that carbon atoms at the end caps of individual nanotubes can be removed via an electric field-induced evaporation process.

Most nanotube-based nanodevices may be subjected to the coupled actions of applied electric field and mechanical loading. An understanding of the coupled effect of mechanical deformation and electric field on the failure of nanotubes is thus indispensable. Guo and Guo [76] conducted theoretical studies on the coupled mechanical and electrical behavior of single-walled carbon nanotubes under an applied electric field and tensile loading by quantum molecular dynamics simulations. They concluded that the nanotube strength under electric field is much lower than under mechanical loading alone and may decrease significantly while increasing the intensity of electric field. But, overall, theoretical and modeling studies on this coupling effect are still limited.

5.4 Constitutive Modeling of Carbon Nanotube-Based Composites

In the last two sections, we focused mainly on the properties and behaviors of individual carbon nanotubes. However, considerable interest has been generated in utilizing carbon nanotubes as nanoscale reinforcement in composites due to their high stiffness, unusual tensile strength, low density, and exceptionally large aspect ratio; advances in scaling up production techniques for carbon nanotubes have enabled their applications. To realize the potential for carbon nanotubes as reinforcement in composites, we must have a fundamental understanding of how the nanoscale material structure influences the properties of the nanotube as well as how nanotubes interact within a composite.

In the literature on nanotube-based composites, there is large variation in the reported elastic properties, but little information is supplied concerning the nanocomposite structure. Reported improvements in elastic modulus are lower than expected if the nanotube is assumed to act as a solid fiber with an elastic modulus of 1.0 TPa. Discrepancies are often explained by assuming that there is insufficient load transfer at the nanotube–matrix interface. Thus, many researchers have focused efforts on functionalizing the nanotube surface to promote adhesion between the nanotube and the polymer matrix. While a strong interface between the reinforcement and matrix is important in terms of load transfer efficiency, improvements in interface properties have a far greater influence on the composite's ultimate strength and fracture behavior than the elastic modulus at low strain. At the nanoscale, the structure of the nanotube plays a significant role in the transfer of load within the composite, and variations in reported elastic properties are probably a consequence of (1) differences in the structure and size of the nanotubes used as reinforcement and (2) variations and lack of control of the nanocomposite structure.

To meet the need of elucidating the fundamental reinforcement mechanisms in nanotube-based composites and developing tools to relate the nanoscale structure to the properties of nanotube-based composites, Thostenson and Chou [83] took into account the nanoscale features of a carbon nanotube and developed a constitutive modeling technique to determine the composite elastic properties based on the properties of the constituent materials, as well as the structure of the carbon nanotube. The technique was then applied to a model system of multiwalled carbon nanotubes embedded in a polystyrene polymer matrix. Their modeling results show that the properties of nanotube composites are strongly influenced by the nanotube diameter. For multiwalled nanotubes, there is typically a distribution of diameters, and modeling the diameter distribution of the reinforcement is essential for an accurate modeling of overall nanotube composite elastic properties. In the following section, we will introduce this constitutive modeling technique.

5.4.1 Micromechanics of Nanotube-Based Composites

For traditional fibrous composites, a wide variety of models have been developed to predict the elastic properties of the composites based on the properties of the constituent materials and the geometry of the reinforcement. This approach is generally termed *micromechanics* because the models deal with the geometric heterogeneity at the microscopic level (microstructure). For traditional composites, where the reinforcement phase diameter is measured in microns, the fiber can be treated as a continuum with isotropic or anisotropic material properties. But at the nanometer-scale, continuum assumptions are no longer valid, and the structure and bonding of the carbon nanotube must be considered.

To take into account the structure of the nanotube, the properties of an "effective fiber" are defined. The definition of effective fiber properties is then used to determine the elastic properties of the composite based on a micromechanics approach. The micromechanical model of Halpin and Tsai [84] for discontinuous fiber composites is adopted for determining the properties of a unidirectional discontinuous fiber composite.

5.4.1.1 Nanotube Load Transfer Mechanism: Effective Fiber Properties

To model the properties of the nanotube-based composite it is important to consider the nanoscale structure of carbon nanotubes and how the nanotube interacts with the polymer matrix. For fiber-like materials, load is transferred to the reinforcement through shear stresses at the fiber–matrix interface. For a single-walled carbon nanotube, the nanotube will carry the entire load transferred at the nanotube–matrix interface. For a multiwalled carbon nanotube, the bonding between the walls of the nanotube is from van der Waals interactions. The relatively weak bonding between the layers of the nanotube results in minimal load transfer between the layers of the nanotube. As a consequence of this weak layer-to-layer interaction in the nanotube, the outermost layer of the multiwalled tube will carry almost the entire load transferred at the nanotube–matrix interface. For modeling of composite elastic properties at low strain, it can be assumed that interfacial stresses do not result in debonding of the reinforcement–matrix interface.

To determine the effective elastic modulus of the nanotube embedded in a composite, the load carrying capability of the outermost layer of the nanotube must be applied to the entire cross-section. The elastic modulus of the nanotube is modeled by considering that the outer wall of the nanotube acts as an effective solid fiber with the same deformation behavior and same diameter (d) and length (l) as shown in Figure 5.34. An applied external force on the nanotube and the fiber will result in an isostrain condition:

$$\varepsilon_{NT} = \varepsilon_{eff}, \qquad (5.64)$$

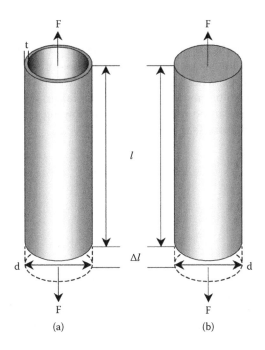

FIGURE 5.34
Schematic of (a) nanotube and (b) effective fiber used to model the elastic properties of an individual carbon nanotube embedded in a composite. (From E. T. Thostenson and T. W. Chou, *J. Phys. D: Appl. Phys.* 36 [2003] 573–582.)

where the subscripts NT and eff refer to the nanotube and effective fiber, respectively. Using Equation 5.64, we can relate the elastic properties of the nanotube to that of an effective fiber:

$$E_{eff} = \frac{\sigma_{eff}}{\sigma_{NT}} E_{NT}. \tag{5.65}$$

Because the applied external force is the same, the effective moduli can be expressed in terms of the ratio of their cross-sectional areas.

$$E_{eff} = \frac{A_{NT}}{A_{eff}} E_{NT}. \tag{5.66}$$

After substituting, the modulus of the effective fiber can be expressed in terms of the elastic modulus of the nanotube, the nanotube outer layer thickness ($t = 0.34$ nm) and the nanotube diameter (d).

$$E_{eff} = \frac{4t}{d} E_{NT}. \tag{5.67}$$

It is understood that the above expression is valid for $(t/d) < 0.25$.

5.4.1.2 Structure–Property Relationship

A wide variety of models have been developed to predict the elastic properties of fiber composites in terms of the properties of the constituent materials. In certain cases where specific fiber arrangements are considered, closed-form elastic solutions can be obtained. Halpin and Tsai [84] showed that many solutions can be reduced to the following general form, widely referred to as the Halpin–Tsai equations:

$$E_c = E_m \left(\frac{1 + \zeta \eta V_f}{1 - \eta V_f} \right) \tag{5.68}$$

$$\eta = \left(\frac{E_f}{E_m} - 1 \right) \bigg/ \left(\frac{E_f}{E_m} + \zeta \right), \tag{5.69}$$

where E_c is the composite elastic modulus, V_f is the fiber volume fraction, and E_f and Em are the fiber and matrix modulus, respectively. In Equations 5.68 and 5.69, the parameter ζ is dependent on the geometry and boundary conditions of the reinforcement phase. For an aligned short fiber composite, this parameter can be expressed as:

$$\zeta = 2\frac{l}{d} + 40V_f, \tag{5.70}$$

and for low volume fractions:

$$\zeta = 2\frac{l}{d}. \tag{5.71}$$

By substituting Equations 5.67, 5.69, and 5.71 into Equation 5.68, we can express the nanocomposite elastic modulus in terms of the properties of the polymer matrix and the nanotube reinforcement:

$$E_{11} = E_m \left(1 + 2\left(\frac{l}{d} \right) \left(\frac{\frac{E_{NT}}{E_m} - \frac{d}{4t}}{\frac{E_{NT}}{E_m} - \frac{l}{2t}} \right) V_{NT} \right) \left(1 - \left(\frac{\frac{E_{NT}}{E_m} - \frac{d}{4t}}{\frac{E_{NT}}{E_m} - \frac{l}{2t}} \right) V_{NT} \right)^{-1} \tag{5.72}$$

where, following standard notation used for traditional fibrous composites, E_{11} is the elastic modulus in the principal material direction, which is the direction of nanotube orientation. Equation 5.72 is valid for $l > d > 4t$. Unlike traditional fibrous composites, where the aspect ratio completely describes the reinforcement in dimensionless terms, the nanotube diameter must be known, since the reinforcement efficiency of the nanotube changes with diameter.

5.4.1.3 Nanotube Structure Influence: Volume Fraction and Density

For carbon nanotubes, there will usually be a distribution of nanotube diameters in a given sample. In addition to the influence of nanotube diameter on the elastic properties of the composite, the volume fraction of the reinforcement is also influenced by the nanotube diameter. The volume occupied by a large-diameter nanotube can be many times higher than the volume occupied by a small-diameter nanotube because the volume occupied by a given nanotube is proportional to d^2. Experimental data for nanocomposites are typically expressed in terms of the weight fraction of reinforcement, since the weight fraction is easily calculated based on the mass of the constituents added during processing. The nanotube weight fraction (W_{NT}) does not explicitly describe the content of reinforcement because it depends on the relative densities of the matrix and the nanotube. Furthermore, the nanotube diameter and wall structure will significantly influence the nanotube density. As a consequence, it is critical to have knowledge of the size and structure of the carbon nanotubes used in processing of the composite system.

The distribution of nanotube diameters for a specific nanotube sample can be determined by measuring the outside diameter of a statistically large sample of nanotubes and then using the experimental data to determine the probability distribution of nanotubes $\xi(d)$. For the purpose of modeling the composite elastic properties, we are interested in the volume fraction of carbon nanotubes within the composite. From the diameter distribution we can then define the volume distribution of nanotubes per unit length $\psi(d)$:

$$\psi(d) = \frac{d^2\xi(d)}{\int\limits_0^\infty (d^2\xi(d))d(d)}. \tag{5.73}$$

The above volume distribution will need to be considered when calculating the overall nanocomposite properties.

For the conversion of weight fraction, measured when processing the nanocomposite, to volume fraction needed for predicting the elastic properties (Equation 5.72), we must know the density of the nanotubes and the polymer matrix. For fibrous composites, the volume fraction of fibers can be calculated based on the density of the constituents:

$$V_f = \frac{\rho_c}{\rho_f} W_f, \tag{5.74}$$

where

$$\rho_c = \rho_f V_f + \rho_m V_m, \tag{5.75}$$

where the ρ is density, and the subscripts f, m, and c refer to the fiber, matrix, and composite, respectively. Substituting Equation 5.75 into Equation 5.74, the volume fraction can be calculated from:

$$V_f = \frac{W_f}{W_f + \frac{\rho_f}{\rho_m} - \frac{\rho_f}{\rho_m} W_f}. \tag{5.76}$$

To gain some insight into the influence of nanotube structure on density, we can calculate the nanotube density per unit length by assuming the graphitic layers of the tube shell to have the density of fully dense graphite ($\rho_g = 2.25$ g/cm³). From the measurements of inside (d_i) and outside diameter (d), the nanotube density can be calculated:

$$\rho_{NT} = \frac{\rho_g (d^2 - d_i^2)}{d^2}. \tag{5.77}$$

Obviously, the density of a multiwalled nanotube will increase with the number of walls.

5.4.1.4 Calculation of Nanocomposite Elastic Modulus

Equation 5.72 expresses the diameter dependence of the carbon nanotube reinforcement on the nanocomposite properties. However, with a distribution of nanotube diameters, Equation 5.72 cannot be applied directly to calculate the nanocomposite elastic modulus. Accurate modeling of the composite elastic modulus requires knowledge of the distribution of nanotube diameters and the volume fraction that tubes of a specific diameter occupy within the composite. If nanotubes are uniformly dispersed and aligned throughout the matrix phase, the contribution of each nanotube diameter can be considered to act in parallel. Therefore, the elastic modulus of the composite can be calculated as a summation of parallel composites over the range of nanotube diameters.

The concept of parallel composites is illustrated in Figure 5.35. Within the entire volume of the composite, the volume can be divided into N individual composites containing a specific nanotube diameter. Each of the N individual composites will have a specific elastic modulus that depends on the local volume fraction of nanotubes at a given diameter. With the assumption of isostrain, the elastic modulus of the composite can be expressed as a summation of the moduli scaled by the partial volume of each nth composite:

$$E_c = \sum_{n=1}^{N} v_n E_n \Big|_{d_n} \tag{5.78}$$

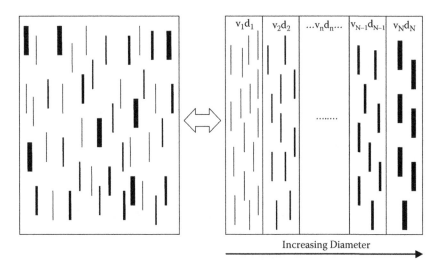

FIGURE 5.35
Illustration of the equivalence between a dispersed composite and N composites each with a specific nanotube diameter and partial volume acting in parallel. (From E. T. Thostenson and T. W. Chou, *J. Phys. D: Appl. Phys.* 36 [2003] 573–582.)

where $E_n|_{d_n}$ is the elastic modulus of the composite calculated from Equation 5.72 at the nanotube diameter included in the nth segment, and v_n is the partial volume of the nth composite:

$$v_n = \frac{V_n}{V}, \quad \sum_{n=1}^{\infty} v_n = 1 \qquad (5.79)$$

where V_n is the volume of the nth composite and V is the overall composite volume.

To calculate the composite elastic modulus at a given nanotube diameter, E_n in Equation 5.78, the local volume fraction at a given nanotube diameter, $V_{NT}|_d$, can be calculated from the volume distribution of nanotubes (Equation 5.73).

$$V_{NT}|_{d_n} = \frac{1}{v_n} \int_{d_n}^{d_n + \, d_n} (V_{NT} \psi(d))d(d) \qquad (5.80)$$

where V_{NT} is the total volume fraction of tubes in the composite calculated from Equation 5.76 and the limits of the integral are the range of diameters included in the nth composite.

5.4.2 Model Nanocomposite System

To predict the elastic modulus of a nanotube composite system, information on the structure of the nanotubes, as well as the structure of the nanocomposite, is required. A model nanocomposite system of aligned multiwalled carbon nanotubes embedded in a polystyrene matrix was produced, and the structures of both the nanotube reinforcement and the nanocomposite were quantified using electron microscopy. The nanocomposite elastic properties were characterized using a dynamic mechanical analyzer (DMA), and the mechanical characterization results are then compared with the structure and property modeling approach developed in Section 5.4.1.

5.4.2.1 Structural Characterization of Model Nanotube-Based Composites

Thostenson and Chou used a microscale twin-screw extruder to obtain the high-shear mixing necessary to untangle the CVD-grown multiwalled nanotubes and disperse them uniformly in a polystyrene thermoplastic matrix. In addition, to create an aligned model composite system, they extruded the polymer melt through a rectangular die, drawing it under tension prior to solidification. The process of extruding the nanocomposite through the die and the subsequent drawing resulted in a continuous ribbon of aligned nanocomposite.

To quantify the structure for the nanotubes used in the model composite systems, high-resolution TEM micrographs were taken of the CVD-grown tubes, and image analysis software was utilized to measure the structural dimensions to quantify both the distribution of nanotube diameters and the nanotube wall structure. Measurements were taken of the outside diameters of nearly 700 nanotubes to obtain a statistically meaningful distribution of diameters. Figure 5.36 shows the resulting histogram for the nanotube diameter distribution. The nanotube diameters show a bimodal distribution with peaks near 18 and 30 nm. To obtain a probability density function for the nanotube diameter distribution, Levenberg-Marquardt nonlinear regression was used to fit the data to a double Lorentzian distribution and a double Gaussian distribution, and the curves were normalized such that the area under the curve is unity. Equations 5.81 and 5.82 show the general forms for the double Lorentz and Gauss equations, respectively.

$$\xi(d) = \frac{a_1}{\left(1 + \dfrac{d - a_2}{a_3}\right)^2} + \frac{a_4}{\left(1 + \dfrac{d - a_5}{a_6}\right)^2}, \tag{5.81}$$

$$\xi(d) = a_1 e^{\left(-\left(\frac{d-a_2}{a_3}\right)^2\right)} + a_4 e^{\left(-\left(\frac{d-a_5}{a_6}\right)^2\right)}. \tag{5.82}$$

The curve fit parameters for the nanotube diameter distributions are shown in Table 5.1 where the units for nanotube diameter are expressed in nanometers.

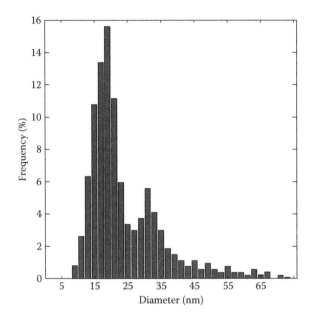

FIGURE 5.36
Experimentally obtained distribution of carbon nanotube diameters. (From E. T. Thostenson and T. W. Chou, *J. Phys. D: Appl. Phys.* 36 [2003] 573–582.)

The Lorentzian and Gaussian probability distributions obtained from the experimental data are shown in Figure 5.37. For small diameter nanotubes, the Gaussian curve most accurately fits the data, but for large-diameter nanotubes, the Gaussian curve underestimates the number of nanotubes. As discussed previously, accurate modeling of the distribution at large nanotube diameters is crucial because the volume occupied by a given nanotube in the composite varies with d^2.

Figure 5.38 shows plots of the volume for both the Lorentzian and Gaussian distributions obtained from the experimental data. In the volume distribution, the relative area under the curve shifts to the larger diameters. Although the height of the peak at 18 nm is three times the height of the peak at 30 nm in the diameter distribution, the two peaks are almost equal in the volume distribution. Of particular importance is the difference in the tail of the curves at large nanotube diameter. The Gaussian curve significantly underestimates the large percentage of volume occupied by large nanotube diameters. Although

TABLE 5.1

Curve Fit Parameters for the Diameter Distribution Functions

Distribution	a_1	a_2	a_3	a_4	a_5	a_6
Lorentz	0.8025	18.23	−3.56	0.02149	31.84	2.946
Gauss	0.0234	31.78	5.84	0.0758	18.03	5.1176

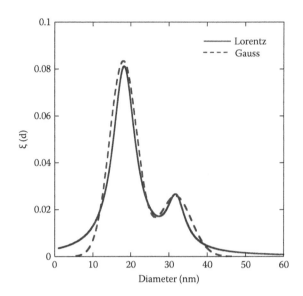

FIGURE 5.37
Diameter distributions of carbon nanotubes. (From E. T. Thostenson and T. W. Chou, *J. Phys. D: Appl. Phys.* 36 [2003] 573–582.)

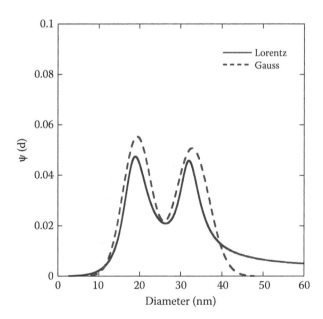

FIGURE 5.38
Volume distributions of carbon nanotubes. (From E. T. Thostenson and T. W. Chou, *J. Phys. D: Appl. Phys.* 36 [2003] 573–582.)

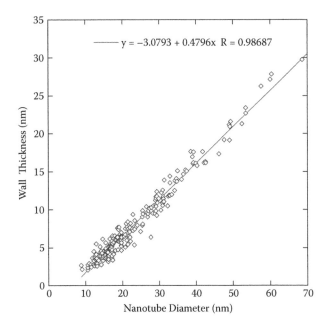

y = −3.0793 + 0.4796x R = 0.98687

FIGURE 5.39
Linear relationship between wall thickness and nanotube diameter for multiwalled nanotubes used in the model composites. (From E. T. Thostenson and T. W. Chou, *J. Phys. D: Appl. Phys.* 36 [2003] 573–582.)

the large diameter nanotubes represent a relatively small percentage of the total number of nanotubes, they occupy a significant percentage of volume within the composite. The Lorentzian curve fit overestimates the number of small diameter nanotubes present, but the difference in the volume distribution for the Gauss and Lorentz curves at small nanotube diameter is insignificant.

The nanoscale tubular structure of the carbon nanotube also results in a distribution of nanotube density. To calculate the density of nanotubes as a function of nanotube diameter, the outside and inside diameters were measured from TEM micrographs. Figure 5.39 shows a plot of the experimental data and indicates a strong linear relationship between the nanotube outside diameter and the wall thickness. At smaller nanotube diameters, the relationship between nanotube wall thickness and outside diameter begins to deviate from the linear curve fit.

Using Equation 5.77, the density of the nanotubes can be calculated from the experimental data. Figure 5.40 shows the calculated nanotube density as a function of the outside diameter, where the curved line is obtained directly from the straight line in Figure 5.39. At larger nanotube diameters, the density of the nanotubes approaches the value for fully dense for graphite (ρ_g = 2.25 g/cm³). Figure 5.41 shows the histogram of calculated nanotube density. For the nanotubes used in the model composite system, the mean density is 1.9 g/cm³.

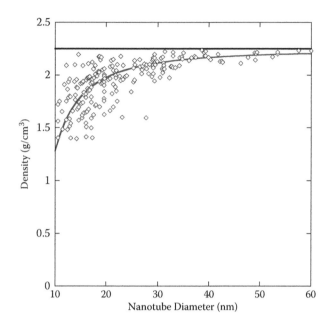

FIGURE 5.40
Variation in calculated nanotube density with outside diameter. (From E. T. Thostenson and T. W. Chou, *J. Phys. D: Appl. Phys.* 36 [2003] 573–582.)

The structure and size of the carbon nanotube expert a significant influence on the properties of the nanocomposite because the nanotube structure influences both the effective elastic modulus and the density of the carbon nanotube. Accurate modeling of the diameter distribution of carbon nanotubes is crucial because large-diameter carbon nanotubes can occupy a significant volume fraction within the composite, as illustrated in Figure 5.38. In addition to occupying a larger relative volume fraction within the composite, large-diameter nanotubes show lower effective elastic moduli (Equation 5.67) and, for the nanotubes used in the model composite system, have higher densities (Figure 5.40).

In addition to the influence of the diameter on the effective elastic modulus of the carbon nanotube and the volume distribution of nanotubes in the composite, the nanotube length is also an important parameter for modeling elastic properties. Variation in nanotube length is difficult to quantify from TEM analysis, since it is clear that a large number of nanotubes are severed by cutting the specimen with a microtome. From measurements on nanotubes that do not appear to be distorted near the ends, it is believed that the majority of the nanotubes in the as-processed composite range between 500 nm and 2 μm, with the average length above 1 μm.

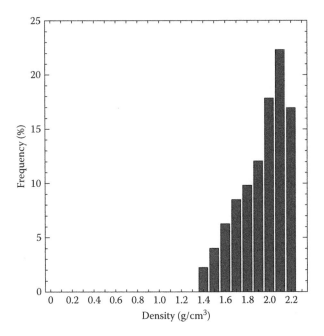

FIGURE 5.41
Histogram showing the distribution of nanotube density. (From E. T. Thostenson and T. W. Chou, *J. Phys. D: Appl. Phys.* 36 [2003] 573–582.)

5.4.2.2 Modeling of Model Nanocomposite Properties

With knowledge of both the nanotube and nanocomposite structures, the micromechanical model developed in Section 5.4.1 can be used to predict the properties of the model nanocomposite system. For input into the microme-chanical model, the modulus of the nanotube, E_{NT}, is assumed to be 1.0 TPa, and the modulus of the matrix, from the characterization results for unrein-forced polystyrene, is 2.4 GPa.

Figure 5.42 shows the influence of nanotube diameter, length, and volume fraction on the composite elastic modulus as predicted by Equation 5.72. While there is a slight increase in elastic modulus at a given nanotube diam-eter and volume fraction with increasing nanotube length, the diameter of the nanotubes plays the most significant role in the composite elastic modu-lus. Figure 5.43 highlights the influence of nanotube length on the composite elastic modulus for a nanocomposite with a diameter of 15 nm and a vol-ume fraction of 0.05 [85]. The increase in composite elastic modulus with nanotube length is most significant for nanotubes with lengths below 1 μm (1000 nm). As the nanotube length increases beyond 1 μm the nanocomposite elastic modulus increases slowly as the modulus approaches the limit for an infinitely long nanotube. The strong diameter dependence of the composite

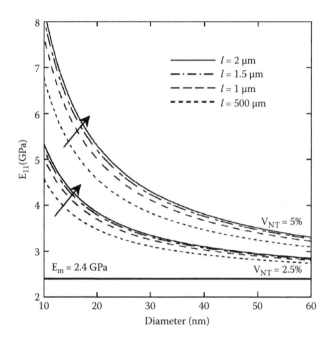

FIGURE 5.42
Influence of nanotube diameter, volume fraction and length on the elastic properties of an aligned nanocomposite modeling. (From Thostenson, E.T., Ph.D. dissertation, University of Delaware, Newark, DE [2004].)

elastic modulus highlights the need to accurately model the dispersion of nanotube diameters in the composite.

To illustrate the importance of modeling the nanotube diameter distribution, the methodology outlined in Section 5.4.1, was used in combination with the structural characterization of the model composite to predict the elastic properties of the composite as a function of the nanotube weight percentage. For conversion of weight loading of nanotubes to volume loading, the density of the matrix was assumed to be $1 \, g/cm^3$. Figure 5.44 shows a direct comparison of the calculated nanotube elastic modulus of varying length nanotubes with the experimental results. For the Lorentz distributions, the calculated elastic modulus compares quite well with the results from the experimental characterization. The Gauss distribution, which ignores the contribution of the larger diameter nanotubes, results in an overestimation of the composite elastic modulus, particularly at higher loading fractions.

For multiwalled nanotubes, there is typically a distribution of diameters, and modeling the diameter distribution of the reinforcement is essential to accurate modeling of overall nanotube composite elastic properties. The nanocomposite elastic properties are particularly sensitive to the nanotube diameter, since larger diameter nanotubes show lower effective moduli and

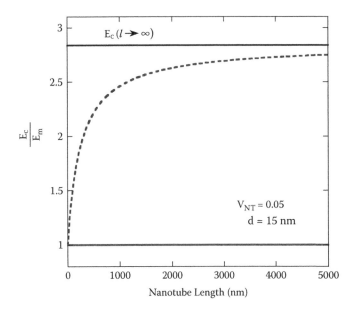

FIGURE 5.43
Influence of increasing nanotube length on the composite elastic modulus. (From Thostenson, E.T., Ph.D. dissertation, University of Delaware, Newark, DE [2004].)

occupy a greater volume fraction in the composite relative to smaller diameter nanotubes.

5.5. Multiscale Modeling of Carbon-Nanotube-Reinforced Composites

For the effective utilization of nanotubes as reinforcements in composites, various attempts have been made regarding the dispersion and alignment of nanotubes, and well-dispersed and well-aligned nanotube reinforced composites are now feasible. On the other hand, the interfacial bonding issue appears to be equally challenging. The major mechanisms of load transfer for traditional fiber-reinforced polymer composites include mechanical interlocking, chemical bonding, and weak van der Waals bonding between the fiber and the matrix. These mechanisms may also be active in nanotube-reinforced polymer composites. Some efforts have been devoted to the study of the load transfer between nanotubes and the matrix. In the study of carbon nanotube–polymethyl methacrylate composites, Jia et al. [86] showed the possibility of existence of carbon–carbon bonds between the nanotube and the matrix. Lordi and Yao [87] examined the factors governing interfacial

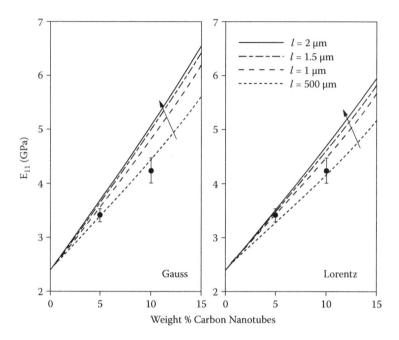

FIGURE 5.44
Influence of nanotube weight %, length and diameter distribution on the elastic moduli of nanotube composites. (From E. T. Thostenson and T. W. Chou, *J. Phys. D: Appl. Phys.* 36 [2003] 573–582.)

adhesion in nanotube-based composites from the molecular mechanics viewpoint. The binding energies and frictional forces were found to play only a minor role in determining the strength of the interface. The key factor in forming a strong bond at the interface is having a helical conformation of the polymer chain around the nanotube. Wagner et al. [88] examined stress-induced fragmentation of multiwalled carbon nanotubes in polymer films. From estimated values of nanotube axial normal stress and elastic modulus, they concluded that the nanotube–polymer interfacial shear stress is on the order of 500 MPa and higher. This value is an order of magnitude higher than the stress transfer ability of current advanced composites. It is well known that the efficiency of load transfer depends on the interfacial shear stress between the reinforcement phase and the matrix. A high interfacial shear stress will transfer the applied load to the fiber over a short distance, whereas a low interfacial shear stress will require a longer transfer length. Covalent bonding can result in more efficient load transfer than van der Waals interactions. This is one of the reasons that some research efforts have been devoted to the functionalization of carbon nanotubes to facilitate the formation of covalent bonds between nanotubes and the matrix.

Despite the various studies conducted on interfacial bonding in nanotube–matrix composites, there is a lack of quantitative understanding of the load transfer efficiencies. The fundamental issues include the interfacial shear stress distribution, the stress concentration in the matrix in the vicinity of nanotube ends, the axial stress profile in the nanotube, and the effect of the nanotube aspect ratio on load transfer. Due to the difficulty in modeling nanotube reinforced composites, analytical studies on the mechanisms of load transfer between the matrix and nanotubes are still very limited. Among the available literature, Lordi and Yao [87] used force-field-based molecular mechanics to model the interactions between nanotubes and several different kind of polymers. Wise and Hinkley [89] used molecular dynamics simulation for addressing the local changes in the interface of a single-walled nanotube surrounded by polyethylene molecules. Odegard et al. [90] studied the effect of chemical functionalization on the mechanical properties of nanotube–polymer composites by using an equivalent continuum modeling technique.

In the study of traditional fiber composites, the approach is to first examine the properties of the constituent phases separately and then combine them to assess the synergy. The fiber and matrix materials are often assumed to be isotropic and homogeneous, and the formation of interphase is taken into account in a multiphase model. However, in a nanotube–polymer composite, the characteristic of the reinforcement phase, unlike the usual fiber material, is highly size and structure dependent. Thostenson and Chou (see Section 5.4.1) characterized extensively multiwalled carbon nanotubes for reinforcing a polystyrene matrix material. Both random and aligned nanotube composites were investigated. It was concluded that the nanotube reinforcement efficiency is highly sensitive to the diameter, wall thickness, and density of the nanotubes, and there is a broad distribution of the structures of the nanotubes in the composite. The findings of Thostenson and Chou and many other studies pertaining to the mechanical properties of carbon nanotubes underscore the necessity to account for the atomistic structures of nanotubes in the modeling of their composites.

In a nanotube-reinforced composite, the volume fraction of polymeric matrix is usually much larger than the volume fraction of nanotubes. Although with current computing power the simulation of both the nanotube and the matrix at the atomistic scale has been attempted, this kind of simulation is very time-consuming because the volume of the matrix is much greater than that of the reinforcement. As a sensible alternative, the polymer matrix can be treated as a continuous medium. Therefore, it is necessary to perform multiscale modeling and to establish a linkage between modeling at the nanoscale and the macroscale.

Among the modeling techniques for nanostructured materials, the molecular dynamics approach has been most frequently adopted by researchers. However, for stress analysis of nanotube–polymer composites, the application of molecular dynamics for modeling of the nanotube is unnecessary because the task is essentially a static problem. To facilitate the simulation of the elastic behavior of nanocomposites, it is desirable to employ computationally efficient techniques to account for the atomistic structures of nanotubes, but also to be able to bridge length scales from atomistic to continuum.

In this section, we mainly introduce the multiscale modeling technique for nanotube-based composites developed by Li and Chou [91]. They analyzed the stress distributions in carbon nanotube–polymer composites under tension by combining the atomistic molecular structural mechanics approach and the continuum finite element method.

5.5.1 Modeling of Nanotube and Matrix

The modeling of nanotubes is conducted at the atomistic scale. The molecular structural mechanics approach is adopted for modeling of the nanotubes because of its computational efficiency. Basically, a single-walled carbon nanotube is simulated as a space frame structure, where the covalent bonds in the nanotube are treated as connecting beams and the carbon atoms as joint nodes in the frame structure. If the equivalent beam is assumed to be of round section, only three stiffness parameters (i.e., the tensile resistance EA, the flexural rigidity EI, and the torsional stiffness GJ) need to be determined for the analysis. Then, the concept of energy equivalence between local potential energies in molecular mechanics and elemental strain energies in structural mechanics is adopted. After a direct relationship between the structural mechanics parameters and the molecular mechanics force field constants is established, the nanotube deformation under certain loading conditions can be readily solved, following structural mechanics technique.

Because the volume of matrix is usually much greater than that of the nanotube reinforcement, it is formidable to simulate the whole composite by atomistic modeling. Therefore, as a tractable compromise, the matrix is treated as a continuous medium, and the finite element method is adopted for modeling its deformation. The matrix phase in computational models is in three-dimensional space. Two kinds of three dimensional finite elements can be used in the meshing of the matrix. One is the 20-node isoparametric cubic element and the other is the 15-node isoparametric wedge-shaped element. The 20-node elements are used in the circumferential region surrounding the nanotube. The 15-node elements are used in the regions directly above and below the nanotube.

5.5.2 Modeling of Nanotube–Polymer Interface

Since the nanotube is modeled at the atomistic scale and the matrix is treated as a continuum, some difficulties arise in the interfacial transition. While much effort has been devoted to functionalizing the nanotube surface to promote adhesion between the nanotube and the polymeric matrix and improving the composite ultimate strength, the elastic modulus of composites at low strain is not significantly influenced by the strength of the reinforcement–matrix interface. Thus, two limiting cases are possible in interfacial load transfer capability. The case of low interfacial load transfer is approximated

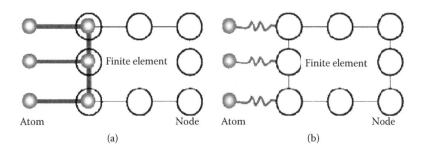

FIGURE 5.45
Nanotube–polymer interface treatment: (a) perfect interface and (b) van der Waals interface.

by the van der Waals interface. The case of high interfacial load transfer is simulated by a perfect interface, which may exist at a covalently bonded nanotube–matrix interface (Figure 5.45).

For simulations of van der Waals interactions at the interface, the truss rod model is adopted, which was first developed for simulating the van der Waals forces between neighboring atomic layers of a multiwalled carbon nanotube. At the nanotube–matrix interface, the activation of a truss rod is determined by the distance between an atom in the nanotube and a node in the finite element. For convenience in computation, the only considerations are the van der Waals interactions between the nanotube and the surface of the polymeric matrix immediately adjacent to the nanotube. This assumption may tend to underestimate the load transfer capability of the nanotube–polymer interface. The center of the atoms in the single-walled nanotube is assumed to be located in the mid-section of the tube thickness, as shown in Figure 5.46(a), which is assumed to be 0.34 nm. The inner surface of the polymer matrix is assumed

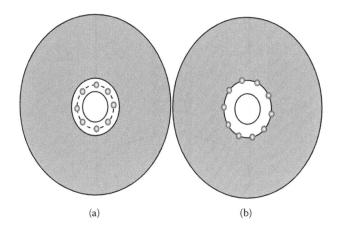

FIGURE 5.46
Top view of nanotube–polymer interface: (a) van der Waals interactions, (b) perfect bonding. (From C. Y. and T. W. Chou, *J. Nanosci. Nanotechnol.* 3 [2003] 1–8.)

to coincide with the outer surface of the nanotube. If the distance between an atom in the nanotube and a node in the inner surface of the matrix is less than 2.5σ, a truss rod is activated.

For a perfectly bonded interface, it is assumed that the outer surface of the nanotube coincides with the inner surface of the polymer matrix. But for matching the atoms in the nanotube and the nodes in the finite elements, the center of an atom in the nanotube is assumed to be located on the outer surface, as shown in Figure 5.46(b), not in the center of the tube wall.

5.5.3 Effective Modulus of Nanotube–Polymer Composites

Combining the molecular structural mechanics approach and finite element method, Li and Chou [91] conducted simulations on a computational model of a continuous nanotube embedded in a polymeric matrix and predicted the effective composite elastic modulus (Figure 5.47). The loading condition was assumed to be tensile in isostrain. The Young's modulus and Poisson ratio of the polymeric matrix were taken as 2.41 GPa and 0.35, respectively, for an epoxy polymer. The nanotube was assumed to be zigzag type. They considered two maximum applied strains, 0.103% and 0.414%.

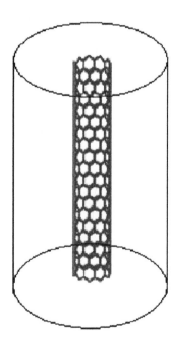

FIGURE 5.47
Model of a composite with a continuous nanotube. (From C. Y. and T. W. Chou, *J. Nanosci. Nanotechnol.* 3 [2003] 1–8.)

TABLE 5.2

Computational Results of Nanotube–Polymer Composites

Applied strain at end	0.103%	0.414%
Average stress in matrix	2.519 MPa	10.102 MPa
Force per bond at axial direction in nanotube	0.086 nN	0.345 nN
Total bonds in cross-section	8	8
Sectional area of nanotube	0.674 nm²	0.674 nm²
Total force in nanotube	0.688 nN	2.760 nN
Sectional area of matrix	19.074 nm²	19.074 nm²
Total force in matrix	0.048 nN	0.193 nN
Effective Young's modulus	36.02 GPa	36.12 GPa

The computational results are listed in Table 5.2. It was found that the majority of the axial force is carried by the nanotube, although the gross volume fraction of nanotube (~3.4%) is much less than that of the matrix (~96.6%). The effective composite axial Young's modulus is calculated by

$$E_c = \frac{N_{nt} + N_m}{(A_{nt} + A_m)\varepsilon_0},$$ (5.83)

where N_m and N_{nt} are the resultant forces in the matrix and the nanotube, respectively, A_m and A_{nt} are the cross-sectional areas of the matrix and the nanotube, respectively, and ε_0 is the applied strain.

For continuous fiber-reinforced composites, the well-known rule of mixtures for the effective axial Young's modulus is

$$E_c = E_f V_f + E_m V_m,$$ (5.84)

where E and V stand for Young's modulus and volume fraction, respectively, and subscripts c, f, and m represent composite, fiber, and matrix, respectively. Using the approximate Young's modulus value of 1000 GPa for carbon nanotubes, the effective Young's modulus of nanotube–polymer composite from Equation 5.84 would be 36.46 GPa. Comparing this value with the computational results in Table 5.2, it can be seen that the stiffness prediction by the rule of mixtures is still valid for nanotube-reinforced nanocomposites.

5.5.4 Stress Distributions in Nanotube–Polymer Composites

The unit-cell model of discontinuous nanotube-based composites under consideration is shown in Figure 5.48 [91]. The elastic behavior and deformation field of this model composite were examined under isostress and isostrain conditions for both the interfacial conditions of van der Waals interactions

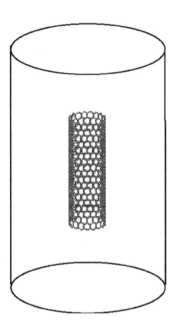

FIGURE 5.48
Model of a composite with discontinuous nanotube. (From C. Y. and T. W. Chou, *J. Nanosci. Nanotechnol.* 3 [2003] 1–8.)

and perfect bonding. Different nanotube aspect ratios, as well as the limiting interfacial conditions, a van der Waals interface and a perfect interface, are examined.

Figure 5.49 displays the distributions of normalized shear stress in a quarter of the matrix material for a perfect interface and a van der Waals interface. The loading is an isostrain case. The stress is normalized by the applied strain multiplied by the Young's modulus of the polymer. The maximum shear stresses occur at the vicinity of nanotube ends. The shapes of shear stress contours for the perfect interface and van der Waals interface cases are similar, but the maximum normalized shear stress in the former case is roughly twice as much as that of the latter case.

The profiles of normalized shear stress for three nanotube aspect ratios with the two interfacial conditions under isostress and isostrain loading conditions indicate only small differences between the maximum stresses for different nanotube aspect ratios for the same loading condition. However, because the region of shear action is larger for a larger aspect ratio, longer nanotubes possess higher load-carrying capability. Under the same loading condition, the shear stresses due to a perfect interface are much larger than those due to a van der Waals interface. Also, under the same interface assumption and the same nanotube aspect ratio, the shear stresses in the

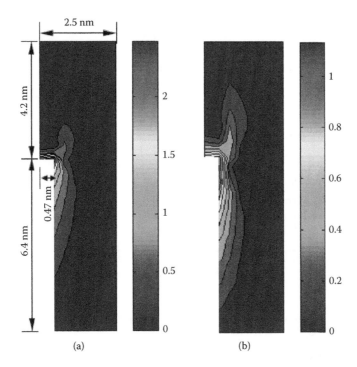

FIGURE 5.49
The shear stress distributions in a polymer matrix (isostrain): (a) Perfect interface, max = 2.78; (b) van der Waals interface, max = 1.13. (From C. Y. and T. W. Chou, *J. Nanosci. Nanotechnol.* 3 [2003] 1–8.)

matrix under the isostrain loading condition are higher than those under the isostress loading condition.

Figure 5.50 displays the axial stress distributions in the nanotube under the isostrain condition. The tensile stress along the tube length (normalized by tube diameter) is symmetrically distributed with the maximum stress appearing at the middle section of the tube. The maximum stress in the nanotube is much higher than that in the polymer matrix. Also, the maximum stress in the case of a perfectly bonded interface is larger than that in a van der Waals interface. Although tensile strength of single-walled nanotubes is reported as high as 63 GPa, the stress concentration may cause failure of nanotubes due to preexisting defects as indicated in some experiments showing that nanotubes in nanotube–polymer composites were broken in tension testing. It should be pointed out that in Li and Chou's computational models the nanotubes are assumed to be without end caps. It was mainly for the convenience in finite element meshing, and it should give the upper bound in stress concentration. In reality, hemispherical end caps often exist in nanotubes, and stress concentrations at the nanotube ends may be less serious.

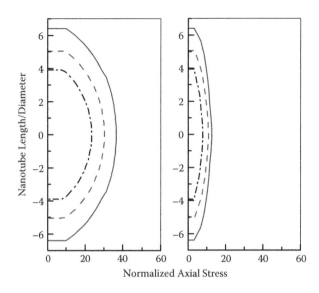

FIGURE 5.50

Axial stress in nanotubes embedded in a matrix (From C. Y. and T. W. Chou, *J. Nanosci. Nanotechnol.* 3 [2003] 1–8.)

5.6 Nanotube-Based Conductive Composites

Although carbon nanotubes possess many potential applications related to their extraordinary mechanical properties, such as stiffness, superlative resilience, tensile strength, and thermal stability, some early applications of carbon nanotubes have utilized their electrical properties for achieving multifunctional carbon nanotube-based composites. In fact, electrically conductive, nanotube-based composites offer the potential of enhancing electrostatic dissipation and electromagnetic interference shielding, as well as corrosion resistance, weight reduction, and processing flexibility. Furthermore, aircraft skins, for instance, require capabilities in impact resistance, structural support, electromagnetic interference shielding, and lightning strike protection, need embedded sensing and actuating networks. These integrated functionalities may be fulfilled by employing multifunctional carbon nanotube-based composites.

Because of their extremely low electric conductivity (on the order of 10^{-12} – 10^{-15} S/m), polymeric and ceramic matrices are usually considered s nonconductive materials. The dispersion of conductive carbon nanotubes into either polymeric or ceramic matrices can result in conductive composites, provided the nanotubes inside the matrices form one or more percolating paths. The electrical conductivity of conductive composites thus formed depends on many factors. First, the intrinsic conductivity of carbon nanotubes plays an important role. Several groups have reported electrical resistivity results for multiwalled nanotubes and single-walled nanotube ropes. Although the differences in electrical properties of different type nanotubes are fairly significant, the electrical conductivities of individual carbon nanotubes are essentially on the order of about 10^4 to 10^7 S/m. With such an exceptionally high conductivity, carbon nanotubes could result in highly conductive composites. Second, size, geometric shape, and even hardness of carbon nanotubes could considerably affect the composite electrical conductivity. Multiwalled carbon nanotubes were reported to have the highest potential for enhancement of electrical conductivity, just because they usually have a better dispersability than single-walled nanotubes. Other factors, such as nanotube distributions, properties of the matrix, and processing techniques, can also have effects on the conductivity. But most importantly, the electrical conductivity is strongly dependent on the nanotube content.

5.6.1 Percolation Threshold

To study electrical conductivity of nanotube-based composites (also particle- or fiber-reinforced composites), the percolation theory has to be employed. Percolation theory is basically a statistical theory describing connections of elements associated spatially. It has lattice (site and bond) and continuum versions. Except for some very simple regular bond or site percolation problems that can be analyzed by analytical methods, the continuum and most lattice percolation problems are simulated by the Monte Carlo method.

An important quantity to describe the onset of percolation is the percolation threshold. Percolation threshold p_c is defined mathematically as a filler volume fraction at which an infinite spanning cluster appears in an infinite system. For all volume fractions $p > p_c$, the probability of finding a spanning cluster extending from one side of the system to the other side is 1, whereas for all volume fractions $p < p_c$, the probability of finding such an infinite cluster is 0. But in the composites literature, the percolation threshold is usually taken as the filler volume fraction at which measurements on specimens or results of numerical simulations begin to show percolation behavior. The percolation threshold may sometimes be surprisingly low for nanotube composites but actually exhibits some degree of probability.

There have been numerous studies on the percolation threshold and electric conductivity of particle- and fiber-reinforced conductive composites. Researchers in recent years have shown increasing interest in carbon nanotube-based conductive nanocomposites. Depending on the polymer matrix, the processing technique, and the nanotube type used, experimental percolation thresholds ranging from 0.001 wt% to more than 10 wt% have been reported [5].

What makes the nanotube percolation more complex is the nanotube waviness. The nanotube wave morphology can be attributed to very large aspect ratio and low bending stiffness, as well as the composite processing

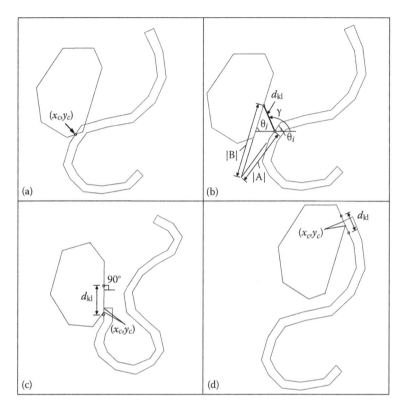

FIGURE 5.51
Contact patterns for fillers with arbitrary shapes. (From C. Y. Li and T. W. Chou, *J. Nanosci. Nanotechnol.* 3 [2003] 1–8.)

techniques. There have been some studies about the effect of nanotube waviness on the percolation threshold and elastic stiffness of composites. For example, Yi et al. [92] and Berhan and Sastry [93] considered the effect of nanotube waviness on percolation onset by approximating nanotubes with a sinusoidal shape. Fisher et al. [94] also assumed nanotubes to be sinusoidal. Shi et al. [95] assumed a helical shape. Dalmas et al. [96] reported their generation of nonstraight nanotubes by using a spline connection. Here, we introduce a versatile percolation simulation method proposed by Li and Chou [97]. The method is applicable for composites containing multiple fillers of arbitrary shapes, including wavy nanotubes.

Consider two fillers with arbitrary shapes, as shown in Figure 5.51. The key issue is to determine the contact status between the two fillers. Because of the irregular shapes, the traditional approach to the cases of circle, square, or ellipse for determining the status of contact is no longer applicable. We propose to approximate the filler shape with straight line segments and replace the filler with a polygon. The lengths and orientations of the sides on the polygon depend on the filler shape and the position of the side. If the filler i has m sides and the filler j has n sides, the length and orientation of these sides can be denoted by L_{ik}, θ_{ik} $(k = 1,2,...,m)$ and $L_{jl}, \theta_{jl}(l = 1, 2,..., n)$, respectively. Here, the orientation angle is defined in the range $0 \leq \theta \leq \pi$.

Assume that the mid-side points of a polygon are represented by x_{ik}, y_{ik} $(k = 1,2,...,m)$ for filler i and x_{jl}, y_{jl} $(l = 1,2,...,n)$ for filler j. In order to determine if the filler i has a contact with the filler j, one side L_{ik}, θ_{ik} in filler i is first selected and sequentially paired with L_{jl}, θ_{jl} $(l = 1,2,...,n)$. The coordinates (x_c, y_c) of the contact point can be given as:

(1) $x_c = x_{ik}, y_c = y_{ik}$, if $x_{ik} = x_{jl}$ and $y_{ik} = y_{jl}$ (Figure 5.51(a)); $\qquad$ (5.85)

(2) $x_c = x_{ik} + A \cos(\theta_{ik}), y_c = y_{ik} + A \sin(\theta_{ik})$, $\qquad$ (5.86)

if $\theta_{ik} \neq \theta_{jl}$ and $L_{ik} \geq 2|A|, L_{jl} \geq 2|B|$ (Figure 5.51(b)), where

$A = d_{kl} \sin(\gamma - \theta_{jl})/\sin(\theta_{ik} - \theta_{jl})$,

$B = d_{kl} \sin(\gamma - \theta_{ik})/\sin(\theta_{ik} - \theta_{jl})$,

$\gamma = \tan^{-1}[(y_{ik} - y_{jl})/(x_{ik} - x_{jl})]$,

$d_{kl} = [(y_{ik} - y_{jl})^2 + (x_{ik} - x_{jl})^2]^{1/2}$,

(3) $x_c = x_{ik}, \quad y_1 + L_1/2 \geq y_c \geq \max(y_1 + d_{kl} - L_2/2, \quad y_1 - L_1/2)$, $\qquad$ (5.87)

if $\theta_{ik} = \theta_{jl} = \pi/2, x_{ik} = x_{jl}$ and $L_{ik} + L_{jl} \geq 2d_{kl}$, where $y_1 = \min(y_{ik}, y_{jl}), L_1 = L_{\min(y_{ik}, y_{jl})}$ and $L_2 = L_{\max(y_{ik}, y_{jl})}$ (Figure 5.51(c));

(4) $x_c \in [x_{ik} + L_{ik} \cos(\theta_{ik})/2, \quad x_{jl} - L_{jl} \cos(\theta_{jl})/2]$,

$\quad y_c \in [y_{ik} + L_{ik} \sin(\theta_{ik})/2, \quad y_{jl} - L_{jl} \sin(\theta_{jl})/2]$, $\qquad$ (5.88)

if $\theta_{ik} = \theta_{jl} \neq \pi/2, y_{ik} = (x_{ik} - x_{jl})\tan(\theta_{ik}) + y_{jl}$ and $L_{ik} + L_{jl} \geq 2d_{kl}$ (Figure 5.51(d)). To simulate the nanotubes in composites in a more realistic manner, the wavy nanotubes are approximated by polygons [98]. First, the nanotube diameter and aspect ratio are given. The nanotube is then divided into N equal-length segments with the segment length always larger than the tube diameter. The profile of the nanotube is formed by setting the starting node at $(0, 0)$ and determining the coordinates of following nodes by the segment length c and an orientation angle $\theta_i(i = 2,..., N+1)$, as shown in Figure 5.52(a). Step increases in θ_i are made by a randomly selected angle

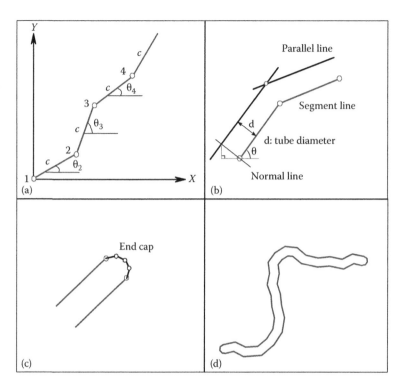

FIGURE 5.52
Steps for generating a wavy nanotube. (From C. Y. Li and T. W. Chou, *Comp. Sci. Technol.* 68 [2008] 3373–3379.)

uniformly distributed between 0 and θ_{max} $(0 < \theta_{max} < \pi/2)$. Second, a profile parallel to the segments established above is formed to take into account the nanotube diameter, as shown in Figure 5.52(b). Corresponding to each end point on segments, a crossing point of a parallel line of a segment line with a normal line of a segment line is produced. Then the positions of the parallel lines are determined. The crossing points of these parallel lines are calculated based on the geometrical relationship. Connecting these crossing points forms the profile parallel to the original segments. Finally, two caps are added to the tube ends (Figure 5.52(c)). An example of a wavy nanotube so produced is shown in Figure 5.52(d). It should be noted that a nanotube with any curvature can be produced by this method, controllably or randomly, through the adjustment of the angle θ_{max}, the number of segments N, and the segment length c.

To simulate the nanotube network in a composite, the wavy nanotube generated above must be put into a random position in a designated area. Suppose the area representing a specimen is a rectangle with side lengths of $L_x \times L_y$. The procedure for producing a nanotube network has five steps. First, a given number of positions are randomly distributed within the rectangular boundary. For each position, there is a corresponding randomly selected orientation. Next, the vertex coordinates (x_0, y_0) of a wavy nanotube are transformed into the physical coordinates (x, y) in the rectangle by

$$\left\{ \begin{array}{c} x \\ y \end{array} \right\} = \left[\begin{array}{cc} \cos\theta & \sin\theta \\ -\sin\theta & \cos\theta \end{array} \right] \left\{ \begin{array}{c} x_0 \\ y_0 \end{array} \right\} + \left\{ \begin{array}{c} x_i \\ y_i \end{array} \right\}, \tag{5.89}$$

where θ is the orientation angle of a filler, and (x_i, y_i) stands for the position where the nanotube would be located. After all the nanotubes are put into the designated positions, a border crossing is then checked. If there are some nanotubes crossing both borders in either x- or y-direction, there is a possibility of a spanning cluster. Otherwise, the simulation returns to the beginning.

If there exists a border crossing, the next step is to find the possible spanning cluster. The process of searching for contact points between two nanotubes follows the method described above (Figure 5.51). Because of the waviness of nanotubes, two nanotubes may have more than one contact point. If the determination of percolation is the only objective of simulations, the process of searching contact points between two nanotubes can be terminated after one of the contact points is identified. However, in the calculation of electric conductivity, all contact points need to be obtained.

Based on the contact points, the nanotube clusters can be identified. The next step is to check if there is any spanning cluster. Sometimes there is one cluster spanning in one direction or one cluster spanning both x- and y-directions. The case of more than one spanning clusters is also not unusual. Figure 5.53 shows an example of a nanotube network having a spanning cluster in only

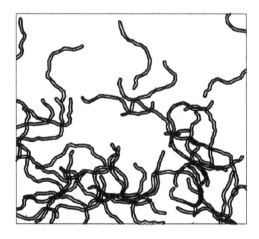

FIGURE 5.53
A network of wavy nanotubes. (From C. Y. Li and T. W. Chou, *Comp. Sci. Technol.* 68 [2008] 3373–3379.)

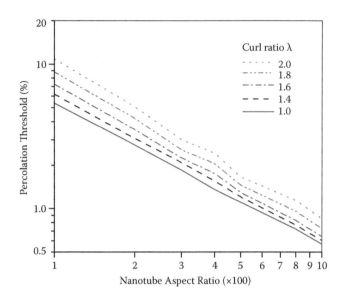

FIGURE 5.54
Percolation thresholds of nanotube-based composites. (From C. Y. Li and T. W. Chou. *Appl. Phys. Lett.* 90 [2007] 174108.)

one direction. Using the same number of nanotubes, running Monte Carlo simulations iteratively for a designated number of times, and summing up the number of times that a spanning cluster occurred, the percolation probability at the specific nanotube volume fraction can be determined.

We obtained the percolation threshold as a function of the nanotube aspect ratio and nanotube curl ratio as shown in Figure 5.54. For wavy nanotubes, we consider an effective nanotube length, which is the maximum distance between a pair of arbitrary points on the nanotube. The curl ratio is then defined as $\lambda = L_{CNT}/L_{effective}$. Our results indicate that increasing nanotube aspect ratio tends to reduce percolation threshold. For straight nanotubes, the relationship between the percolation threshold and the nanotube aspect ratio is logarithmically linear, in agreement with Natsuki et al. [99]. It is also shown that the percolation threshold of wavy nanotubes increases with increasing curl ratio. The relationship between the percolation threshold and the aspect ratio of wavy nanotubes at a certain curl ratio is also roughly logarithmically linear.

5.6.2 Backbone of Percolating Network

Percolation theory has extensive applications in a wide variety of fields, such as materials science (polymers, concrete, composites, and porous media), geophysics (oil exploitation, geothermal power, groundwater pollution control, and earthquake prediction), information technology (Internet, wireless communications), sociophysics (social hierarchies, political persuasion, and marketing), and medical or biological studies (epidemics, species evolution), etc. Except for some very simple regular bond or site percolation problems that can be examined by analytical methods, most complex percolating systems have to be simulated by the Monte Carlo (MC) method.

In the MC simulations, there are two main tasks that are fundamentally important. One is to find the spanning cluster, and the other is to identify the backbone. These tasks are very time consuming for large percolation systems, especially in three dimension (3D) systems. There have been tremendous efforts to discover efficient algorithms for speeding up the MC simulations. The Hoshen–Kopelman cluster labeling algorithm [100] and its modifications [101], as well as the nearest-neighbor search method [102], are popular and efficient algorithms for finding the spanning cluster in 2D and 3D systems. The parallel implementations of the Hoshen–Kopelman algorithm have also been designed for large scale MC simulations. But for backbone identification, the pursuit of efficient algorithm has not reached the same level of success.

A backbone is a subset of the spanning cluster and essentially comprises the biconnected nodes in computer science. It plays a fundamental role in any transport process in percolating systems. The Tarjan's recursive depth-first-search (DFS) algorithm [103] is well-known and the most frequently used for backbone identification. However, its major drawback is the intensive use of the stack. When dealing with a large system, the high number of consecutive recursive calls issued by this algorithm often causes a stack overflow. Thus, the limitation of the recursive algorithm is obvious. The parallel implementation of recursive algorithm also seems to be rather difficult. Other algorithms, such as the burning algorithm, the dual lattice algorithm, the matching algorithm, the hull-generating algorithm, and the algorithm based on the modified Hoshen–Kopelman algorithm are either slower than the Tarjan's and not easy to implement, or a little faster but suitable only for strictly specific networks. Recently, Li and Chou [104] proposed an effective method, termed *direct electrifying algorithm*, for backbone identification by directly employing this definition. The effectiveness of this algorithm has been demonstrated by 2D and 3D lattice site and bond percolation problems. Here, we only introduce the 2D site percolation case.

Let us first consider the site percolation on a square lattice of size $L \times L$. The algorithm for backbone identification requires three steps. The first step consists of randomly generating occupied sites based on a given probability p and determining whether a spanning cluster has formed. The Hoshen–Kopelman algorithm is used for the identification of spanning clusters. Different from the traditional assumption of "bus bar" geometry in which the two opposite sides of a lattice entirely are connected to superconducting electrodes, we only assume that the sites on two opposite sides are individually connected to superconductors. With this modification, the number of

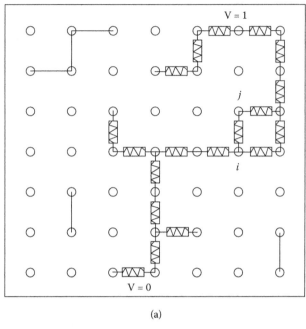

(a)

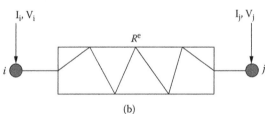

(b)

FIGURE 5.55
(a) Resistor network; (b) isolated resistor element. (From C. Y. Li and T. W. Chou, *J. Phys. A: Math. Theor.* 40 (2007) 14679–86.)

dangling arcs connected with the spanning cluster would be significantly reduced. The other two sides are assumed to be open as usual. The output of the first step includes the information of sites on the spanning cluster and the connectivity between these sites.

The second step is to calculate the electric current flowing through the spanning cluster. We assume that the connection between any pair of two neighboring sites on the spanning cluster represents a resistor with unit resistance, as shown in Figure 5.55(a). A voltage is applied to the superconducting electrodes by assuming the electric potential to be 0 at the bottom and 1 at the top. The voltage distribution at each site of the spanning cluster can be solved by establishing a system of algebraic equations based on the finite element method.

For a typical resistor element *i-j* shown in Figure 5.55(b), the elemental matrix representing the relation between the current (*I*) entering the element at the ends and the end potential (*V*) is

$$\begin{Bmatrix} I_i^e \\ I_j^e \end{Bmatrix} = [K_{ij}^e] \begin{Bmatrix} V_i \\ V_j \end{Bmatrix} = \frac{1}{R^e} \begin{bmatrix} 1 & -1 \\ -1 & 1 \end{bmatrix} \begin{Bmatrix} V_i \\ V_j \end{Bmatrix}. \tag{5.90}$$

According to the Kirchhoff's current law, a system of algebraic equations can be assembled for the entire spanning cluster:

$$\mathbf{I} = \mathbf{KV}, \tag{5.91}$$

where $\mathbf{V} = \{V_1, V_2, \ldots, V_n\}^T$ stands for the nodal potentials at n sites belonging to the spanning cluster, $\mathbf{I} = \{I_1, I_2, \ldots, I_n\}^T$ is the vector of external input current at the n sites (Here $\mathbf{I} = \mathbf{0}$, because we are not inputting any current at any site into the spanning cluster), and the global coefficient matrix is

$$\mathbf{K} = \sum_{e=1}^{m} [K_{ij}^e], \tag{5.92}$$

where m is the number of resistor elements.

After applying the voltage boundary conditions to Equation 5.91, the electric potentials at each site of the spanning cluster can be obtained. The current flowing through each resistor element can then be determined by

$$I^e = (V_i - V_j)/R^e. \tag{5.93}$$

The final step is to extract the backbone from the spanning cluster based on the current-carrying definition. If the current in a resistor element is non-zero, it means this resistor is carrying current and its two ends must belong to the backbone. The backbone can be identified after all the resistors in the spanning cluster are scanned. As shown in Figure 5.56, the dangling ends, loops, and arcs carry no current. For bond percolation problems, the procedure for backbone identification is roughly the same as described above for site percolation.

It should be pointed out that the backbone extracted based on the above algorithm would actually be the effective backbone. There are some so-called perfectly balanced bonds (PBBs) that carry no current as a result of the voltages on their ends being equal. These belong to the geometrical backbone. Although the PBBs are often single bonds, they can be multiple bonds. If the finding of geometrical backbone is the objective, then a small change should be made to the algorithm described above. Batrouni et al. [105] made a systematic study of the density of perfectly balanced bonds in

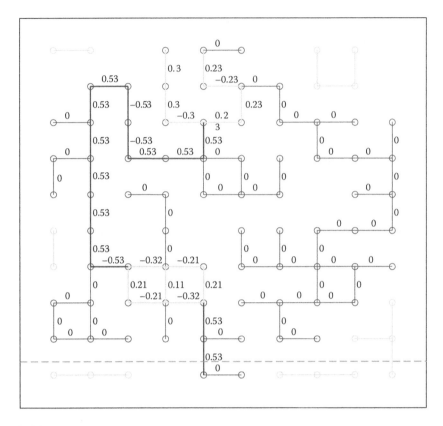

FIGURE 5.56
An example of identified backbone. Electric current value, positive directions are ↓ and ←.
Heavy line = red bonds; medium line = dangling bonds; light line = blobs; dashed line = section for determining the total current. (From C. Y. Li and T. W. Chou, *J. Phys. A: Math. Theor.* 40
[2007] 14679–86.)

the geometrical backbone by using noisy resistors. We adopt the same tactic in our algorithm.

For finding the geometrical backbone, the only need is to add small random noisy resistances (uniformly distributed between −0.001 and 0.001) to the resistors' R^e unit resistance) in the elemental matrix $[K_{ij}^e]$ shown in Equation 5.69. Then the PBBs will also have a current in addition to the effective backbone. Thus, based on the "nonzero-current" criterion, the entire geometrical backbone can be accurately identified. The choice to use uniform unity resistances or noisy resistances depends on the objective of the simulations. Usually, the backbone identification is for calculating transport properties. The PBBs are not really important in such circumstances. If this is the case, then the algorithm without noisy resistances is accurate enough. If the geometrical backbone is the real objective, then the algorithm with noisy resistances should be used.

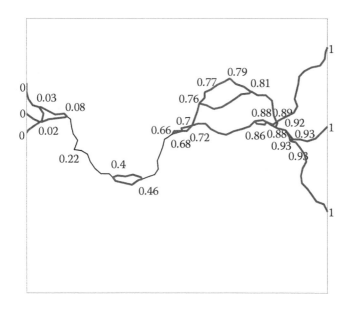

FIGURE 5.57

The backbone of a percolation network. Text: voltage values. Gray line = red bonds; black line = blobs. (From C. Y. Li and T. W. Chou, *Comp. Sci. Technol.* 68 [2008] 3373–3379.)

A network consisting of wavy nanotubes is quite different from the site or bond lattices. But the direct electrifying algorithm is still suitable for extracting backbone from the percolating cluster. Because the objective is to calculate the electric conductivity, the effective backbone is what we need to extract, which is defined as the set of nanotube segments in which a "non-zero" current flows when a voltage is applied across the percolating cluster at the border electrodes. Figure 5.57 shows an example of an effective backbone extracted from a percolating network of wavy nanotubes.

5.6.3 Contact Resistance

For a percolating network of carbon nanotubes, there are two sources of electrical resistance. One is the resistance along the nanotube itself, and the other is the contact resistance between junctional nanotubes. The electrical resistance of carbon nanotubes depends only on the intrinsic conductivity of nanotubes and nanotube size. But the contact resistance is affected by several factors: contact area, contact gap, junction type (metallic/metallic or metallic/semiconducting), and so on. Previous studies indicate that the overall resistances of SWCNT bundle–bundle networks and carbon nanotube-based composites are dominated by the contact resistance [106]. Measurements on crossed SWCNTs gave contact resistance of about 100–400 kΩ for metal–metal or semiconducting–semiconducting SWCNT junctions and two orders higher for metal–semiconducting junctions. Theoretical calculations predicted that the contact resistance between CNTs can vary from 100 kΩ to 3.4 MΩ.

The complexity of contact resistance is further enhanced when CNTs are dispersed in a matrix material. There may be a thin insulating layer formed between the contact points of the junction CNTs. This insulating layer may not prevent the electric tunneling effect if it is sufficiently thin but certainly increases the contact resistance. In their studies on the electrical conductivity of CNT–polymer composite thin films, Kilbride et al. [107] found the electrical conductivity was significantly lower than expected. They suggested that conduction in the composite films is dominated by electric tunneling, and a thick coating of polymer around CNTs results in poor electrical conductivity. By fitting their simulation results to the experimental data of other researchers, Foygel et al. [108] estimated the contact resistance between CNTs in composites to be in the order of 10^{13} Ω and suggested that the high resistance was caused by tunneling-type contacts between the CNTs belonging to the percolation cluster.

In fact, the tunneling percolation model has been extensively used for explaining the nonuniversal behavior of the conductivity of particle- or fiber-reinforced composites. For example, Balberg [109] suggested that the different values of critical exponent of the conductivity resulted from the different average inter-particle (or inter-fiber) distances in different networks. However, many questions—such as what is the maximum possible tunneling gap in composites, what are the tunnel resistances corresponding to the different tunneling gaps, and how does the tunneling resistance affect the critical behavior of the electrical conductivity in composites?—remained unanswered, especially for nanotube-based composites. In the following, we introduce Li and Chou's studies [106] of contact resistance and their investigations of the effect of electrical tunneling on electrical conductivity, with a particular focus on carbon nanotube-based composite films.

The model composite of Li and Chou is a multilayer nanotube-based composite shown in Figure 5.58(a), which is composed of in-plane randomly distributed carbon nanotubes in a polymer or ceramic matrix. Because of the uniform nanostructure in the thickness direction, it is reasonable to consider only a representative layer. The translation of the layer structure in the thickness direction should generate the overall microstructure of the composite. The composite layer is simulated as a pseudo-3D problem in that the nanotubes of diameter d are distributed in two stacking tube layers (Figure 5.58(b)). The smallest spacing between the tube layers could be 0 if the crossing nanotubes contact with each other directly. But because of the thin insulating matrix film between crossing nanotubes, the largest gap will be determined by the maximum possible thickness of insulating film that allows the tunneling penetration of electrons. The thickness of the composite layer is assumed to be twice the tube diameter plus the largest inter-tube spacing (Figure 5.58(c)).

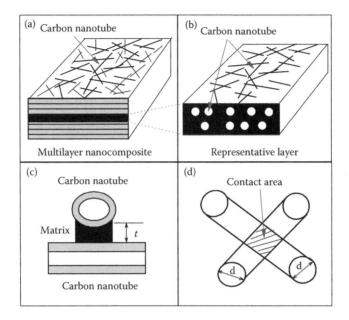

FIGURE 5.58
Computational model of contact resistance in a nanotube composite: (a) multilayer nanotube-based composite; (b) representative layer; (c) inter-tube insulating matrix of thickness t in the contact area; (d) two crossing nanotubes of diameter d. (From C. Y. Li et al. *Appl. Phys. Lett.* 91 [2007] 223114.)

Nanotubes at a contact point in a percolating network are assumed to be overlapping, not penetrating, each other. Although the crossing angles of arbitrary pairs of nanotubes may vary, for simplicity, the contact area at the overlapping position is approximated by $A_C = d^2$ (Figure 5.58(d)). It is understood that A_C so defined corresponds to a right crossing angle. The nanotube segments between contact points in a percolating cluster comprise a conducting network.

The electrical resistance T_{NT} of each nanotube segment in the network is calculated by

$$R_{NT} = \frac{4L_m}{\sigma_{NT} \cdot \pi d^2},$$ (5.94)

where σ_{NT} is the electrical conductivity of carbon nanotubes and L_m is the length of a nanotube segment. The contact resistance R_C is assumed to be the sum of the nanotube contact resistance without an insulating thin film ($R_{directcontact}$), and the resistance from the electric tunneling effect is due to an insulating thin film (R_{tunnel}), that is,

$$R_C = R_{directcontact} + R_{tunnel}.$$ (5.95)

Equation 5.95 ensures that the contact resistance of crossed nanotubes is properly represented when there is no insulating film between junction nanotubes.

The tunneling resistance depends on the thickness and material of the insulating layer. Simmons [110] derived a generalized formula for the electric tunneling effect between similar electrodes separated by a thin insulating film. If we assume the thickness of insulating film in the contact area of crossing nanotubes to be uniform and neglect the variation of barrier height along the thickness, the formula for a rectangular potential barrier can be employed. The current density penetrating the insulating film can be expressed as

$$J = \alpha(\ t)^{-2}[\ \exp(-1.025\ t\ ^{\frac{1}{2}}) - (\ + U)\exp(-1.025\ t(\ + U)^{\frac{1}{2}})] \quad (5.96)$$

where

$$= \ _0 - (U/2t)(t_1 + t_2) - [5.75/K\ t]\ln[t_2(t - t_1)/t_1(t - t_2)], \quad (5.97)$$

$\alpha = 6.2 \times 10^{10}$, t is the thickness of insulating film (in Å), and $\Delta t = t_2 - t_1$ is the difference of the limits of barrier at Fermi level with

$$t_1 = 6/K\ _0, \quad t_2 = t(1 - 46/(3\ _0Kt + 20 - 2UKt)] + 6/K\ _0. \quad (5.98)$$

The symbol φ_0 stands for the height of the rectangular barrier and is approximately taken as the work function of carbon nanotubes (in V); K is the dielectric constant of insulating material. U is the voltage across the film and can be calculated by

$$U = e/C = et/A_C K \varepsilon_0, \quad (5.99)$$

where e is the unit electric charge, C denotes the capacitance and ε_0 is the permittivity of vacuum. The tunneling resistivity (in Ω-cm^2, not to be confused with the volume resistivity that is in Ω-cm) of the insulating film can be given by [94]

$$\rho_{tunnel} = U/J. \quad (5.100)$$

The tunneling resistance is then obtained by

$$R_{tunnel} = \rho_{tunnel}/A_C. \quad (5.101)$$

Figure 5.59 illustrates the tunneling resistance as a function of the thickness of an insulating layer and the nanotube diameter. In our calculations, the work function of carbon nanotubes is 5.0 eV. The dielectric constants are $K = 3.98$ for epoxy and $K = 4.5$ for alumina. From Figure 5.59, it is obvious that the

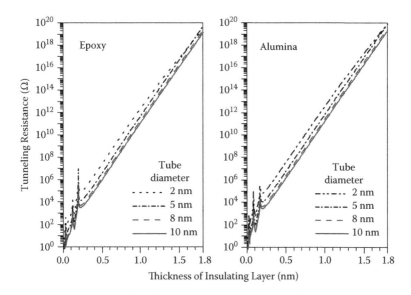

FIGURE 5.59
Contact resistance due to electric tunneling effect in nanotube–epoxy and nanotube–alumina composites. (From C. Y. Li et al. *Appl. Phys. Lett.* 91 [2007] 223114.)

thickness of insulating layer between crossing nanotubes plays a significant role in the tunneling resistance, which increases very rapidly with increasing layer thickness. The use of matrix material, either polymeric or ceramic, appears to be not important because of the small differences in their dielectric constants. But the tunneling resistivity actually increases with increasing dielectric constant of the insulating film. Furthermore, different matrices need different composite processing techniques and may result in different thicknesses of insulating film and, consequently, a difference in the tunneling resistance.

Based on Equation 5.95 and Figure 5.59, the contact resistance between carbon nanotubes could vary in a very wide range. It is affected by a number of factors, such as nanotube type, nanotube diameter, contact area, matrix material, and thickness of the insulating film between nanotubes. Among these factors, the thickness of the insulating film plays the most important role. The insulating film must be sufficiently thin to have an electric tunneling effect. It is essential to determine the maximum possible insulating film thickness that electrons can penetrate. Li and Chou's simulation work on the electrical conductivity of composites at very low nanotube concentrations assumed the same contact resistances at all contact points. Their results indicate that the contact resistance plays a dominant role in the conduction of composites and the conductivity drops below 10^{-12} S/m, which is usually considered the conductivity of an insulator, when the contact resistance is larger than 10^{19} Ω. The thickness of the insulating (epoxy or alumina) film corresponding to this magnitude of resistance is 1.8 nm. This cutoff thickness is consistent with predictions that the tunneling distance is typically on the order of tens of angstroms [109].

5.6.4 Electrical Conductivity of Nanotube-Based Composites

In view of the potential applications, researchers worldwide have been motivated to incorporate carbon nanotubes into polymeric or ceramic matrices for developing carbon nanotube-based composites. There have been numerous reports on the improvement in stiffness, strength, and electric conductivity due to the addition of carbon nanotubes in matrix materials. In terms of electrical conductivity, most of the measured values range from 10^{-5} to 10^{-2} S/m at near or above percolation threshold. But electrical conductivity tailored to the range of 0.01–480 S/m by varying the loading level from 0.11 to 7 wt% has also been reported. This large range of electric conductivity reflects the complex nature of carbon-nanotube-based conductive composites and the effects of multiple factors. It has been difficult to ascertain the full potential of carbon nanotubes due to their often highly entangled and agglomerated morphology, which tends to reduce the effective aspect ratio and hinder the dispersion. Uniform dispersion of nanotubes within a matrix is always critical for enhancing the properties of carbon-nanotube-based composites and considerable efforts have been made in meeting this challenge.

Besides the dispersion of nanotubes, another challenge in the processing of composites is the control of microstructures and nanotube alignment. The subject of reinforcement and filler alignment in conventional composites has been extensively studied. It is well known that unidirectional composites possess the highest strength and stiffness. In carbon-nanotube-based composites, researchers have also used various techniques, such as mechanical stretching, tape casting, air blowing, melt spinning, and electric and magnetic fields, to gain orientation control and/or alignment of nanotubes in the matrix material and have indeed significantly improved the composite mechanical properties. But for electrical conductivity of nanotube composite, there seem to be some inconsistency in the experimental results concerning the effect of nanotube alignment.

5.6.4.1 Effect of Nanotube Content

Like conventional carbon-fiber-reinforced composites, carbon nanotube-based composites exhibit significant electrical percolation behavior. The high aspect ratio of carbon nanotubes enables the development of percolation networks at relatively low nanotube contents. The electrical conductivity of composites strongly depends on the concentration of conductive nanotubes. Figure 5.60 shows the simulation results of the conductivity of nanotube composites at different nanotube concentrations [106]. The influence of contact resistance is also demonstrated. These contact resistance values could

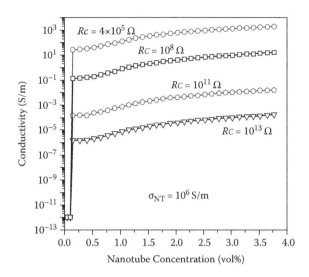

FIGURE 5.60
The effects of contact resistance and nanotube concentration on the conductivity of carbon nanotube-based composites. (From C. Y. Li et al. *Appl. Phys. Lett.* 91 [2007] 223114.)

be considered average contact resistances within the nanotube network. As expected, the conductivity shows a sharp jump at a certain nanotube concentration and then gradually increases with increasing nanotube concentration. For a given concentration of nanotubes, the contact resistance has a significant effect on the conductivity. The nanotube mats or bucky papers have been reported to have conductivities in the range from 10^3 to 10^6 S/m. By contrast, the experiment conductivity results of nanotube-based composites have shown much lower values ranging from 10^{-5} to 10^2 S/m.

Based on Figure 5.60, it is reasonable to believe that the large variation in conductivity reported in the scientific literature is mainly due to the difference of contact resistance in different networks. In nanotube-based composites, tunneling gaps between crossing nanotubes cause large contact resistances and result in a significant decrease of conductivity. In nanotube mats, the relatively smaller contact resistances result in much higher conductivity. It is noted that the extrapolation of the conductivity of nanotube networks without insulating layers (the top curve in Figure 5.60) at 100% concentration gives about 7.2×10^5 S/m, which is very close to the conductivity 10^6 S/m of individual nanotubes we used. This indicates the intrinsic resistance of nanotubes plays a dominant role in nanotube mats, consistent with the conclusion of Hone et al. [111].

The dependence of conductivity of composites σ on the nanotube concentration p above the percolation threshold pc is usually described by a scaling law according to the percolation theory:

$$\sigma = \sigma_0 (p - p_c)^\gamma,$$

(5.102)

where σ_0 is a fitting constant, and γ is the critical exponent. The theoretical predictions and experimental results have shown that the γ value ranges from 1.2 to 3.1. The significant deviation of the critical conductivity exponent from its universal value may be attributed to the tunneling effect.

5.6.4.2 Effect of Nanotube Waviness

Due to the dominant role of contact resistance in the electrical conductivity of nanotube-based composites, the waviness of carbon nanotubes affects not only the percolation threshold but also the electrical conductivity. Two neighboring wavy nanotubes in a percolating cluster tend to have a higher chance of contacting each other with more than one contact point. By contrast, two straight nanotubes can only have one contact point. Therefore, the waviness of nanotubes results in more contact points in the percolation network and thus more contact resistances. The electrical conductivity of composites in this case is expected to be reduced, compared with straight nanotubes.

Figure 5.61 shows the electrical conductivity of composites with increasing nanotube concentration for different curl ratios of carbon nanotubes. The contact resistance in these simulations is assumed to be the same for all contact points. The specimen size is taken as $20 \times 20\ \mu m^2$ and the nanotube aspect ratio is 1000 (diameter 2 nm, length 2 μm). It can be seen that the electrical conductivity of composites increases with increasing nanotube concentration for both straight and wavy nanotubes. The electrical conductivity of composites with wavy nanotubes is lower than that of composites with straight nanotubes.

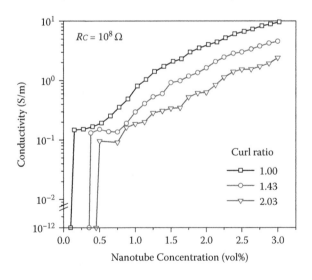

FIGURE 5.61
Effect of nanotube waviness on electrical conductivity. (From C. Y. Li et al. *Comp. Sci. Technol.* 68 [2007] 1445–1452.)

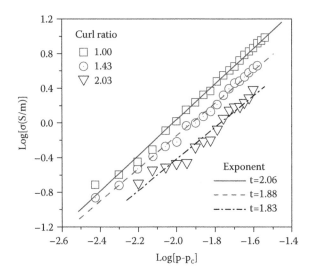

FIGURE 5.62
Critical exponent of electrical conductivity.

Figure 5.62 shows the log–log plot of electrical conductivity versus the nanotube volume concentration. The percolation thresholds are 0.0015, 0.00375, and 0.005 for wavy nanotubes with curl ratios 1.00, 1.43, and 2.03, respectively. The linear relationship indicates that the basic scaling law $\sigma \sim (p - p_c)^\gamma$ in the percolation theory is followed. The fitting of the log–log data gives the critical exponent $\gamma = 2.06$, 1.88, and 1.83 for curl ratios 1.00, 1.43, and 2.03, respectively; the larger the curl ratio, the smaller the critical exponent. These exponent values are in the range of so-called universal values of 1.33 to 2.0 but seem to be closer to 2.0 for 3D systems. The deviation of critical exponent from universal values indicates the significant effect of nanotube waviness.

The electrical conductivity gradually decreased with increasing nanotube curl ratio for a given nanotube concentration. In fact, the reduction of conductivity due to the increase of nanotube waviness can also be concluded from the analysis of the results of Dalmas et al. [96]. The difference of conductivities for composites with different nanotube curl ratios becomes larger when the nanotube concentration increases. It is interesting to note the logarithmically linear relationship between the conductivity and the nanotube curl ratio. This means that the conductivity exhibits an inverse power law dependence on the curl ratio, that is,

$$\sigma \sim \lambda^{-\tau}, \tag{5.103}$$

where τ is a new critical exponent. The linear fitting of the log–log plots for different nanotube concentrations results in the exponent in the range of

2.2–2.6. This power law relationship of the effect of nanotube waviness on electrical conductivity of composites was first reported by Li and Chou [112]. Further studies are needed to confirm this relationship.

5.6.4.3 Effect of Nanotube Alignment

If the nanotubes in a matrix are well dispersed, they can be considered completely random, and the electrical conductivity of the composite should show isotropic behavior. If an alignment method is employed, carbon nanotubes in a matrix will be somehow oriented in a predetermined direction. Depending on the alignment method and other processing conditions, the nanotubes usually show different degrees of alignment. According to the report of Thostenson and Chou [113], in which the distribution of nanotube alignment was obtained using image analysis, nanotube orientations essentially show a normal distribution along the principal nanotube direction.

Choi et al. [114] reported that the nanotube alignment contributed to the enhancement of electrical conductivity of carbon nanotube–polymer composites. The effect was attributed to the more efficient percolation path along the parallel direction and/or the decrease of disorder by the alignment of nanotubes. Haggenmueller et al. [115] also reported single-walled carbon nanotube-polymer composites with enhanced electrical properties by aligning the nanotubes. But the results of Du et al. [116] indicated that the alignment of nanotubes in the polymer matrix significantly lowered the electrical conductivity compared with that of the unaligned composite with the same nanotube content. Du et al. [117] further concluded that the highest conductivity occurs when the nanotubes in the composite are slightly aligned rather than randomly distributed.

The controversial results can hardly be explained by the difference in the orientation controlling techniques used because the results compared were usually obtained by the same processing technique. Motivated by clarifying the inconsistency in interpreting the experimental results, Li and Chou [118] examined the effects of nanotube alignment and nanotube waviness on the electrical conductivity of nanotube-based composites using Monte Carlo simulations. In their study, the orientation of aligned nanotubes was assumed to be in a form of normal distribution. The nanotube alignment angle was θ_a, which can vary from 0° to 90° with respect to the x-direction. When $\theta_a = 90°$, the distribution of nanotubes in the matrix is completely random. When $\theta_a = 0°$, the nanotubes are perfectly aligned in the horizontal direction. For completely random nanotubes, the orientation angle with respect to the horizontal direction is uniformly distributed in the range of $-90° \leq \theta \leq 90°$. For aligned nanotubes, the orientation angle θ with respect to the horizontal direction is selected in the range $-\theta_a \leq \theta \leq \theta_a$ according to a modified normal distribution:

$$\theta = \theta_a \{2[\theta_i - \min(\theta_n)] / [\max(\theta_n) - \min(\theta_n)] - 1\}, (i = 1, 2, ... n), \quad (5.104)$$

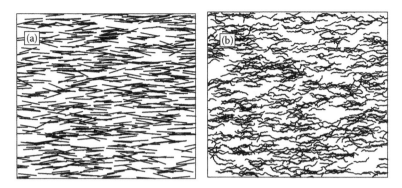

FIGURE 5.63
Examples of simulations of aligned carbon nanotubes in a matrix: (a) straight nanotubes; (b) wavy nanotubes. (From C. Y. Li et al. *J. Nanosci. Nanotechnol.* 68 [2008] 2518–2524.)

where θ_n represents a collection of n random numbers (for n nanotubes) produced from the standard normal distribution, and θ_i is the i-th random number.

In their studies, both straight and wavy nanotubes were simulated. For wavy nanotubes, an effective nanotube length is defined as the maximum distance between a pair of arbitrary points on the nanotube. The curl ratio is then defined as $\lambda = L_{CNT}/L_{\text{effective}}$. The direction of nanotube effective length is taken as the axial direction. For straight nanotubes, the curl ratio is unity and the axial direction is along the nanotube length. Figure 5.63 shows two examples of aligned carbon nanotubes.

Figure 5.64 shows the electrical conductivity in the parallel direction. The general trend is that the electrical conductivity increases with increasing nanotube content, but the nanotube alignment tends to reduce the electrical conductivity. Compared to completely random nanotube distribution ($\theta_a = 90°$), the electrical conductivity of composites with well aligned nanotubes (e.g., $\theta_a = 30°, 20°$) drops several orders in magnitude. For very small percentage nanotube contents (<1.0 vol%), the effect seems to be more obvious when the alignment angle is less than 50°. For large nanotube contents, the alignment effect decreases more gradually with alignment angle (Figure 5.64(a) and (b)). The reason that completely random nanotube distribution results in the highest conductivity may be the multiple paths formed in one percolating cluster and thus reducing total resistance by the parallelism of resistors, and the electric current flowing through the network is increased. In contrast, the decrease of conductivity in well aligned systems may be attributed to the fact that the correct alignment tends to orient nanotubes more along the parallel direction, and thus it is difficult to form multiple percolating clusters with the same number of nanotubes as in the random case. It is also noted that the critical exponent of conductivity, t, increases significantly with increasing alignment (Figure 5.64(c)). This means that higher nanotube

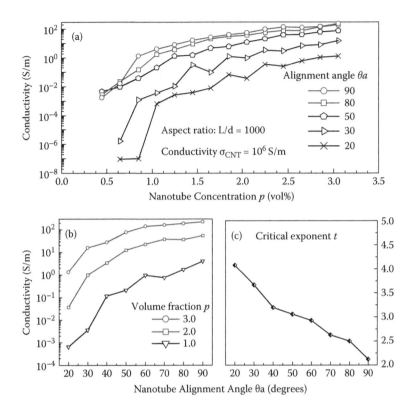

FIGURE 5.64

Electrical conductivity in the parallel direction. (From C. Y. Li et al. *J. Nanosci. Nanotechnol.* 68 [2008] 2518–2524.)

content enhances the conductivity of composites more rapidly in the aligned conditions. The same general trend in the electrical conductivity of composites was also seen in the transverse direction.

The above simulation results establish the general trend of decreasing electrical conductivity of composites with increasing nanotube alignment, and are consistent with the observations of some experiments. However, it is still unclear why some experiments observed the enhancements of conductivity by nanotube alignment. While the above simulations assume straight nanotubes, in practice, carbon nanotubes possess a certain degree of waviness, especially when they are in a nonaligned condition. A percolating network of random wavy nanotubes tends to have a lower conductivity compared with that of straight nanotubes of the same volume content. The alignment processes such as drawing and extrusion can reduce, but not completely eliminate, the nanotube waviness, resulting in higher conductivity.

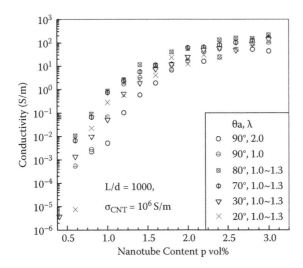

FIGURE 5.65
Effect of nanotube alignment and waviness on composite electrical conductivity (λ: curl ratio; θ_a: alignment angle; σ_{CNT}: intrinsic conductivity). (From C. Y. Li et al. *J. Nanosci. Nanotechnol.* 68 [2008] 2518–2524.)

Figure 5.65 shows the electrical conductivity of composites in the parallel direction with increasing nanotube content for different nanotube curl ratios and alignment angles. The curl ratio for the aligned cases ($\theta_a = 80°, 70°, 30°$, and 20°) is assumed to vary in a small range ($\lambda = 1.0$ to 1.3) instead of having a fixed value to take into account the different degrees of stretching of nanotubes in the alignment processing. It can be seen that the conductivities in the parallel direction are generally higher than or in the same order of magnitude as that of the completely random system ($\theta_a = 90°$). Unlike the results for straight nanotubes where their alignment tends to reduce conductivity, the trend now is that the conductivity is enhanced by nanotube alignment. This confirms the fact that nanotube waviness plays a role in the increase of conductivity. It is believed that the conductivity increase occurs because the wavy nanotubes are stretched and their curl ratios are reduced in the alignment process. The final conductivity depends on the combination of the alignment and waviness effects. For example, the conductivity seems to be of the highest value at the alignment angles of 70° to 80°, and further alignment tends to reduce the conductivity. Thus, the best result in terms of high conductivity can actually be obtained when the nanotubes in the composite possess some degree of waviness, and this conclusion is consistent with the experimental observation of Du et al. [117]. Figure 5.65 indicates that the effect of alignment becomes weaker for larger nanotube contents.

5.7 Defect Sensing Using Nanotube Networks

Electrically conductive nanotube-based composites offer the potential of added multifunctionality, including structural support, electromagnetic interference shielding, lightning strike protection, and embedded sensing and actuating networks. There has been considerable interest in making macroscopic engineering materials that can exploit the multifunctionality of the nanotube properties. Wood et al. [119] and Zhao et al. [120] examined the Raman spectral shifts of carbon nanotubes embedded in a polymer matrix due to elastic strain in the nanotubes and proposed the use of nanotubes as microscale sensors for measuring the strain fields around defects or fibers. Dharap et al. [121] investigated resistance-based CNT strain sensors by using thin films of randomly oriented CNTs. Zhang and coworkers [122] reported that multiwalled CNT-reinforced composites can be utilized as strain sensors and suggested that the instantaneous change in resistance with strain can be utilized for self-diagnostics and real-time health monitoring. Kang et al. [123,124] utilized single-walled CNT–polymer composite films for strain sensing and suggested that a sensor network attached to the surface of a structural component could enable structural health monitoring. Some researchers [125–127] have also proposed using CNT composites as chemical sensors, based on their changes in electrical resistivity. The principle behind these examples of CNT-based composite sensors is the sensing of the change in volume resulted from chemical, thermal, or mechanical loading.

Fiedler and coworkers [128] were the first to propose the concept of conductive modification with nanotubes for both strain and damage sensing. Because CNTs possess higher electrical conductivity than carbon fibers, it is expected that the sensitivity of changes in electrical resistance resulting from strain or damage could be enhanced. Thostenson and Chou [129] demonstrated that the change in the size of reinforcements, from conventional micron-sized fiber reinforcement to carbon nanotubes with nanometer-level diameters, enables a unique opportunity for the creation of multifunctional in situ sensing capability. They found that the percolating networks of CNTs are remarkably sensitive to the onset of matrix-dominated failure, and can detect the progression of damage. More recently, Park et al. [130] evaluated the inherent sensing of CNT–epoxy composites using an electro-mechanical testing technique and concluded that uniform dispersion and interfacial adhesion are key factors for improving sensing performance.

Although some modeling work has been carried out on the electrical resistance-based damage detection in carbon fiber composites [131,132], there have been few reports on the modeling of nanotube networks for damage sensing in fiber composites. Li and Chou [133] reported their studies on the computational modeling of a carbon nanotube network embedded in a glass fiber composite with a particular focus on the variation of electrical resistance with damage evolution.

Given the potential applications of multifunctional nanocomposites, an in-depth understanding of the key factors controlling the effectiveness of carbon nanotube network for damage sensing is indispensable. This section gives a overview of recent advances in the studies on defect sensing of nano and micro hybrid composites by using nanotube networks.

5.7.1 Processing of Nanotube–Fiber Composites

Over the past several years there have been extensive investigations on the development of processing techniques for producing nanotube-reinforced polymer composites. These techniques have been highlighted in a review by Thostenson et al. [4]. For effective reinforcement, nanotubes must be uniformly dispersed within the matrix. Van der Waals interactions between small-diameter nanotubes result in aggregates of ropes. Also, agglomeration is significant in chemical vapor deposition (CVD)-grown multiwalled nanotubes because of nanoscale entanglement. For processing nanotube composites, many approaches involve several steps that may include high speed mixing, high-energy sonication, and solution evaporation processing, surfactant-assisted processing through formation of a colloidal intermediate, functionalization of nanotubes with a polymer matrix, and high-shear mixing.

Owing to the difference in reinforcement scale between conventional micrometer-sized fiber reinforcement and carbon nanotubes with nanometer-sized diameters (the diameters of carbon nanotubes are three orders of magnitude smaller than those of traditional advanced fibers such as glass, carbon, or aramid fibers), it is possible to have carbon nanotube reinforcement in the matrix-rich areas between fibers in an individual bundle as well as between adjacent plies, as illustrated in Figure 5.66.

In Thostenson and Chou's study [129], multiwalled carbon nanotubes grown using chemical vapor deposition techniques with purity higher than 95% (Iljin Nanotech) were used to create the percolating network of sensors. Scanning electron microscopy (SEM) images of as-grown carbon

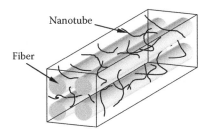

FIGURE 5.66
The penetration of carbon nanotubes in a fiber bundle. (From E. T. Thostenson and T. W. Chou. *Adv Mater* 18 [2006] 2837–2841.)

nanotubes show they are highly entangled, and have diameters on the order of 20–30 nm and lengths of several micrometers. The epoxy polymer matrix used is a bisphenol-F epoxy resin cured with an aromatic amine curing agent (Epikote Resin 862/Epikure W Curing Agent; Hexion Specialty Chemicals).

To fabricate the nanotube–epoxy fiber composites, the carbon nanotubes were first dispersed in the epoxy resin using a calendering approach [134]. A laboratory-scale three-roll mill (Exakt-80E; Exakt Technologies) was utilized to impart high-shear mixing to the nanotube–epoxy mixture to untangle and disperse the nanotubes. Nanotubes were weighed out for a final concentration of 0.5 wt%, added to the epoxy resin by hand mixing and then processed using a three-roll mill. The mixture was processed by passing through the mill at progressively decreasing gap settings down to 5 μm. The evolution of nanoscale composite structure during the process was evaluated using transmission electron microscopy (JEOL 2000FX; 200 kV) to ensure a high degree of dispersion. Vacuum-assisted resin transfer molding (Figure 5.67) was used to fabricate the fiber–epoxy composites with embedded carbon nanotubes. Unidirectional nonwoven glass fiber mats were stacked in preferred orientations. For cross-ply composites $[0/90]_s$, the laminate consisted of four layers with two layers on the top and bottom with a zero-degree orientation to the load direction and two layers in the middle orthogonal to the top and bottom plies. Unidirectional composites were fabricated from five layers of fabric oriented in the same direction $[0]_5$, and the center ply was cut to initiate delamination during tensile loading. Composites tested in flexure were eight layers of unidirectional fabric $[0]_8$. The layups were then compacted under vacuum using a vacuum bag. After dispersing nanotubes in the epoxy resin it was heated to 50°C to reduce the viscosity and then mixed with the curing agent at a ratio of 26.4:100 curing agent to epoxy. The mixture was mechanically stirred and degassed in a vacuum oven for 15 min at 50°C. Composites were fabricated by drawing the resin mixture through the glass fibers under vacuum and then curing for 6 h at 130°C.

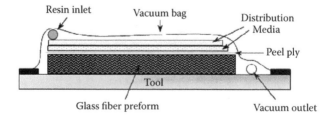

FIGURE 5.67
Processing of nanotube–fiber composites.

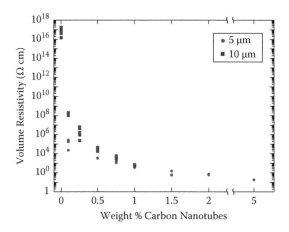

FIGURE 5.68
Electrical percolation behavior in nanotube–epoxy composites. (From E. T. Thostenson and T. W. Chou. *Adv Mater* 18 [2006] 2837–2841.)

5.7.2 Piezoresistive Properties of Nanocomposites

If the high aspect ratio of carbon nanotubes is preserved while processing a polymer nanocomposite, the nanotubes can form a conductive percolating network throughout the polymer matrix at relatively low concentrations. In their research on nanotube–fiber composites, Thostenson and Chou processed polymer nanocomposites using a high-shear stress field [134] and found that the electrical percolation threshold occurred at a concentration below 0.1 wt% for the highly dispersed structure (5 μm) and the partially agglomerated structure (10 μm) (Figure 5.68).

The percolation threshold for conductive particles embedded in an insulating polymer matrix is sensitive to the structure of the reinforcement, and the decrease in electrical resistivity with an increase in reinforcement content is attributed to the probability of reinforcement contact. The low percolation threshold of the nanocomposites indicates that the relatively large aspect ratios of the carbon nanotubes are likely maintained during processing, and the nanotubes form a percolating network throughout the matrix. Electrical properties are sensitive to local statistical perturbations in the microstructure that create a conducting path for transport. For the partially agglomerated structure, the mean values of electrical resistivity are higher than for the more highly dispersed structure, particularly at lower fractions of carbon nanotubes. Because of local agglomeration of nanotubes, the statistical fraction of nanotubes participating in conductive percolation is lower, resulting in higher resistivity.

Thostenson and Chou performed a tensile test on a nanotube–epoxy specimen and found a highly linear relationship between the specimen deformation and electrical resistance (Figure 5.69).

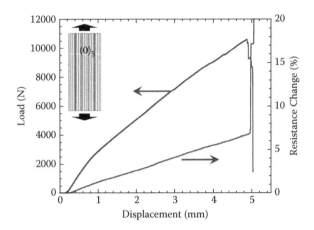

FIGURE 5.69
Resistance change with deformation for a 0.5 wt% nanotube–epoxy composite loaded in tension. (From E. T. Thostenson and T. W. Chou. *Adv Mater* 18 [2006] 2837–2841.)

5.7.3 Structure/Deformation Response to Specific Failure Modes

In Thostenson and Chou's research, specific mechanical tests were designed to promote distinct failure modes. Key experiments in tension were to evaluate interlaminar delamination in unidirectional composites and transverse microcracking in cross-ply laminates. When subject to bending, unidirectional specimens were tested with different spans to promote different failure modes. The specimen strain was not recorded using an external extensometer or bonded strain gages in order to avoid any influence on the resistance measurements. The deformation data is instead presented in terms of the cross-head displacement of the testing machine and is proportional to strain.

5.7.3.1 *In Situ Sensing of Delamination*

A five-ply unidirectional composite was fabricated and the center ply of the laminate was cut in the middle of the specimen to promote ply delamination during tensile loading. The discontinuity at the center ply of the laminate resulted in the accumulation of shear stresses at the ends of the ply and these shear stresses initiated delamination of the center and adjacent plies. Figure 5.70 shows the results of the tensile test.

The specimen resistance increases linearly with initial deformation and is consistent with our earlier observation of linear increase in resistance with deformation in nanotube–epoxy specimens. A sharp increase in resistance occurs when the ply delamination is initiated. As the delamination grows with increasing load, there is a large increase in resistance marked primarily

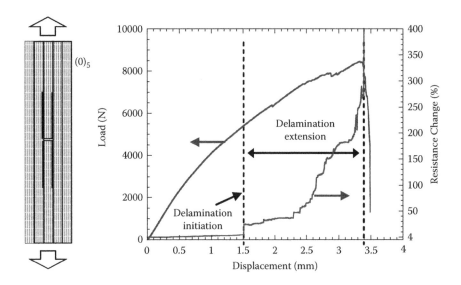

FIGURE 5.70
Load-displacement and resistance curves for the [0]₅ specimen with center ply cut to initiate delamination. (From E. T. Thostenson and T. W. Chou. *Adv Mater* 18 [2006] 2837–2841.)

by a progressive increase in the slope of the resistance curve with extension of the ply delamination.

5.7.3.2 In Situ Sensing of Transverse Microcracking

It is widely known in cross-ply laminates loaded in tension that failure initiates in the 90° plies by the formation of microcracks that are oriented normal to the direction of applied load. As the laminate is further strained, more cracks are formed until the ply becomes saturated with a regularly spaced array of cracks; the crack density is related to laminate configuration. The influence of transverse microcrack development in plies oriented normal to the direction of loading was investigated using a [0/90]ₛ cross-ply specimen with plies oriented along the loading axis (0°) on the outside of the laminate and plies oriented at 90° in the center. As observed in the unidirectional specimen, there is a linear increase in resistance with initial loading in the cross-ply specimens (Figure 5.71).

Upon the initiation of microcracking in the 90° plies, there is a sharp change in the resistance. In the progression from the first initiation of cracking to ultimate failure of the composite laminate, the resistance changes are marked by step increases corresponding to the accumulation of microcracks and linear increases with deformation between the step increases. By comparing resistance curves in Figures 5.70 and 5.71, it is possible to identify the nature and progression of damage using nanotubes. For both the [0] and [0/90]ₛ specimens, there is a slight degradation in the stiffness with

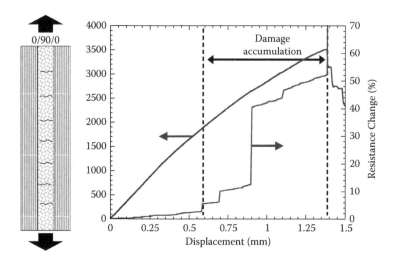

FIGURE 5.71
Load-displacement and resistance curves for the [0/90]$_s$ specimen. (From E. T. Thostenson and T. W. Chou. *Adv Mater* 18 [2006] 2837–2841.)

increasing damage accumulation, but the change in slope is small relative to the overall specimen stiffness.

5.7.3.3 Damage Accumulation in Flexure

For laminated composites the reinforcement is primarily oriented in-plane. When subject to bending, the layers of the composite transfer the load through shear stresses at the ply interfaces. Composites are particularly susceptible to failure at the ply interfaces because the polymer matrix must transfer the entire shear load from layer to layer. When subjected to out-of-plane loads, such as low velocity impact, the local shear stress can result in delamination of the plies. In order to assess the capability of nanotubes to sense through-thickness interlaminar fracture, unidirectional specimens were tested in three-point bending at varying spans. As the span is increased, the moment is higher and will produce stresses high enough to promote fiber fracture as in the initial deformation mechanism. Specimens were tested at span-to-thickness ratios of four and eight to promote the different types of failure. Figure 5.72 shows the load versus midspan displacement results for the short and long spans, respectively. With initial loading, both show a slight decrease in the resistance as the beam is deformed in both tension and compression. At the point of failure for the beam with the short span, the resistance increases by several orders of magnitude indicating that the specimen has delaminated completely.

For the long-span specimen there are slight decreases in load that correspond to a flattening of the resistance curve. These load drops are likely due to fiber breakage and crushing under the load nose; observation of the

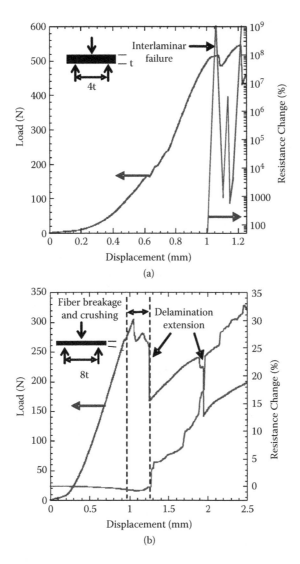

FIGURE 5.72
Flexural test results showing (a) short span to promote interlaminar failure and (b) long span to promote fiber failure. (From E. T. Thostenson and T. W. Chou. *Adv Mater* 18(2006) 2837–2841.)

specimen shows fiber and matrix damage directly below the midspan load point. Subsequent sharp decreases in load are accompanied by step increases in resistance and likely correspond to matrix cracking and the extension of local delamination. Unlike the specimen that fails in interlaminar shear, where the resistance changes several orders of magnitude corresponding to complete ply separation, the more incremental changes in resistance correspond to damage.

5.7.4 Sensing of Damage in Joint

Because of manufacturing requirements and design specifications, large and complex sections of fiber-reinforced composites often need to be joined together to form the final structures. It is important to understand the failure behavior of these joints under a variety of static and dynamic loading conditions. Since fiber-reinforced composites have low bearing strength as well as low in-plane shear strength, mechanical fastening of composites through bolting or riveting often results in incipient damage around the fastener under in-service loads. A variety of complex failure modes can occur, depending on the joint construction as well as the local micro-scale structure of the composite surrounding the joint. Bearing, shear-out, cleavage, and tensile failure are commonly encountered failure modes. The growth of damage around the bolt hole can result in significant reduction in overall structural load-carrying ability.

Nondestructive techniques such as x-ray or ultrasonic inspection can shed some light on local damage, but frequently bolted connections may be disassembled for inspection and then reassembled. These techniques are not capable of real-time monitoring of damage evolution. Acoustic emission is often utilized during testing to detect the occurrence of damage connections, but interpretation of the results is often qualitative or the analysis complex. Among the variety techniques that are currently employed to evaluate damage in joint connections, there is no technique that can accurately identify the individual mechanisms and subsequent progression of damage in real-time during mechanical characterization.

Thostenson and Chou [135] extended the approach of a nanotube-based in situ sensing to detect localized damage in glass fiber composites due to the deformation of mechanically fastened joints. Single and double-lap joint configurations tested in tension are investigated under monotonic and cyclic loading conditions, and specimens were designed to promote shear-out as a dominant failure mode. Because typical mechanical fasteners are metallic and electrically conductive, the physical contact between the materials was considered when interpreting the data.

The geometric configurations of the test set-up for single and double-lap joints are shown in Figure 5.73. The alignment fixture and end tabs applied to the specimens were fabricated from electrically insulating glass–epoxy laminates to isolate the electrical resistance measurement from the metallic specimen grips on the load frame. End tabs were bonded to one end of the nanotube–fiber composite specimen, and test specimens were machined into strips such that the orientation of the fiber ran along the axis of the specimen (0_ orientation). Electrodes were applied to the specimens using conductive silver paint. The electrodes were applied along the axis of the test specimen on either side of the bolt, and slots were machined in the alignment fixture to accommodate the attachment of electrode wires.

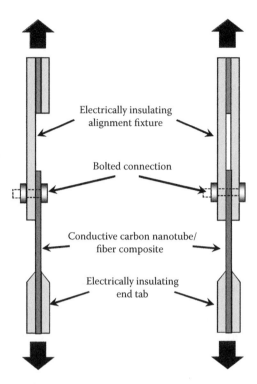

FIGURE 5.73
Diagram showing the configuration and test set-up of specimens in single (left) and double (right) lap geometries. (From E. T. Thostenson and T. W. Chou, *Comp. Sci. Technol.* 68 [2008] 2557–2561.)

Figure 5.74 shows a load deformation and electrical resistance response of a single-lap joint configuration where the bolt threads have been insulated with PTFE and tightened to an applied torque of 14.1 Nm (125 in lb). With initial loading, there is a linear change in electrical resistance. At approximately 60% of the ultimate load, the resistance response begins to deviate from the linear response, and more noise in the measurement is noted. This likely corresponds to the initial stages of bearing damage in the composite and the subsequent formation of longitudinal cracks. At peak load there is a sharp knee in the resistance response, followed by further increases in resistance as the material is sheared out. The resistance signature is promising in that the deviation from the linear response likely corresponds to accumulation of damage, but the overall change in resistance is relatively small, less than 3% of the total resistance. This is likely due to the conductive bolt head in direct contact with the composite laminate. Due to the contact, a conducting path is created near the microscale cracks formed under the bolt head, effectively shorting out the resistance response. Furthermore, it has been observed that specimen bending results in an overall decrease in resistance due to the

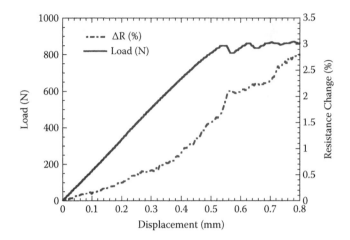

FIGURE 5.74
Load-displacement and resistance curves for a single-lap joint configuration. (From E. T. Thostenson and T. W. Chou, *Comp. Sci. Technol.* 68 [2008] 2557–2561.)

decrease in resistance of the side of the specimen, which undergoes compressive deformation. The eccentric loading about the connection results in an observable bending of the joint that varies, depending on the length of the relatively thin composite specimen.

The double-lap configuration overcomes the problems associated with bending and also prevents contact of the bolt with the surface of the laminate. Figure 5.75 shows a load deformation and resistance signature response of a double-lap specimen configuration, where the bolt threads have been

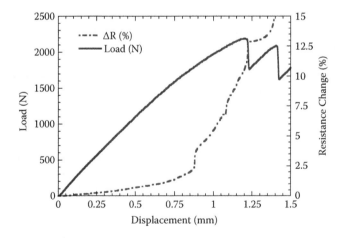

FIGURE 5.75
Load-displacement and resistance curves for a double-lap joint configuration. (From E. T. Thostenson and T. W. Chou, *Comp. Sci. Technol.* 68 [2008] 2557–2561.)

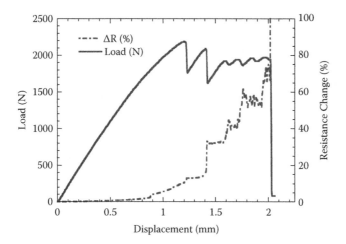

FIGURE 5.76
Load-displacement and resistance curves for a double-lap joint, showing the complete resistance response after initial failure to ultimate fracture. (From E. T. Thostenson and T. W. Chou, *Comp. Sci. Technol.* 68 [2008] 2557–2561.)

insulated with PTFE and the bolt tightened to an applied torque of 14.1 Nm (125 in lb). Similar to the resistance signature of the single-lap connection, the response appears to deviate from linearity at approximately 60% of the ultimate load of the joint. There is a sharp jump in the resistance curve, likely indicating the initiation of the longitudinal cracks, followed by a progressive increase in the slope of the curve until the joint reaches peak load. This observed response is very similar to the response observed in our previous research [129] for delamination initiation and extension during tensile loading of a composite, as the propagation of longitudinal cracks around the joint is analogous to a delamination failure.

Figure 5.76 shows the entire load displacement curve up to the final fracture of the specimen shown in Figure 5.75. Beyond the initial load drop there is a stick/slip nature to the fracture. With each subsequent load drop, there are substantial jumps in electrical resistance that correspond to continued damage. It is clear from the above graphs that the technique of in situ sensing holds great promise for examining the initiation and evolution of damage within the composite substrate in mechanically fastened connections. The observed resistance response for the double-lap joint is also considerably higher than that of the single-lap connection at the peak load. It should be noted that the ultimate failure loads for all the double-lap specimens are consistently higher than for the single-lap specimens. This is due to the shear load transfer at the interfaces due to friction between the composite and the alignment fixture because of clamping of the substrates on either side of the composite.

It has been demonstrated that carbon nanotube networks can be utilized as in situ sensors for sensing of local composite damage accumulation and also detection of fastener loosening in bolted joints. It is important to consider both the loading condition and also the physical contact of constituents when designing specimens to detect the onset and progression of damage in mechanically fastened connections. The research work of Thostenson and Chou [135] encompassed the first step toward the development of hierarchical strategies for damage sensing and health monitoring of joints where the nature and extent of damage can be monitored in situ and in real time.

5.7.5 Damage Sensing Simulation of Nano/Micro Hybrid Composites

In the experimental work of Thostenson and Chou [129], some of the typical fiber composite failure modes were activated and were correlated to the electrical resistance measurements from distributed carbon nanotube sensor networks. Among the patterns of damage evolution in unidirectional and cross-ply composites, the matrix cracking in the 90° layer of the [0/90]$_s$ laminate exhibited a progressive accumulation of damage and is particularly interesting from both the experimental and analytical modeling point of view. Li and Chou [133] thus chose the cross-ply configuration as the model system for their simulation work.

The focus of the simulation was on the demonstration of the interaction between the nanotube network and transverse cracks in the 90° plies. For simplicity, a two-dimensional model, as shown in Figure 5.77, was adopted. This two-dimensional model may inevitably increase the contact between nanotubes comparing with an actual nanocomposite, in which the conductivity

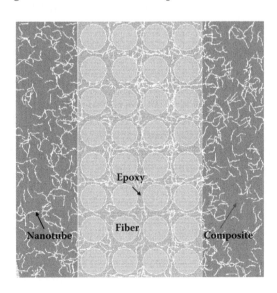

FIGURE 5.77
Nanotube network in the [0/90]$_s$ fiber-composite. (From C. Y. Li and T. W. Chou, *Comp. Sci. Technol.* 68 [2008] 3373–3379.)

network is three-dimensional in nature and nanotubes are not limited to a planar distribution. But this undesirable effect was somewhat reduced in the simulations by assuming that the gaps of inter-nanotube contacts are statistically distributed in a rather large range. The middle section shows two layers of 90° fibers, while the two sections on the sides represent the 0° fiber layers. The carbon nanotubes are allowed to penetrate into the inter-fiber matrix region in the 90° layers. To avoid the complexity in dealing with overlapping nanotubes and fibers, the two 0° layers are replaced by their effective medium and consequently the 0° fibers are not shown in Figure 5.77. The simulation was just to demonstrate the effect of the essential factors contributing to the nanocomposite electromechanical behavior in damage sensing.

5.7.5.1 Percolation of Nanotube Network

In a nanotube–fiber hybrid composite, where the fibers occupy a significant portion of the volume, the nanotubes are infused into the gaps between neighboring fibers and may assume a larger degree of waviness. Wavy nanotubes dispersed in a matrix also tend to have more contact points than straight nanotubes, which could have a considerable effect on the electrical conductivity due to the dominant role of contact resistance [106]. The orientation of a nanotube in the 90° ply also needs to be adjusted to avoid overlapping with the fiber cross-section. The contact points between two neighboring nanotubes are determined by following the method described in Chapter 6, Section 6.1.

Based on the knowledge of contact points, the nanotube clusters can be identified. The next step is to check for the existence of spanning clusters. In some cases, there is no spanning cluster, while in others more than one spanning cluster has formed. The percolation probability at a specific nanotube volume fraction can be determined by the percentage of the number of times that at least one spanning cluster has occurred out of the total number of Monte Carlo simulations. Here, we only need to select one percolation nanotube network for studying the effect of damage on electrical conductivity of the network. Figure 5.78 gives the percolating nanotube network in the composite shown in Figure 5.77.

5.7.5.2 Electrical Resistance of Percolating Nanotube Network

The electrical resistance of a percolating nanotube network comes from two sources (i.e., the intrinsic resistance of nanotubes and the contact resistance at nanotube junctions). Thus, the electrical conductivity of the nanotube network strongly depends on the morphology of the nanotube network and the number of contact points. Depending on a number of contributing factors, the contact resistance between carbon nanotubes in composites could vary in a wide range, from 10^2 kΩ to 10^{16} kΩ. Li and Chou [133] adopted different approaches for dealing with the two possible nanotube contact configurations. One configuration is the overlapping contact. In this case, it is often

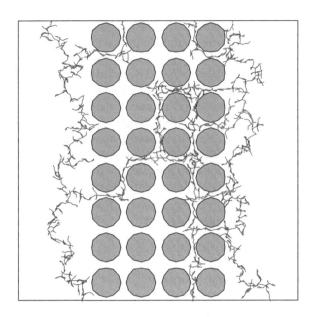

FIGURE 5.78
A percolating nanotube network in a fiber composite. (From C. Y. Li and T. W. Chou, *Comp. Sci. Technol.* 68 [2008] 3373–3379.)

difficult to determine the thickness of an insulating film and, hence, the precise value of the contact resistance. It was assumed that the thicknesses of insulating films follow a normal distribution in the range of 0–1.8 nm, and the corresponding distribution of contact resistances is calculated using the method introduced in Section 6.4. The distribution of contact resistance resulting from the normal distribution of insulating films is shown in Figure 5.79. The lower bound of contact resistance is taken as 100 kΩ, which is the lowest contact resistance between nanotubes, assuming no insulating film. The other nanotube contact configuration is the in-plane contact, where two neighboring nanotubes are not overlapping but are situated close enough to permit electrical tunneling. In this case, based on the tunneling gap size, the contact resistance at a specific contact point is approximately calculated using Equation 5.95.

The percolating nanotube network can then be replaced by a resistor network that includes nanotube resistors and contact resistors. The calculation of the electrical current flowing through the percolating nanotube network can be carried out following the method described in Chapter 6, Section 6.2.

5.7.5.3 Damage Evolution in Composites

As indicated in the model composite of Figure 5.77, the 0° plies were replaced by their effective media, of which the elastic modulus and tensile strength

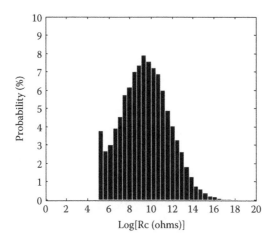

FIGURE 5.79
The distribution of contact resistance *Rc*. (From C. Y. Li and T. W. Chou, *Comp. Sci. Technol.* 68 [2008] 3373–3379.)

are computed from the rule of mixtures. The damage evolution of the model composite was simulated by fixing the lower boundary while applying a uniform displacement on the upper end. This configuration of deformation is equivalent to an isostrain axial loading condition of a specimen twice the length of the specimen in Figure 5.77.

To simulate the damage evolution of fiber-reinforced composites with randomly distributed CNTs, the commercial finite element method software ANSYS [136] was employed. The coordinates of wavy nanotubes, glass fibers, and matrix are generated by in-house software and then exported into ANSYS. The nanotubes with overlapping contacts are assumed to be directly connected because of the two-dimensional nature of the model. The criterion for damage initiation is set by the use of principal tensile stress. For identifying the location of potential damage, the maximum principal stress in each element is traced in every load step in the simulation process. If the maximum principal stress of an element is equal to or larger than the tensile strength of the corresponding material of that element, the critical element is identified and assigned as a death element and deactivated in the next load step. The death elements give rise to stress redistribution, which may result in significant stress concentrations. These stress concentrations contribute to the deterioration of the load-carrying capacity as well as the damage progression of the composite.

Figure 5.80 displays the overall view of damage evolution in the composite with increasing applied strain. Multiple damage spots are visible at the strain level of about 0.433%. These damage spots are located in the interfiber matrix region where the strain concentration is high, as shown by the strain

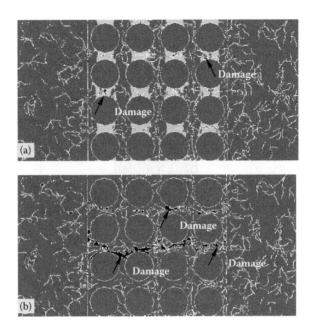

FIGURE 5.80
Damage evolution in the composite under different imposed strains: (a) $\varepsilon_0 = 0.433\%$, (b) $\varepsilon_0 = 1.167\%$. Contours show the first principal strain in the upper half of the model configuration. (From C. Y. Li and T. W. Chou, *Comp. Sci. Technol.* 68 [2008] 3373–3379.)

contours in Figure 5.80(a). With increasing tensile strain, the number of damage spots also increases, and the size of damage spots expands. When the applied strain reaches 1.167% (Figure 5.80(b)), damage spots merge to form microscopic cracks, which extend through the entire width of 90° plies. These cracks could further propagate into the 0° plies, eventually resulting in the failure of the cross-ply composite.

5.7.5.4 Damage Sensing by Electrical Resistance Method

The effective resistance of a composite can be changed when it is deformed under applied loading. Several factors may contribute to the electrical resistance change. First, when a fiber–nanotube–polymer hybrid composite is deformed, the nanotube length and diameter will alter, resulting in the change of the nanotube's intrinsic resistance, and hence, the effective resistance of the nanotube network. However, this resistance change is expected to be negligible because of the extremely small elastic deformation in nanotubes.

The second and more important factor contributing to the resistance change of the composite is the contact resistance. Under applied load, the thickness of the insulating matrix film between adjacent nanotubes may be changed considerably. The contact resistance increases dramatically with the increase in insulating film thickness. The matrix damage also contributes to the change of contact resistance. Figure 5.81 illustrates the effect of damage evolution on the change of contact resistance. The damage spot assumes the form of a nanoscopic void, and its formation give rise to the increase of the opening gap at the contact area where electrical tunneling takes place and thus increases the contact resistance. With the evolution of damage, the

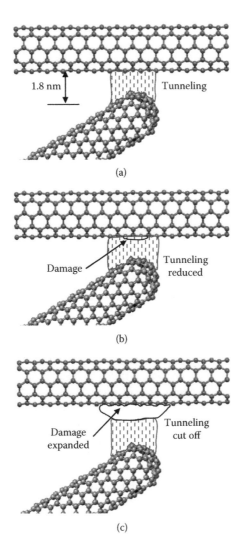

FIGURE 5.81
An illustration of electrical tunneling affected by the damage evolution. (From C. Y. Li and T. W. Chou, *Comp. Sci. Technol.* 68 [2008] 3373–3379.)

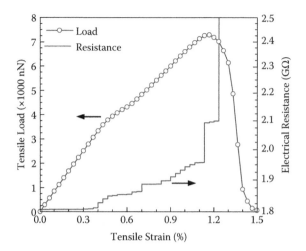

FIGURE 5.82
Resistance changes versus tensile load applied on a cross-ply composite. (From C. Y. Li and T. W. Chou, *Comp. Sci. Technol.* 68 [2008] 3373–3379.)

electrical tunneling at a damage area can be eventually cut off, resulting in significant change of the effective resistance of the entire composite. The real effect of the damage is in changing the contact resistance.

Figure 5.82 displays the load–strain relationship of the cross-ply composite. A "knee" is visible in the curve at about 0.433% strain, which is an indication of significant damage accumulation in the composite. This is evident from the multiple damage spots in the 90° plies. The ultimate load level of the composite occurs at about 1.2% applied tensile strain, when multiple microcracks have formed in the 90° plies and also extended into the 0° plies. The load-carrying capacity of the composite drops rapidly after 1.2% applied strain, and eventually the composite fails at about 1.5% applied strain when microcracks propagate through the 0° plies.

The change of electrical resistance with applied tensile strain is also shown in Figure 5.82. It can be seen that the resistance change is very small in the beginning. But significant change occurs around 0.4% strain, which roughly corresponds to the knee point in the load–strain curve. A careful browsing of the image of strain distribution at this stage reveals that some damage spots are located right in the percolation path, and two of the conducting branches have been cut off. The resistance change reflects the cut-off of the current paths due to matrix cracking. It should be noted that several small damage areas actually appeared earlier when the loading level was lower, but the resulting resistance change is insignificant because the locations of these damages are not on the percolation path.

5.8 Concluding Remarks

Nanocomposite is one of the evolving areas of composite research, and it was motivated by the recent enthusiasm about nanotechnology. A unique feature in nano-structured material development is the so-called "bottom-up" approach, which has been practiced by composites researchers and technologists in processing and manufacturing for decades. The capability to tailor materials at the nanoscale has opened a new field of research work aimed at tailoring their mechanical and physical properties and developing intelligent systems that are capable of sensing and actuation. Recent advances in producing carbon nanotubes in large scale with relatively low costs have stimulated research to create multifunctional nanotube-based composite materials.

In such a multifunctional multiphysical material system, the characteristic length scales span several orders of magnitude and highlight the need to develop both experimental and analytical techniques to bridge these scales toward the optimization of systems. From the electromechanical coupling of individual carbon nanotubes to the global deformation of a polymer–nanotube composite, the behavior of carbon nanotube-based composites presents a lot of challenges for research. But at the same time, it also brings tremendous opportunites for innovative approaches in the processing, characterization, analysis and modeling of the new generalization of composite materials. It is our desire that the content in this chapter could provide some basic knowledge and stimulate the interest of young researchers.

References

1. S. Iijima, *Nature* 354 (1991) 56–58.
2. P. G. Collins and P. Avouris. *Scientific American* 2000; 283(6): 62–69.
3. E. T. Thostenson, Z. F. Ren, and T. W. Chou, *Comp. Sci. Technol.* 61 (2001) 1899–1912.
4. E. T. Thostenson, C. Y. Li, and T. W. Chou, *Comp. Sci. Technol.* 65(2005) 491–516.
5. C. Y. Li, E. T. Thostenson, and T. W. Chou, *Comp. Sci. Technol.* 68 (2007) 1445–1452.
6. C. Y. Li and T. W. Chou. *Int. J. Solids Struct.* 40 (2003) 2487–2499.
7. M. Monthioux and V. L. Kuznetsov, *Carbon* 44 (2006) 1621–1623.
8. A. K. Rappe, C. J. Casewit, K. S. Colwell et al. *J. Am. Chem. Soc.* 114(1992) 10024–10035.
9. D. W. Brenner, *Phys. Rev.* B 42 (1990), 9458.
10. S. L. Mayo, B. D. Olafson, and W. A. Goddard, *J. Phys. Chem.* 94 (1990) 8897–8909.
11. W. D. Cornell, P. Cieplak, C. I. Bayly et al., *J. Am. Chem. Soc.* 117 (1995) 5179–5197.

12. B. R. Gelin, *Molecular Modeling of Polymer Structures and Properties*, Hanser/ Gardner Publishers, Cincinnati, OH, 1994.
13. W. Weaver Jr. and J. M. Gere, *Matrix Analysis of Framed Structures*, 3rd edition, Van Nostrand Reinhold Co. Ltd., New York, 1990.
14. C. Y. Li and T. W. Chou, *J. Nanosci. Nanotechnol.* 6 (2006) 54.
15. E. Riks, *Int. J. Solids Struct.* 15 (1979) 529–551.
16. Y.B. Yang and M. McGuire, *Proc. 1985 Int. Conf. Num. Meth. Engng*, J. Middleton and G. N. Pande (eds.), University College Swansea, Wales, UK, 913–921.
17. Y. B. Yang and M. S. Shieh, *J. AIAA* 28 (1990) 2110–2116.
18. V. N. Popov, V. E. Van Doren, and M. Balkanski, *Phys. Rev.* B 61 (2000) 3078–3084.
19. E. Hernandez, C. Goze, P. Bernier, and A. Rubio, *Phys. Rev. Lett.* 80 (1998) 4502–4505.
20. B.T. Kelly, *Physics of Graphite*. Applied Science, London. 1981.
21. E.W. Wong, P.E. Sheehan, and C.M. Lieber, *Science* 277 (1997) 1971–1975.
22. J. P. Salvetat, J. M. Bonard, and N. H. Thomson, *Appl. Phys.* A 69 (1999) 255–260.
23. C.Y. Li and T.W. Chou, *Comp. Sci. Technol.* 63 (2003) 1517–1524.
24. J. P. Lu, *Phys. Rev. Lett.* 79 (1997) 1297–1300.
25. R. S. Ruoff, J. Tersoff, D. C. Lorents et al., *Nature* 364 (1993) 514.
26. T. Hertel, R. Martel, and P. Avouris, *J. Phys. Chem.* B 102 (1998) 910.
27. N. Lordi and N. Yao, *J. Chem. Phys.* 109 (1998) 2509.
28. W. Shen, B. Jiang, B. S. Han, and S. S. Xie, *Phys. Rev. Lett.*, 84 (2000) 3634.
29. M. F. Yu, T. Kowalewski, and R. S. Ruoff, *Phys. Rev. Lett.* 85 (2000) 1456.
30. S. Iijima, C. Brabec, A. Maiti, and J. Bernholc, *J. Chem. Phys.* 104 (1996) 2089.
31. J. Tang, L.-C. Qin, T. Sasaki et al., *Phys. Rev. Lett.*, 85 (2000) 1887.
32. C. Y. Li and T. W. Chou, *Phys. Rev.* B 69 (2004) 073401.
33. H. Reich, C. Thomsen, and P. Ordejon, *Phys. Rev.* B 65 (2002) 153407.
34. M. M. J. Treacy, T. W. Ebbesen, and T. M. Gibson, *Nature* 381 (1996) 680–687.
35. A. Krishnan, E. Dujardin, T.W. Ebbesen et al., *Phys. Rev.* B 58 (1998) 14013–14019.
36. K. Sohlberg, B. G. Sumpter, R. E. Tuzun, and D. W. Noid, *Nanotechnology* 9 (1998) 30.
37. J. Yoon, C. Q. Ru, and A. Mioduchowski, *Phys. Rev.* B 66 (2002) 233402.
38. E. S. Snow, P. M. Campbell, and J. P. Novak, *Appl. Phys. Lett.* 80 (2002) 2002.
39. C. Y. Li and T. W. Chou, *Phys. Rev.* B 68 (2003) 073405.
40. C. Y. Li and T. W. Chou, *Appl. Phys. Let.* 84 (2004) 121–123.
41. L. X. Benedict, S. G. Louie, and M. L. Cohen, *Solid State Comm.*, 100 (1996) 177.
42. W. Yi, L. Lu, D.-L. Zhang, Z. W. Pan, and S. S. Xie, *Phys. Rev* B, 59 (1999) R9015.
43. A. Mizel, L. X. Benedict, M. L. Cohen et al., *Phys. Rev.* B, 60 (1999) 3264.
44. J. Hone, B. Batlogg, Z. Benes, A. T. Johnson, and J. E. Fischer, *Science*, 289 (2000) 1730.
45. J. C. Lasjaunias, K. Biljakovic, Z. Benes et al., *Phys. Rev.* B, 65 (2002) 113409.
46. V. N. Popov, *Phys. Rev.* B, 66 (2002) 153408.
47. J. X. Cao, X. H. Yan, Y. Xiao, Y. Tang, and J. W. Ding, *Phys. Rev.* B, 67 (2003) 045413.
48. E. Dobardzic, I. Milosevic, B. Nikolic, T. Vukovic, and M. Damnjanovic, *Phys. Rev.* B, 68 (2003) 045408.
49. S. L. Zhang, M. G. Xia, S. M. Zhao, T. Xu, and E. H. Zhang, *Phys. Rev.* B, 68 (2003) 075415.
50. C. Y. Li and T.W. Chou, *Phys. Rev.* B, 71(2005) 075409.
51. C. Y. Li and T.W. Chou, *Mater. Sci. Eng.* A 409 (2005) 140–144.

52. R. S. Ruoff, and D. C. Lorents, *Carbon*, 33(1995) 925–930.
53. S. Bandow, *Jpn. J. Appl. Phys.*, Part 2, 36 (1997) L1403–L1405.
54. Y. Maniwa, R. Fujiwara, H. Kira et al., *Phys. Rev. B*, 64(2001) 073105.
55. Y. Yosida, *J. Appl. Phys.*, 87 (2000) 3338–3341.
56. Y. Maniwa, R. Fujiwara, H. Kira et al., *Phys. Rev. B*, 64(2001) 241402R.
57. N. R. Raravikar, P. Keblinski, A. M. Rao et al., *Phys. Rev. B*, 66 (2002) 235424.
58. P. K. Schelling and P. Keblinski, *Phys. Rev. B*, 68 (2003) 035425.
59. Y. K. Kwon, S. Berber, and D. Tomanek, *Phys. Rev. Lett.*, 92 (2004) 015901.
60. H. Jiang, B. Liu, and Y. Huang, *J. Eng. Mater. Tech.*, 126 (2004) 265–270.
61. C. Y. Li and T. W. Chou, *Phys. Rev. B* 71 (2005) 235414.
62. L. Lou, P. Nordlander, and R. E. Smalley, *Phys. Rev. B*, 52 (1995) 1429–1432.
63. R. V. Rotkin, V. Shrivastava, K. A. Bulashevich, and N. R. Aluru, *Int. J. Nanoscience*, 1 (2002) 337–346.
64. J. Luo, L.-M. Peng, Z. Q. Xue, and J. L. Wu, *Phys. Rev. B* 66 (2002) 115415.
65. P. Keblinski, S. K. Nayak, P. Zapol, and P. M. Ajayan, *Phys. Rev. Lett.* 89 (2002) 255503.
66. X. Zheng, G. H. Chen, Z. Li, S. Deng, and N. Xu, *Phys. Rev. Lett.* 92 (2004) 106803.
67. C. Y. Li and T. W. Chou, *Appl. Phys. Lett.* 89 (2006) 063103.
68. C. Ke and H. D. Espinosa, *J. Appl. Mech.* 72 (2005) 721–725.
69. G. G. Chen, S. Bandow, E. R. Margine et al., *Phys. Rev. Lett.* 90(2003) 257403.
70. P. S. Dorozhkin and Z. C. Dong. *Appl. Phys. Lett.* 85 (2004) 4490.
71. M. Paillet, P. Poncharal, and A. Zahab, *Phys. Rev. Lett.* 94 (2005) 186801.
72. G. Y. Sun, J. Kurti, M. Kertesz, and R. H. Baughman, *J. Am. Chem. Soc.* 124 (2002) 15076–15080.
73. Y. N. Gartstein, A. A. Zakhidov, and R. H. Baughman, *Phys. Rev. Lett.* 89 (2002) 045503.
74. A. Z. Hartman, M. Jouzi, R. L. Barnett, and J. M. Xu, *Phys. Rev. Lett.* 92 (2004) 236804.
75. M. Verissimo-Alves, B. Koiller, H. Chacham, and R. B. Capaz, *Phys. Rev. B*, 67 (2003) 161401.
76. W. L. Guo and Y. F. Guo. *Phys. Rev. Lett.* 91 (2003) 115501.
77. C. Y. Li and T. W. Chou, *Nanotechnology* 17 (2006) 4624–8.
78. C. Y. Li and T. W. Chou, *Carbon* 45 (2007) 922–30.
79. J. Luo and J. L. Wu. *Sci. China* G 47 (2004) 685–93.
80. K. Raghavachari and J. S. Binkley. *J. Chem. Phys.* 87 (1987) 2191.
81. Y. H. Lee, S. G. Kim, and D. Tomfinek, *Chem Phys. Lett.* 265 (1997) 667–72.
82. Z. L. Wang, R. P. Gao, W. A .de Heer, and P. Poncharal, *Appl. Phys. Lett.* 80 (2002) 856.
83. E. T. Thostenson and T. W. Chou, *J. Phys. D: Appl. Phys.* 36 (2003) 573–582.
84. J. C. Halpin, *Primer on Composite Materials: Analysis*, Technomic Publishing Company, Lancaster, PA (1984).
85. E. T. Thostenson, Carbon nanotube-reinforced composites: processing, characterization and modeling. Ph.D. Dissertation, University of Delaware, Newark, DE (2004).
86. Z. Jia, Z. Wang, C. Xu et al., *Mater. Sci. Eng. A* 271, 395 (1999).
87. V. Lordi and N. Yao, *J. Mater. Res.* 15 (2000) 2770.
88. H. D. Wagner, O. Lourie, Y. Feldman, and R. Tenne, *Appl. Phys. Lett.* 72(1998) 188.
89. K. Wise and J. Hinkley, Molecular dynamics simulations of nanotube-polymer composites. American Physical Society Spring Meeting (2001) Seattle, WA.

90. G. M. Odegard, S. J. V. Frankland, and T. S. Gates, *44th AIAA Structure, Structures Dynamics, and Materials Conference*, Norfolk, VA, 2003–1701.
91. C. Y. and T. W. Chou, *J. Nanosci. Nanotechnol.* 3 (2003) 1–8.
92. Y. B. Yi, L. Berhan, and A. M. Sastry, *J. Appl. Phys.* 96 (2004)1318–27.
93. L. Berhan, and A.M. Sastry. *Phys. Rev.* E 75 (2007) 041121.
94. F. T. Fisher, R.D. Bradshaw, and L.C. Brinson. *Comp. Sci. Technol.* 63 (2003) 1689–1703.
95. D. L. Shi, X.Q. Feng, Y. Huang et al., *J. Eng. Mater. Tech.* 126 (2004) 250–257.
96. F. Dalmas, R. Dendievel, L. Chazeau et al., *Acta Mater.* 54 (2006) 2923–31.
97. C. Y. Li and T. W. Chou. *Appl. Phys. Lett.* 90 (2007) 174108.
98. C. Y. Li and T. W. Chou, *Comp. Sci. Technol.* 68 (2008) 3373–3379.
99. T. Natsuki, M. Endo, and T. Takahashi, *Phys A: Stat. Mech. Appl.* 352 (2005) 498–508.
100. J. Hoshen and R. Kopelman, *Phys. Rev.* B 14 (1976) 3438–3445.
101. F. Babalievski, *Int. J. Modern Phys.* C 9 (1998) 43–60.
102. M. T. Orchard. 1991 *Int. Conf. on Acoustics, Speech and Signal Processing*, 4 (1991) 2297–3000.
103. R. E. Tarjan, *SIAM J. Comput.* 1, 146–160 (1972).
104. C. Y. Li and T. W. Chou, *J. Phys. A: Math. Theor.* 40 (2007) 14679–86.
105. G. G. Batrouni, A. Hansen, and S. Roux, *Phys. Rev.* A, 38, 3820 (1988).
106. C. Y. Li, E. T. Thostenson, and T. W. Chou. *Appl. Phys. Lett.* 91 (2007) 223114.
107. B. E. Kilbride, J. N. Coleman, J. Fraysse et al. *J. Appl. Phys.* 92 (2002) 4024–30.
108. M. Foygel, R. D. Morris, D. Anez et al., *Phys. Rev.* B 71 (2005) 104201.
109. I. Balberg, *Carbon* 40 (2002) 139–143.
110. J. G. Simmons, *J. Appl. Phys.* 34, 1793–1803 (1963).
111. J. Hone, M. C. Llaguno, N. M. Nemes et al., *Appl. Phys. Lett.* 77 (2000) 666–668.
112. C. Y. Li, E. T. Thostenson and T. W. Chou, *Comp. Sci. Technol.* 68 (2007) 1445–1452.
113. E. T. Thostenson and, T.W. Chou. *J. Phys. D: Appl. Phys.* 35 (2002) L77–80.
114. E. S. Choi, J. S. Brooks, D. L. Eaton et al., *J. Appl. Phys.* 94 (2003) 6034–9.
115. R. Haggenmueller, H. H. Gommans, A. G. Rinzler et al., *Chem. Phys. Lett.* 330 (2000) 219–225.
116. F. M. Du, J. E. Fischer, and K. I. Winey, *J. Polym. Sci. Part B–Polym. Phys.* 41, 3333 (2003).
117. F. M. Du, J. E. Fischer, and K. I. Winey, *Phys. Rev.* B 72, 121404 (2005).
118. C. Y. Li, E. T. Thostenson, and T. W. Chou, *J. Nanosci. Nanotechnol* 68 (2008) 2518–2524.
119. J. R. Wood, Q. Zhao, M. D. Frogley et al., *Phys. Rev.* B, 62, 7571(2000).
120. Q. Zhao, J. R. Wood and H. D. Wagner, *Appl. Phys. Lett.* 78(2001) 1748.
121. P. Dharap, Z. L. Li, S. Nagarajaiah, and E. V. Barrera, *Nanotechnology* 15(2004) 379–382.
122. W. Zhang, J. Suhr, and N. Koratkar, *J. Nanosci. Nanotechnol.* 6(2006)960–964.
123. I. P. Kang, M. J. Schulz, J. H. Kim et al., *Smart Mater. Struct.* 15(2006) 737–748.
124. I. Kang, J. W. Lee, G. R. Choi et al., *Key Engineering Materials: Advanced Nondestructive Evaluation.* 321–323(2006) 140–145.
125. H. Yoon, J. N. Xie, J. K .Abraham et al., *Smart Mater. Struct.* 15(2006) S14–S20.
126. B. Zhang, R.W. Fu, M.Q. Zhang et al., *Sens. Actuat. B-Chem.* 109(2005) 323–328.
127. C. Wei, L. M. Dai, A. Roy, and T. B. Tolle, *J. Am. Chem. Soc.* 128 (2006)1412–1413.
128. B. Fiedler, F.H. Gojny, M. H. G. Wichmann, W. Bauhofer, and K. Schulte, *Ann. De Chim.-Sci. Des Materiaux* 29(2004)81–94.

129. E. T. Thostenson and T. W. Chou, *Adv. Mater.* 18(2006) 2837–2841.
130. J.M. Park, D.S. Kim, S.J. Kim et al., *Composites*: Part B 38(2007) 847–861.
131. W. A. Curtin, *Adv. Appl. Mech.* 36(1999) 163–253.
132. Z. H. Xia and W. A. Curtin, *Comp. Sci. Technol.* 67(2007)1518–1529.
133. C. Y. Li and T. W. Chou, *Comp. Sci. Technol.* 68 (2008) 3373–3379.
134. E. T. Thostenson and T. W. Chou, *Carbon* 44 (2006) 3022–3029.
135. E. T. Thostenson and T. W. Chou, *Comp. Sci. Technol.* 68 (2008) 2557–2561.
136. S. Moaveni, Finite *Element Analysis: Theory and Applications with ANSYS,* 2nd edition. Prentice Hall, Upper Saddle River, NJ (2003).

6

Natural Fiber Composites in Biomedical and Bioengineering Applications

Karen Hoi-yan Cheung

Hong Kong Polytechnic University

CONTENTS

6.1 Development of Biomaterials

Bioengineering refers to the application of concepts and methods of the physical sciences and mathematics in an engineering approach towards solving problems in repair and reconstruction of lost, damaged, or deceased tissues [1]. Any material that is used for this purpose can be regarded as a biomaterial. According to Williams (1986), biomaterial is a material used in implants or medical devices and intended to interact with biological systems [2]. The most common types of medical devices include substitute heart valves and artificial hearts, artificial hip and knee joints, dental implants, internal and external fracture fixators and skin repair templates. The development of biomaterial that is suitable to be used in a human body with such a complicated biological and sensitive system is a great challenge to all scientists. There are various kinds of biomaterials with different characteristics, structures, and material properties, but all these materials must fulfill the requirements of being a viable implant. They

(a)

(b)

FIGURE 6.1
Different types of metallic and ceramic biomaterials.

must be biocompatible (i.e., they must not elicit an unresolvable inflammatory response nor demonstrate immunogenicity or cytotoxicity to the host).
Moreover, they should have the desired shape, structure, and interconnected porosity to suit different types of bioengineering applications for
cell growth and proliferation, appropriate mechanical properties, and the
ability to retain their integrity at the implantation site until their desired
functions are fulfilled. Apart from the above criteria, they should be biodegradable and bioresorbable; that is, they can be metabolized or excreted
by the host. Extra operations for removal of the implants will then not be
required, and the pain of patients can be alleviated. Last but not least, they
have to be easily sterilized to prevent infection.

Metallic and ceramic materials, including stainless steel, titanium alloys,
bio-glass, hydroxyapatite, etc., are widely accepted as biomaterials due to
their bioinert characteristics. They are usually used in the clinical applications of artificial joints, dental implants, femoral stems, etc., as shown in
Figure 6.1. Generally, they are implanted inside the host body without generating any adverse response and interaction with surrounding tissues.

For biomedical applications, most of the current clinical applications
involve complex structural and chemical functions; therefore, ceramic
and metallic materials have been widely accepted for the development of
implants. Different types of applications of ceramic and metallic implants

TABLE 6.1

Examples and Applications of Ceramic and Metallic Implant Materials

Biomaterials	Examples	Applications
Metallic materials	Stainless steel, cobalt-based alloys, amalgams, titanium and titanium alloys, nickel–titanium alloy (SMA), gold, platinum, silver	Artificial joints (hip ball and sockets), femoral stem, dental implants, tooth fixation (crowns and permanent bridges), artificial hearts, endovascular therapy, bone plates, staples, wires, pins, stents, pacemaker electrodes, artificial inner ears, intramedullary nail, disc prostheses, internal and external fixators (plates, washers, and screws), spinal fusion, medical devices, surgical instruments (chisels, scalpels, pliers, forceps, etc.), hypodermic needles, craniofacial and maxillofacial treatments, self-expanding cardiovascular stents (SMA), guide wires for introduction of therapeutic and diagnosis devices (SMA), snare wires (SMA), vena cave filters (SMA), orthodontic arch wires (SMA), studs of earrings (silver), sanitizing agent (silver), burn therapy (silver), wound dressing (silver), urinary bladder catheters, and stethoscope diaphragms (silver).
Ceramic materials	Alumina, zirconia, silicon nitride, hydroxyapatite (HA), bio-glass, glass-ceramics, crystalline or glassy forms of carbon and its compounds	Orthopedic and dental implants, hip and knee replacements, porous coating for femoral stems, porous alumina spacers, scaffolds, artificial teeth, bone filler, orbital implants within eye socket, crows, shoulder reconstruction surgery, coating for dental implants, knee joints, and spinal implants.

Source: Bartolo, P. and Bidanda, B. (Ed.), *Bio-Materials* and *Prototyping Applications* in *Medicine.* Springer Science + Business Media, LLC, USA, 2008; Teoh, S.H. (Ed.), *Engineering Materials for Biomedical Applications.* World Scientific Publishing Co. Pvt. Ltd., Singapore, 2004; Ramakrishna et al., *An Introduction to Biocomposites. Vol. 1 Series on Biomaterials and Bioengineering.* Imperial College Press, U.K., 2004.

TABLE 6.2

Mechanical Properties of Typical Implant Materials

	Elastic Modulus (GPa)	Yield Strength (MPa)	Tensile Strength (MPa)	Elongation at Break (%)
Al_2O_3	350	—	1000–10000	0
CoCr alloy	225	525	735	10
Stainless steel	210	240	600	55
316	120	830	900	18
Ti-6Al-4V	15–30	30–70	70–150	0–8

Source: Bartolo, P. and Bidanda, B. (Ed.), *Bio-Materials and Prototyping Applications in Medicine.* Springer Science + Business Media, LLC, USA, 2008; Teoh, S.H. (Ed.), *Engineering Materials for Biomedical Applications.* World Scientific Publishing Co. Pvt. Ltd., Singapore, 2004; Ramakrishna, S., Huang, Z.M., Kumar, G.V., Batchelor, A.W., and Mayer, J., *An Introduction to Biocomposites. Vol. 1 Series on Biomaterials and Bioengineering.* Imperial College Press, U.K., 2004.

are as listed in Table 6.1. For bone implants, some requirements such as increase in surface area and porosity are essential for bone growth. A large surface area may characterize porous materials, resulting in a high tendency for bioresorbability, which induces high bioactivity. Interconnected pores permit tissue attachment and in-growth, thus anchoring the prosthesis with the surrounding bone, and preventing the loosening of implants [3].

However, there are still various problems related to the metallic and ceramic biomaterials. For example, (1) mismatches between the stiffness of implants and tissue as listed in Table 6.2, (2) the existence of uneven or excess neo-cell growth around the implants and porosis under the implants, (3) corrosion and bioinertness of the implants, and (4) the necessity of several subsequent surgical operations for the removal of the implants after completion of their intended functions. These drawbacks are of major concern since they may significantly affect mechanical properties, such as lowering the density and altering the architecture of bone after recovery, which may diminish the bone strength and lead to refracture of bones after implant removal. On the other hand, patients may suffer great discomfort, inconvenience, and pain because of these materials, which induce extra surgical operations and treatments. In this respect, polymers have been extensively used in various medical and pharmaceutical applications recently, which is mainly due to their availability in a broad diversity of compositions, properties, and forms (solids, fibers, fabrics, films, and gels), and their ease of fabrication into different complex shapes and structures. Nevertheless, their low mechanical properties and brittle nature limit the range of applications. The other drawbacks of conventional materials for medical applications as listed in Table 6.3 provide some insights into the development of advanced composite materials as an alternative choice in bioengineering applications.

In order to solve the above problems, the development of a novel type of biomaterials with biocompatible, biodegradable, and bioresorbable

TABLE 6.3

Disadvantages of Conventional Implant Materials

Biomaterials	Disadvantages
Metallic materials	Low biocompatibility, relatively heavy in weight, poor durability, chromium releases, toxicity of corrosion products, production of polyethylene wear, excessively high rigidity, stiffness mismatch to tissues, high specific gravity, fracture due to corrosion fatigue, lack of biocompatibility, inadequate affinity for cells and tissues integration, shielding of x-rays, intrinsically soft and ductile, high processing cost
Ceramic materials	Produced only by high temperature sintering, poor tensile properties, brittle, low fracture strength, difficulty in fabrication, low mechanical reliability, lack of resilience, production of polyethylene wear
Polymeric materials	Low mechanical properties, limitations to moderate load-bearing applications, poor durability, shrinkage, difficulty to anchor to bone, abrasion and wear of prostheses. Strength deterioration in long term (fatigue), generation of particulate matter may cause synovitis and inflammation. Sensitivity to hydrolytic and stress-induced degradation. Lack anisotropy and nonlinear compliance for vascular prostheses, too flexible, absorb liquid and swell. Properties may be affected by sterilization processes.

characteristics is necessary. The advantages of using bio-polymers and composite materials over the traditional metallic and ceramic materials for implant development include:

- A wide range of properties can be tailor-made to suit different final product design and applications.
- Uniformity and mechanical properties are controllable.
- They possess good biocompatibility, biodegradability, and bioresorbability.
- The degradation rate of biodegradable implants can be engineered to match a tissue regeneration rate and to reduce stress-bearing capability over time since the polymer will be degraded naturally (i.e., the load can be transferred to the fractured bone gradually).
- The mechanical properties adequately address short-term function without interfering with long-term function.
- Toxicity of degradation products is low or negligible, in terms of both local tissue response and systemic response.
- The histological response is generally predictable.
- They are approved for numerous usages in human medical and pharmaceutical applications by the U.S. Food and Drug Administration (FDA).

- Drug delivery compatibility exists in applications that release or attach active compounds.
- Risks associated with the presence of nonbiodegradable foreign materials remaining permanently in the human body can be eliminated.
- Pain, both physically and physiologically, due to the need of secondary surgical operation for removing ceramic and metallic implants can be alleviated.

6.2 Biodegradable Polymers

Biodegradable materials like polymers can be decomposed naturally but their degraded products will remain inside the human body. Bioresorbable materials will degrade after a certain period of implantation time, and nontoxic products will be produced in the ways of elimination with time and/or metabolism. For the chemical degradation, two different modes are defined: (1) hydrolytic degradation or hydrolysis, which is mediated simply by water, and (2) enzymatic degradation, which is mainly mediated by biological agents such as enzymes. The field of biodegradable polymers is a fast-growing area of polymer science because of the interest in such compounds for temporary surgical and pharmacological applications. There are two main types of biodegradable polymers, including natural polymers and synthetic polymers.

Natural polymers used in tissue engineering applications include collagen [7–9], alginate [10–13], agarose [14,15], chitosan [16,17], fibrin [18,19], and hyaluronic acid-based materials [12,18,20,21]. They often possess highly organized structures and may contain an extracellular substance called ligand that can be bound to cell receptors. Although they are of known biocompatibility, lack of supply in large quantities and difficulty in processing into scaffolds have limited their use in clinical applications. Moreover, as natural polymers can guide cells to grow at various stages of development, they may stimulate an immune response at the same time. This leads to concerns over antigenic and delivery of diseases for allograft. Since the degradation of these polymers depends on the enzymatic processes, the degradation rate may vary from patient to patient.

Synthetic polymers are man-made polymers, which have advantages over the use of natural origin polymers as they are more flexible, predictable and able to be processed into different size and shapes. The physical and chemical properties of a polymer can be easily modified, and the mechanical and degradation characteristics can be altered by their chemical composition of the macromolecule. The functional groups and side chains of these polymers can be incorporated (i.e., the synthetic polymers can be self-cross-linked or cross-linked with peptides or other bioactive molecules, which may be desirable biomaterials for cartilage tissue engineering). Additionally, synthetic polymers are generally degraded by simple hydrolysis that is desirable as the degradation rate does not have variations from host to host, unless there are inflammations and implant degradation, etc., to affect the local pH variations. The most extensively used synthetic polymers in biomedical applications belong to the polyester family, which are poly(glycolic acid) (PGA), poly(lactic acid) (PLA), and their copolymers [22,23]. Other biodegradable synthetic polymers include polycaprolactone (PCL) [23–27], polyhydroxybutyrate (PHB) [22,28–30] and poly(propylene fumarate) (PPF) [31–33], and their market is expanding rapidly worldwide.

6.2.1 PGA, PLA, and Their Copolymers

Physical and chemical properties of polymers such as the molecular weight and the polydispersity index affect the mechanical properties of the polymer as well as its ability to be formulated as orthopedic devices. Different factors like the molecular weight, copolymer composition, crystallinity, and geometric regularity of individual chains significantly affect the mechanical strength of the polymer. PGA, PLA, and their copolymers are usually characterized in terms of intrinsic viscosity, which is directly related to their molecular weights. In general, high molecular weight is needed to produce devices of high mechanical strength. PGA is a highly crystalline thermoplastic while PLA is a semicrystalline thermoplastic, and they are insoluble in most types of organic solvents. They can be fabricated into a diversity of complex shapes for clinical applications by common processing techniques including extrusion, injection, and compression molding, solvent casting, etc. Thus, PGA, PLA, and their copolymers are considered as biodegradable and compostable thermoplastics, with high strength and modulus.

PLA, PGA, and their copolymers can generally be formed into film, tubes, matrices, and other forms by using standard processing techniques such as injection molding, extrusion molding, solvent casting, and spin casting to fulfill the surface area and cellular requirements of a variety of tissue engineering constructs. The properties of these biodegradable polymers make them the preferred materials for a variety of medical devices and pharmaceutical applications like tissue fixation, tissue regeneration, sutures, stents, wound dressing, dialysis, antitissue adhesion, and drug delivery system, etc. Table 6.3 shows the potential applications for PGA, PLA, and their copolymers.

Although many researchers have tried to investigate porous PLA scaffolds for use in orthopedic applications, PLA is primarily used as a nonwoven mesh for tissue engineering applications. Ishaug-Riley et al. (1999) have shown that fewer chondrocytes were attached to PLLA than to PGA at the initial stage, but both surfaces allow extensive cell proliferation, giving a similar total number of cells at confluence. Taboas et al. (2003), Liao et al. (2004), and Montjovent et al. (2005) have developed a conceptual framework for 3-D scaffolds with controlled scaffold material composition, and architecture (local and global porosity), using polymer–ceramic (PLA/HA ceramic) composites (Figure 6.2)

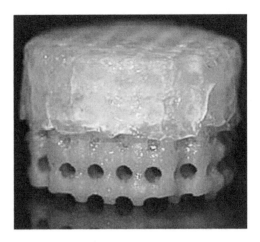

FIGURE 6.2
3-D scaffold made by PLA/HA ceramic. (From Taboas et al. *Biomaterials* 2003; 24: 181–194.)

and nano-HA/Collagen/PLA composite, respectively, for cell culture and proliferation. It was found that the 3-D scaffold provided better mechanical performance, depending on porosity and lay-up orientations, for bone tissue engineering (biomimetic scaffolds engineering and scaffolds for complex biomechanical applications) than the woven type 2-D scaffold materials.

At temperature below the glass transition temperature (Tg), polymer acts more like a glass, and at temperatures above Tg, the polymer acts more like a rubber. Tg of PGA and PLA is above the physiological temperature of 37°C, and hence they are glassy in nature. Thus, they have a rigid structure to give suitable mechanical strength to be formulated as medical devices. In general, Tg increases with an increase in the lactide content ratio in the copolymer and with an increase in their molecular weight. The crystallinity of PGA, PLA, and their copolymers directly affects their mechanical strength, the capacity to undergo hydrolysis, and the degradation rate.

Hydrolytic degradation behavior of polyester polymers is mainly affected by (1) the chemical composition, copolymer ratio, morphology (crystalline or amorphous), glass transition temperature, molecular weight and molecular weight distribution of the polymers, the concentration of additives including solvent, monomers and catalyst, etc.; (2) the processing methodologies and dimensions (size, shape, surface properties, and porosity, etc.) of the end devices; (3) the physical and chemical factors, including pH value, ionic strength, moisture content, and temperature, etc., at the implantation site, and (4) the mechanisms of hydrolysis, including autocatalytic, noncatalytic, and enzymatic. Degradation of PGA and PLA begins with random hydrolysis in an aqueous environment through cleavage of its backbone ester linkages by purely hydrolytic (nonenzymatic) mechanism. In vivo, enzymes are considered to enhance the initial degradation. PGA breaks down by certain enzymes into glycolic acid during biodegradation while PLA biodegrades into lactic acid, which enters the tricarboxylic acid (TCA) cycle, and is metabolized and excreted from the human body as energy, carbon dioxide, and water.

PLA degrades slower than PGA due to its hydrophobic characteristic, which limits the water absorption of thin films and slows down the back-bone hydrolysis rate. Based on the available data, the duration of degrada-tion of PLA can range from 12 months to over 2 years, while the duration of degradation of PGA can be within 12 months. The duration of degradation of the copolymers of PGA and PLA ranges from 1 month to more than 1 year, depending on the copolymer ratio of the final material. It means that if more PLA is blended with PGA, the complete degradation period will be longer due to the hydrophobic characteristic of PLA. On the other hand, if more PGA is blended with PLA, it will take a shorter time to degrade the material completely due to the hydrophilic characteristic of PGA. However, since they are still sensitive to hydrolytic degradation, special control of operation tem-peratures and moisture content during storage and fabrication environment of these polymers are necessary to prevent degradation.

PLA alone can have a relatively low tensile strength and modulus of elasticity, depending on which isomer is used in the polymer. Lactic acid exists in two ste-reoisomerism forms, which can be separated into four morphologically distinct polymers, namely, D-PLA (PDLA), L-PLA (PLLA), D,L-PLA (PDLLA), and meso-PLA. Degradation products of these materials reduce the local pH, accelerate the polyester degradation rate, and induce inflammatory reaction. Figure 6.3a–c shows the general polymerization method of this class of polymers.

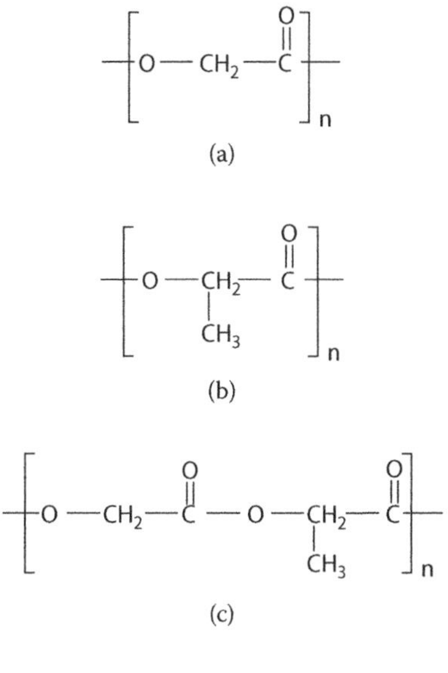

FIGURE 6.3
Ring-opening polymerization of (a) PGA, (b) PLA, and (c) their copolymer.

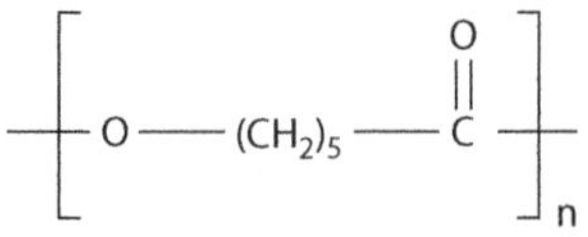

FIGURE 6.4
Polycaprolactone (PCL).

6.2.2 Other Biodegradable Polymers (PCL, PHB, and PPF)

Other biodegradable polymers, including PCL, PHB, and PPF, are widely studied as biomaterials for clinical applications. PCL can be prepared by ring-opening polymerization as shown in Figure 6.4. It is a semicrystalline thermoplastic with low glass transition temperature (Tg) and melting temperature (Tm). The mechanical properties, processability, and biodegradability of PCL can be adjusted to expand its utility by blending or copolymerizing with other polyesters [24,25]. PCL is more stable in ambient conditions; it is also cheaper and readily available commercially in large quantities when compared to PLA. It involves similar degradation mechanisms as PLA, that is, (1) random hydrolytic ester cleavage, and (2) weight loss through diffusion of oligometric species from the bulk. The products generated during degradation are either metabolized through the TCA cycle or eliminated by direct renal secretion. PCL has an even lower degradation rate than PLA, and takes about 3 years for complete degradation. PCL is regarded as hard and soft tissue-compatible biomaterial. It is investigated in bone and cartilage tissue engineering applications as a scaffold for supporting cell growth [26,27]. On the other hand, PCL can be effectively used to entrap antibiotic drugs; hence, it is suitable for biomedical application as a long-term implantable drug delivery system [24]. However, its slow degradation and resorption mechanisms due to its hydrophobic and high crystallinity characteristics limit its applications [27].

Poly(hydroxyalkonate)s (PHAs) are biocompatible and biodegradable thermoplastic polyesters produced by various microorganisms. Their mechanical properties and biocompatibility can be varied by blending and surface modification or composition with other polymers, enzymes, or inorganic materials for different clinical applications. Nevertheless, commercial availability and the time-consuming extraction of these types of bacterial culture polymers limit their applications. The most extensively studied PHA for bone tissue engineering is poly(3-hydroxybutyrate) (PHB), which can be produced in high yield by fermentation of a variety of bacteria strains as shown in Figure 6.5. PHB serves as an intracellular energy and carbon storage product in microorganisms. It degrades into D-3-hydroxybutyrate, which is a normal constituent of human blood. Its copolymers with different ratios of hydroxyvalerate (PHBV) are the most widely used in industry due to their flexibility and processability compared to pure PHB [29].

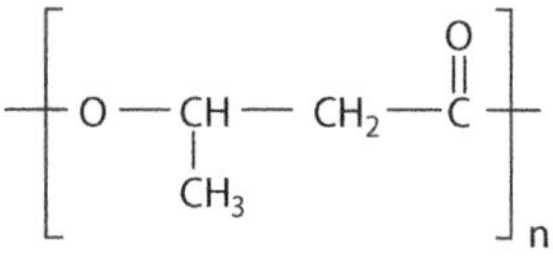

FIGURE 6.5
Polyhydroxybutyrate (PHB).

Other polymers in the group consist of poly-4-hydroxybutyrate (P4HB), co-polymers of 3-hydroxybutyrate and 3-hydroxyhexanoate (PHBHHx), and poly-3-hydroxyoctanoate (PHO); they have been shown to be suitable for tissue engineering applications [22,30]. The degradation rate of PHBV mainly depends on the ratio of copolymer composition, and it degrades faster in aqueous medium. Apart from this, PHBV is suitable for diversified clinical applications because of its natural originality, thermoplasticity, biocompatibility, biodegradability, piezoelectricity, optical activity, and stereospecificity [28]. It is now being intensely studied as a tissue engineering substrate since it maintains its integrity during the cell culture period and is applicable to cell culture and proliferation.

Another type of commonly used biodegradable polymer is PPF, which is a copolyester based on fumaric acid. PPF is unsaturated linear polyester as shown in Figure 6.6. It contains two ester groups and one unsaturated carbon–carbon double bond in its backbone for subsequent cross-linking reactions (i.e., thermal-crosslinking and photo-crosslinking, etc.) [35]. The presence of the backbone double bond leads to a difficulty in achieving high molecular weight PPF. In order to obtain better mechanical strength, extra reactions to form cross-linkage networks and the incorporation of reinforcements such as ceramic materials are required. The mechanical properties of PPF can vary greatly, depending on the method of synthesis and the

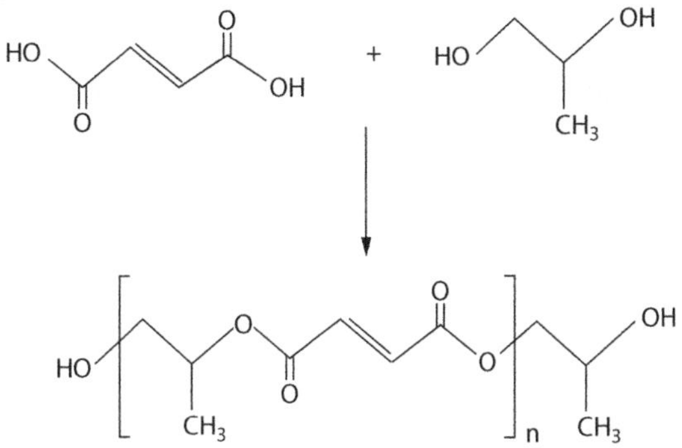

FIGURE 6.6
Poly(propylene fumarate) (PPF).

cross-linking agent used. PPF undergoes hydrolytic degradation to fumaric acid and propylene glycol in several months, and its degradation time and mechanical properties mainly depend on varying the PPF molecular weight and other components for PPF-based composites [36,37]. Therefore, preservation of the double bonds and control of molecular weight are critical for PPF. Fumaric acid and propylene glycol are biocompatible and can be removed from the host body. Fumaric acid is a naturally occurring substance, which is commonly used as a diluent in drug formulations. Other clinical applications for PPF include 3-D scaffolds for tissue regeneration and substrate for osteoblast cell culture.

6.3 Natural Fiber Reinforcements

Biofibers or natural fibers are naturally renewable resources. When they are imbedded into biodegradable polymeric materials, a novel type of bio-composites can be produced. These fibers represent alternatives to conventional reinforcing fibers including glass, carbon, Kevlar®, etc., as they have the advantages of low cost, low density, high toughness, good specific strength properties, biodegradability, and tool wear reduction. They can be well recognized as potential micro-reinforcements for the enhancement of the mechanical, thermal, and structural properties of biodegradable polymer composites, without generating any harmful by-products and adverse effects during degrading process to the patients. They are subdivided into groups based on their origins and sources from plants, animals, or minerals. Generally, plant-based natural fibers are lignocelluloses in nature and are composed of cellulose, hemi-cellulose and lignin, like flax, jute, sisal, kenaf, etc. These fibers are extracted from seeds (cotton and coir), fruits (oil palm and coconut), leaves (sisal and pineapple leaf), stem and skin (flax, jute, kenaf) of plants. The most commonly used plant-based natural fibers in composite applications are extracted from bast and leaf (hard fiber), including hemp, jute, sisal, kenaf, flax, etc. Animal-based natural fibers, like wool, mohair, spider and silkworm silks, etc., contain proteins. These fibers are obtained from hairy mammals (wool, cashmere), dried saliva of silkworms during preparation of cocoon (silkworm silk), and bird feather (avian fiber). Mineral fibers are naturally occurring fibers like asbestos [38,39]. Figure 6.7 shows various types of natural fibers.

These fibers are extracted from plants and animals before they can be used as reinforcement. Plant-based natural fibers are usually extracted by microbiological retting (e.g., sisal and coir), hand scraping (e.g., sisal, banana fiber, and flax), processing with a raspador machine (e.g., sisal and banana fiber), boiling leaf sheaths in sodium hydroxide solution (e.g., banana fiber) or a cotton gin (cotton), and decorticating (e.g., coir, flax), etc. For animal-based fibers

FIGURE 6.7
Various types of natural fibers.

they are usually extracted from animals (wool and chicken feathers) and insects (spider silk and silkworm) or by soaking in boiling water (silkworm silk).

Natural fibers exhibit wide variation in diameter along the lengths of individual filaments inside as shown in the SEM micrographs (Figure 6.8). The quality and material properties of these fibers depend on the size, maturity, and the extraction methods. The mechanical properties of these natural fibers depend on the origins of the plants and animals including their growing and living environment and conditions, the quality of fibers, their chemical composition and dimensions, and even their ages. The modulus of fiber decreases with an increase in diameter, whereas the density, ultimate tensile strength, initial modulus and electrical resistivity are related to the internal structural and chemical composition of the fiber. Moreover, the strength and stiffness of these fibers depend on the

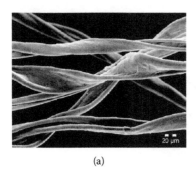
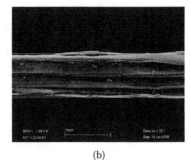

(a)   (b)

FIGURES 6.8
SEM micrographs of (a) plant-based (cotton) and (b) animal-based (silkworm silk) natural fibers.

chemical constituents and chemical structure, as well as the angle between axis and fibril of the fiber (i.e., the decrease in this angle will lead to better mechanical properties) [40]. Therefore, the reproducibility of the mechanical properties of natural fibers is difficult to maintain. Table 6.4 lists the mechanical properties of widely used natural fibers. However, some of the natural fibers have appreciable mechanical properties compared to synthetic fibers. They have high stiffness, strength, and ductility, which are mainly due to the crystallinity index of the fibers.

In the literature, there is formulation describing the relationship for bast and fruit fibers between Young's modulus and microfibrillar angle, as well as the modified equation by McLaughlin and Tait for simulating the relationship

TABLE 6.4

Potential Applications for PGA, PLA, and Their Copolymers

Products	Functions	Examples
Fixation device	Fix bone fragments, stabilize small joints, repair soft tissue tears, reattach ligaments and tendons	Screw, plate, pin, rod, anchor, staple and tack
Bone substitute	Fill voids fracture defects, spinal defects or fusion	Resorbable bone substitute
Ligaments and tendons	Repair, augment and/or replace ligaments or tendons	Resorbable ligaments or augmentation devices, suture
Cartilage implant	Replace and regenerate cartilage	Resorbable cartilage implants
Drug delivery system	Carrying a variety of drug classes to provide prolonged medication	Microspheres, microcapsules, membrane, tablet, and other implants

between the effective modulus of fiber and microfibrillar for various natural fibers [70]. For bast and fruit fibers

$$Y_c = Y_f x \cos^2\theta,$$

where Y_c and Y_f are the axial Young's modulus of fiber and microfibril, and θ is the microfibrillar angle. For all types of natural fibers

$$Y_f = W_c Y_c \cos^2\theta + W_{nc} Y_{nc},$$

where Y_f is the effective modulus, Y_c is the modulus of crystalline region, and W_c and W_{nc} are the weight fractions of the crystalline region and noncrystalline region, respectively.

On the other hand, disadvantages and limitations of natural fibers used as reinforcement in composites are their lack of interfacial adhesion, moisture absorption due to their hydrophilic characteristic, low processing temperature below 200°C, low reproducibility, and low dimensional stability. It is obvious that the interfacial properties of fiber and matrix play an important role in the physical and mechanical properties of the composites; therefore, surface modification is a critical key to enhance the interfacial bonding between natural fibers and polymers. Surface modifications of natural fibers can be done chemically and physically in order to control their susceptibility to moisture or modify the structure and properties. Chemical surface modifications include silane treatment, acetylation, isocyanate treatment, stearic acid treatment, graft copolymerization, duralin treatment, permanganate treatment, peroxide treatment, benzoylation treatment, meleated coupling. These modifications can be defined as chemical reactions between reactive constituents of the natural fibers and chemical reagents with or without a catalyst, to form a covalent bond between them. In addition, surface modifications with chemicals can change the fundamental properties of a fiber. Physical surface modifications include mercerization for disruption of hydrogen bonding in the network structure, in order to increase surface roughness and number of possible reaction sites for better mechanical interlocking; corona and cold plasma treatment for surface oxidation activation by different gases to increase surface crosslinking, surface energy, and reactive free radicals and groups [39,41].

One of the animal-based natural fibers made of silkworm silk has a long medical history of impacting bioengineering applications. Silk fibers have been used in biomedical applications particularly as sutures in which the silk fibroin filaments are usually coated with waxes or silicone to enhance material properties and reduce fraying. But, in fact, there remains much confusion about the usage of these fibers due to the absence of detailed characterization of the fibers, including the extent of extraction of the sericin coating, the chemical nature of wax-like coatings sometimes used, and many related processing factors. For example, the sericin glue-like

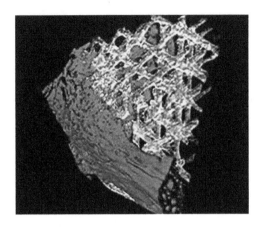

FIGURE 6.9
3-D tissue scaffold. (From www.eng.uab.edu/polymers.)

proteins are the major cause of adverse problems with biocompatibility and hypersensitivity to the silk fiber. The variability of source materials has raised potential concerns with this class of fibrous protein. Yet, silk fiber's knot strength, handling characteristics, and ability to lay on the tissue surface make it a popular suture in cardiovascular applications where bland tissue reactions are desirable for the coherence of the sutured structures [42].

A 3-D scaffold permits the in vitro cultivation of cell-polymer constructs that can be readily manipulated, shaped, and fixed to the defect site as shown in Figure 6.9 [43]. The matrix acts as the translator between the local environment (either in vitro or in vivo) and the developing tissue, aiding in the development of biologically viable functional tissue. However, during the 1960s to the early 1980s, the use of virgin silk had a negative impact on the general acceptance of this biomaterial from the surgical practitioner's perspective. For example, the reaction of the host tissue to silk fiber and the inflammatory potential are main concerns [44–46]. Recently, silk matrices are being rediscovered and reconsidered as potentially useful biomaterials for a range of applications in clinical repairs and in vitro as scaffolds for tissue engineering [47].

Silk fibers are susceptible to proteolytic degradation in vivo; silk-based implants like 3-D scaffolds can retain their integrity at implantation sites over a long period of time until the regenerated tissues fulfill the desired functions. Degradation rates are mainly dependent on health and physiological status of patients, mechanical environment of the implantation site, and types and dimensions of the silk fiber. The slow degradation rate of the silk fiber in vitro and in vivo makes it useful in biodegradable scaffolds for slow tissue ingrowths, since the biodegradable scaffolds must be able to retain their mechanical properties and support the growth of cells at the implantation site until the regenerated tissue is fully capable. The degradation rate should be matched with the rate of neo-tissue formation so as to compromise with the load-bearing capabilities of the tissue. Additionally, scaffold structures, including size and connectivity of pores, determine the transport of nutrients, metabolites, and regulatory molecules to and from cells. The matrix must support cell attachment, spreading, growth, and differentiation. Meinel et al. (2004) concentrated on cartilage tissue engineering with silk protein scaffolds, and the authors have identified and reported that silk scaffolds were particularly suitable for tissue engineering of cartilage, beginning with human mesenchymal stem cells (hMSCs), which are derived from bone marrow, mainly due to their high porosity, slow degradation, and structural integrity.

Recent research on silk fibers has focused on the evolution of a wire rope matrix for the development of autologous tissues-engineered anterior cruciate ligaments (ACLs) using the patient's own adult stem cells [50]. Silk fibroin offers versatility in matrix scaffold design for a number of tissue engineering needs in which the mechanical performance and biological interactions are the major factors for success with bone, ligaments, tendons, blood vessels, and cartilage. Silk fibroin can also be processed into foams, films, fibers, and meshes.

6.4 Biocomposites for Biomedical and Bioengineering Applications

In the past few decades, research and engineering interests have undergone a major transition from focusing on traditional monolithic materials to fiber-reinforced polymer-based materials because of their high strength-to-weight ratio, noncorrosive property, and high fracture toughness. The matrix material surrounds and supports the reinforcements by maintaining their relative positions to transfer loading. The reinforcements impart their unique and desired mechanical and physical properties to enhance the properties of the matrix. A conglomeration produces material properties that are unavailable from individual constituent materials, while a wide variety of matrices and reinforcements allow designers and engineers to select an optimum combination. Among all reinforcements, glass fiber is the most widely used for composite materials due to their low cost and fairly good mechanical properties. However, this fiber has serious drawbacks including nonrenewable, nonrecyclable, high energy consumption in the manufacturing process, health risks when inhaled, and its nonbiodegradable nature.

Within the past few years, there has been a dramatic increase in the use of natural fibers for composites. Recent advances in natural fiber development, genetic engineering, and composite science offer significant opportunities

TABLE 6.5

Mechanical Properties of Natural Fibers

Natural fibers	UTS (MPa)	Elongation at Break (%)	E (GPa)
Flax	300–1500	1.3–10	24–80
Jute	200–800	1.16–8	10–55
Sisal	80–840	2–25	9–38
Kenaf	295–1191	3.5	2.86
Abaca	980		7.31×10^{-4}
Pineapple	170–1627	2.4	60–82
Banana	529–914	3	27–32
Coir	106–175	14.21–49	4–6
Oil palm (empty fruit)	130–248	9.7–14	3.58
Oil palm (fruit)	80	17	
Ramie	348–938	1.2–8	44–128
Hemp	310–900	1.6–6	30–70
Wool	120–174	25–35	2.3–3.4
Spider silk	875–972	17–18	11–13
Cotton	264–800	3–8	5–12.6
Silkworm silk	500–740	4–20	5–17

for improved materials from renewable resources with an enhanced support for global sustainability [40]. A material that can be used for medical application must possess a number of specific characteristics. The most fundamental requirement related to biocompatibility is the absence of adverse effect to the host tissues. Therefore, those traditional composite structures with a nonbiocompatible matrix and reinforcements are substituted with bioengineered composites for biomedical applications. Tables 6.5 and 6.6 summarize the mechanical properties of hard and soft tissues [6], and there are several important factors that should be considered in selecting a material for biomedical applications [51] (Table 6.7).

6.4.1 Plant-Based Natural Fiber Biocomposites

Huda et al. (2006) have compared the properties of conventional reinforcement (chopped glass fibers) and natural reinforcement (recycled newspaper cellulose fiber) PLA composites—glass/PLA and RNCF/PLA composites, respectively. The reinforcing effect of RNCF on PLA matrix was demonstrated. The mechanical and thermal properties of PLA were enhanced with the reinforcement of cellulose fibers, which showed RNCF-reinforced PLA biocomposite can be used as substitute glass-reinforced composite in applications.

Other recent studies have been conducted on thermal and mechanical properties, biodegradability and different applications of natural- or biofiber-reinforced PLA biocomposites including abaca fiber–PLA biocomposite

TABLE 6.6

Mechanical Properties of Hard and Soft Tissues

Hard tissue	Young's Modulus (GPa)	Tensile Strength (MPa)
Cortical bone (longitudinal direction)	17.7	133.0
Cortical bone (transverse direction)	12.8	52.0
Cancellous bone	0.4	7.4
Enamel	84.3	10.0
Dentine	11.0	39.3

Soft tissue	Young's Modulus (MPa)	Tensile Strength (MPa)
Articular cartilage	10.5	27.5
Fibrocartilage	159.1	10.4
Ligament	303.0	29.5
Tendon	401.5	46.5
Skin	0.1–0.2	7.6
Arterial tissue (longitudinal direction)		0.1
Arterial tissue (transverse direction)		1.1
Intraocular lens	5.6	2.3

Note: Tissues show broad variation.

Source: Ramakrishna et al., *An Introduction to Biocomposites. Vol. 1 Series on Biomaterials and Bioengineering.* Imperial College Press, U.K., 2004.

[53,54], bamboo fiber–PLA biocomposite [55], lyocell fabric–PLA biocomposite [56], kenaf fiber–PLA biocomposite [57], PLA–corn starch (CS) biocomposite [58], microcrystalline cellulose (MCC)–PLA biocomposite [59], and collagen–PLA biocomposite [60]. These show the popularity of PLA and plant-based natural fibers used as biomaterials for tissue engineering applications.

6.4.2 Animal-Based Natural Fiber Biocomposites

In the biocomposite field, some researchers have investigated different combinations of silk-based composites. Annamaria et al. (1998) have discovered in their study that environmentally-friendly biodegradable polymers can be produced by blending silk sericin with other resins. Nomura et al. (1995) have noted that polyurethane foams incorporating sericin coating have excellent moisture-absorbing and -desorbing properties. Hatakeyama (1996) has also reported on the production of sericin-containing polyurethane with excellent mechanical and thermal properties. Sericin blends well with water-soluble polymers, especially with polyvinyl alcohol (PVA). Ishikawa et al. (1987) have investigated the fine structure and the physical properties of blended films made by sericin and PVA. Moreover, a recent patent has reported that a PVA/sericin cross-linked hydrogel membrane produced by using dimethyl

TABLE 6.7

Key Factors for the Selection of Materials for Biomedical Applications

	Descriptions		
	Chemical/ biological Characteristics	Physical Characteristics	Mechanical/ Structural Characteristics
1st Level material properties	Chemical composition (bulk and surface)	Density	Elastic modulus Shear modulus Poisson's ratio Yield strength Compressive strength
2nd Level material properties	Adhesion	Surface topology Texture Roughness	Hardness Flexural modulus Flexural strength
Specific functional requirements (based on applications)	Biofunctionality Bioinertia Bioactivity Biostability Biodegradability	Form and geometry Coefficient of thermal expansion Electrical conductivity Color, aesthetics Refractive index Opacity or translucency	Stiffness or rigidity Fracture toughness Fatigue strength Creep resistance Friction and wear resistance Adhesion strength Impact strength Proof stress Abrasion resistance
Processing and fabrication	Reproducibility, quality, sterilizability, packaging, secondary processability		
Characteristics of host	Tissue, organ, species, age, sex, race, health condition, activity, systemic response		
	Medical/surgical procedure, period of application/usage		
	Cost		

Source: Ramakrishna et al. Biomedical applications of polymer-composite materials: A review. *Composites Science and Technology.* 2001; 61(9): 1189–1224.

urea as the cross-linking agent has high strength, high moisture content, and durability for usage as a functional film [65].

Silk fibroin film has good dissolved oxygen permeability in wet state, but it is too brittle to be used on its own when in a dry state; whereas chitosan is a biocompatible and biodegradable material that can be easily shaped into films and fibers. Park et al. (1999) and Kweon et al. (2001) have introduced the innovation of silk fibroin/chitosan blends as potential biomedical composites since the crystallinity and mechanical properties of silk fibroin can be greatly enhanced with increasing chitosan content.

Another type of biodegradable composites is silk fibroin/alginate blend sponges [68]. For the biotechnological and biomedical fields, silk fibroin filament's reproducibility, environmental and biological compatibility, and nontoxicity benefit in many different clinical applications. As the collective properties, especially the mechanical properties of silk fibroin sponges in a dry state, are too weak to be suitable for wound dressing, they can be enhanced by blending silk fibroin films with other synthetic or natural polymers (e.g., polysaccharide–sodium alginate).

Furthermore, Katori and Kimura (2002) and Lee et al. (2005) have examined the effect of silk–poly(butylene succinate) (PBS) biocomposites. They found that the mechanical properties, including tensile strength, fracture toughness and impact resistance, and thermal stability of biocomposites can be greatly affected by their manufacturing processes. A good adhesion between the silk fibers and PBS matrix was also found through the observation and analysis by scanning electron microscopy (SEM) imaging.

6.5 Summary

Since the last decade, interest has been generated in the impact of natural fiber-reinforced biodegradable polymer biocomposites upon the applications of environmentally-friendly products, clinical implants, and the related equipment. This chapter explores the idea and develops the conceptual framework for the evolution of novel types of biocomposites. Data provided by preceding studies tend to support the potential reinforcing effect of natural fibers, both plant-based and animal-based ones, and this topic is in need of further study and clarification in different aspects.

Future research on biocomposites may be focused on the improvement for the interfacial bonding properties and stress transfer properties between natural fibers and polymers with the help of different surface modification treatments to produce high performance biocomposites. In addition, by mimicking the natural fibers with the use of various laboratory techniques, more reliable and reproducible material properties from these renewable resources can be obtained. Physical properties (appearance, size, shape, etc.), chemical structures (crystalline and amorphous proportions, etc.), and performance (mechanical and thermal properties, etc.) of silk fiber can be altered or tailor-made to obtain a more advanced biomaterial to suit different field of bioengineering applications. Biomaterial implants are developed at laboratory-evolution stage nowadays. However, the disparities among these biomaterials and conventional composite materials, the selection of fabrication methodologies including resin transfer molding (RTM), hand lay-up, autoclave molding, extrusion, and injection molding must be addressed to produce biocomposites with the desired material properties for various types of implants applicable to different body sites. Moreover, natural fiber-reinforced biodegradable polymer biocomposites are potential candidates to be developed as completely biodegradable and resorbable implants in the human body, both in vitro and in vivo. Bioresorbability, biodegradability, and biocompatibility tests are necessary for determining the potential of these types of biocomposites to be used for clinical applications as implants. Last but not least, variability of natural fibers in structures, dimensions, and material properties are in need of further study and clarification, both experimentally and theoretically, and even by making use of numerical modeling in order to provide a broader conceptualization of the use of natural fibers in bioengineering applications.

In conclusion, natural fibers are valuable and commercially available natural resources. With their supreme material and mechanical properties, the physical, mechanical, and thermal properties of biodegradable polymers can be enhanced to form reliable and potential biocomposites for biomedical and bioengineering applications.

Acknowledgment

This project was supported by The Hong Kong Polytechnic University Grant and Research Grant Council (PolyU G-U688).

References

1. Berger, S.A., Goldsmith, W., and Lewis, E.R., Eds. *Introduction to Bioengineering*, Oxford University Press, Oxford, 1996.
2. Williams, D.F. Definitions in biomaterials. *Proceedings of a Consensus Conference of the European Society for Biomaterials*, Chester, England, 1986; 4, Elsevier, New York.
3. Guelcher, S.A. and Hollinger, J.O. *An Introduction to Biomaterials*. CRC Press, Boca Raton, FL, 2006.
4. Bartolo, P. and Bidanda, B. (Ed.), *Bio-Materials and Prototyping Applications in Medicine*. Springer Science + Business Media, LLC, USA, 2008.
5. Teoh, S.H. (Ed.), *Engineering Materials for Biomedical Applications*. World Scientific Publishing Co. Pte. Ltd., Singapore, 2004.
6. Ramakrishna, S., Huang, Z.M., Kumar, G.V., Batchelor, A.W., and Mayer, J., *An Introduction to Biocomposites. Vol. 1 Series on Biomaterials and Bioengineering*. Imperial College Press, U.K., 2004.
7. Frenkel, S.R., Toolan, B.C., Menche, D., Pitman, M., and Pachence, J.M. Chondrocyte transplantation using a collagen bilayer matrix for cartilage repair. *Journal of Bone and Joint Surgery* 1997; 79B: 831–836.
8. Pachence, J.M. Collagen-based devices for soft tissue repair. *Journal of Biomaterials Research* 1996; 33: 35–40.
9. Pieper, J.S., Van der Kraan, P.M., Hafmans, T., Kamp, J., Buma, P., Van Susante, J.L.C., Van den Berg, W.B., Veerkamp, J.H., and Van Kuppevelt, T.H.. Crosslinked type II collagen matrices: preparation, characterization and potential for cartilage engineering. *Biomaterials* 2002; 23: 3183–3192.

10. Dausse, Y., Grossin, L., Miralles, G., Pelletier, S., Mainard, D., Hubert, P., Baptiste, D., Gillet, P., Dellacherie, E., Netter, P., and Payan, E.. Cartilage repair using new polysaccharidic biomaterials: macroscopic, histological and biochemical approaches in a rat model of cartilage defect. *Osteoarthritis Cartilage* 2003; 11: 16–28.

11. Hauselmann, H.J., Fernandes, R.J., Mok, S.S., Schmid, T.M., Block, J.A., Aydelotte, M.B., Kuettner, K.E., and Thonar, E.J. Phenotypic stability of bovine articular chondrocyte after long term culture in alginate beads. *Journal of Cell Science.* 1994; 107: 17–27.

12. Murphy, C.L. and Sambanis, A.. Effect of oxygen tension and alginate encapsulation on restoration of the differentiated phenotype of passaged chondrocytes. *Tissue Engineering.* 2001; 7: 791–803.

13. Paige, K.T., Cima, L.G., Yaremchuk, M.J., Schloo, B.L., Vacanti, J.P., and Vacanti, C.A.. De novo cartilage generation using calcium alginate-chondrocyte constructs. *Plastic and Reconstructive Surgery.* 1996; 97: 168–178.

14. Mauck, R.L., Soltz, M.A., Wang, C.C.B., Wong, D.D., Chao, P.H.G., Valhmu, W.B., Hung, C.T., and Ateshian, G.A. Functional tissue engineering of articular cartilage through dynamic loading of chondrocyte-seeded agarose gels. *Journal of Biomechanical Engineering* 2000; 122: 252–260.

15. Weisser, J., Rahfoth, B., Timmermann, A., Aigner, T., Brauer, R., and Von der Mark, K. Role of growth factors in rabbit articular cartilage repair by chondrocytes in agarose. *Osteoarthritis Cartilage.* 2001; 9A: 48–54.

16. Chenite, A., Chaput, C., Wang, D., Combes, C., Buschmann, M.D., Hoemann, C.D., Leroux, J.C., Atkinson, B.L., Binette, F., and Selmani, A. Novel injectable neutral solutions of chitosan form biodegradable gels in situ. *Biomaterials.* 2000; 21: 2155–2161.

17. Madihally, S.V. and Matthew, H.W.. Porous chitosan scaffolds for tissue engineering. *Biomaterials.* 1999; 20: 1133–1142.

18. Hendrickson, D.A., Nixon, A.J., Grande, D.A., Todhunter, R.J., Minor, R.M., Erb, H., and Lust, G. Chondrocyte-fibrin matrix transplant for resurfacing extensive articular cartilage defects. *Journal of Orthopaedic Research.* 1994; 12: 485–497.

19. Sims, C.D., Butler, P.E.M., Cao, Y.L., Casanova, R., Randolph, M.A., Black, A., Vacanti, C.A., and Yaremchuk, M.J. Tissue engineered neo-cartilage using plasma derived polymer substrates and chondrocytes. *Plastic and Reconstructive Surgery.* 1998; 101: 1580–1585.

20. Hollister, S.J., Maddox, R.D., and Taboas, J.M. Optimal design and fabrication of scaffolds to mimic tissue properties and satisfy biological constraints. *Biomaterials* 2002; 23(20): 4095–4103.

21. Katstra, W.E., Palazzolo, R.D., Rowe, C.E., Giritlioglu, B., Teung, P., and Cima, M.J. Oral dosage forms fabricated by three dimensional printing. *Journal of Control Release* 2000; 66: 1–9.

22. Gunatillake, P.A. and Adhikari, R. Biodegradable synthetic polymers for tissue engineering. *European Cells and Materials.* 2003; 5: 1–16.

23. Yang, S., Leong, K.F., Du, Z., and Chua, C.K.. The design of scaffolds for use in tissue engineering. Part I. Traditional factors. *Tissue Engineering.* 2001; 7(6): 679–689.

24. Li, W.J., Danielson, K.G., Alexander, P.G., and Tuan, R.S. Biological response of chondrocytes cultured in three-dimensional nanofibrous poly(e-caprolactone) scaffolds. *Journal of Biomedical Materials Research Part A.* 2003; 67A(4): 1105–1114.

25. Coombes, A.G.A., Rizzi, S.C., Williamson, M., Barralet, J.E., Downes, S., and Wallace, W.A.. Precipitation casting of polycaprolactone for applications in tissue engineering and drug delivery. *Biomaterials*. 2004; 25: 315–325.

26. Williams, J.M., Adewunmi, A., Schek, R.M., Flanagan, C.L., Krebsbach, P.H., Feinberg, S.E., Hollister, S.J., and Das, S. Bone tissue engineering using polycaprolactone scaffolds fabricated via selective laser sintering. *Biomaterials*. 2005; 26: 4817–4827.

27. Kweon, H.Y., Yoo, M.K., Park, I.K., Kim, T.H., Lee, H.C., Lee, H.S., Oh, J.S., Akaike, T., and Cho, C.S. A novel degradable polycaprolactone networks for tissue engineering. *Biomaterials*. 2003; 24: 801–808.

28. Kose, G.T., Kenar, H., Hasirci, N., and Hasirci, V. Macroporous poly(3-hydroxybutyrate-co-3-hydroxyvalerate) matrices for bone tissue engineering. *Biomaterials*. 2003; 24: 1949–1958.

29. Ito, Y., Hasuda, H., Kamitakahara, M., Ohtsuki, C., Tanihara, M., Kang, I.K., and Kwon, O.H. A composite of hydroxyapatite with electrospun biodegradable nanofibers as a tissue engineering material. *Journal of Bioscience and Bioengineering*. 2005; 100(1): 43–49.

30. Deng, Y., Lin, X.S., Zheng, Z., Deng, J.G., Chen, J.C., Ma, H., and Chen, G.Q. Poly(hydroxybutyrate-co-hydroxyhexanoate) promoted production of extracellular matrix of articular cartilage chondrocytes in vitro. *Biomaterials*. 2003; 24: 4273–4281.

31. Ishaug-Riley, S.L., Okun, L.E., Prado, G., Applegate, M.A., and Ratcliffe, A. Human articular chondrocyte adhesion and proliferation on synthetic biodegradable polymer films. *Biomaterials* 1999; 20(23–24): 2245–2256.

32. Taboas, J.M., Maddox, R.D., Krebsbach, P.H., and Hollister, S.J. Indirect solid free form fabrication of local and global porous, biomimetic and composite 3D polymer-ceramic scaffolds. *Biomaterials* 2003; 24: 181–194.

33. Liao, S.S., Cui, F.Z., Zhang, W., and Feng, Q.L. Hierarchically biomimetic bone scaffold materials: nano-HA/collagen/PLA composite. *Journal of Biomedical Materials Research. Part B, Applied Biomaterials* 2004; 69(2): 158–165.

34. Montjovent, M.O., Mathieu, L., Hinz, B., Applegate, L.L., Bourban, P.E., Zambelli, P.Y., Manson, J.A., and Pioletti, D. Biocompatibility of bioresorbable poly(L-lactic acid) composite scaffolds obtained by supercritical gas foaming with human fetal bone cells. *Tissue Engineering* 2005; 11(11/12): 1640–1649.

35. Fisher, J.P., Vehof, J.W.M., Dean, D., van der Waerden, J.P.C.M., Holland, T.A., Mikos, A.G., and Jansen, J.A. Soft and hard tissue response to photocrosslinked poly(propylene fumarate) scaffolds in a rabbit model. *Journal of Biomedical Materials Research Part A*. 2001; 59(3): 547–556.

36. Temenoff, J.S. and Mikos, A.G. Injectable biodegradable materials for orthopedic tissue engineering. *Biomaterials*. 2000; 21: 2405–2412.

37. Wang, S., Lu, L., and Yaszemski, M.J. Bone-tissue-engineering material poly(propylene fumarate): correlation between molecular weight, chain dimensions, and physical properties. *Biomacromolecules*. 2006; 7: 1976–1982.

38. Lee, S.M., Cho, D.H., Park, W.H., Lee, S.G., Han, S.O., and Drzal, L.T. Novel silk/poly(butylenes succinate) biocomposites: the effect of short fiber content on their mechanical and thermal properties. *Composites Science and Technology*. 2005; 65: 647–657.

39. Chand, N. and Fahim, M. *Tribology of Natural Fiber Polymer Composites*. Woodhead Publishing Limited, Cambridge, U.K., 2008.

40. Mohanty, A.K., Misra, M., and Hinrichsen, G. Biofibers, biodegradable polymers and biocomposites: an overview. *Macromolecular Materials and Engineering.* 2000; 276/277: 1–24.
41. Bogoeva-Gaceva, G., Avella, M., Malinconico, M., Buzarovska, A., Grozdanov, A., Gentile, G., and Errico, M.E. Natural fiber eco-composites. *Polymer Composites.* 2007; 28: 98–107.
42. Postlethwait, R.W. Tissue reaction to surgical sutures. In: Dumphy, J.E., Van Winkle, W., Eds. *Repair and Regeneration.* New York: McGraw-Hill, 1969; p. 263–285.
43. Freed, L.E., Grande, D.A., Emmanual, J., Marquis, J.C., Lingbin, Z., and Langer, R. Joint resurfacing using allograft chondrocytes and synthetic biodegradable polymer scaffolds. *Journal of Biomedical Materials Research* 1994; 28: 891–899.
44. Morrow, F.A., Kogan, S.J., Freed, S.Z., and Laufman, H. In vivo comparison of polyglycolic acid, chromic catgut and silk in tissue of the genitourinary tract: an experimental study of tissue retrieval and calculogenesis. *Journal of Urology* 1974; 112: 655–658.
45. Nebel, L., Rosenberg, G., Tobias, B., and Nathan, H. Autograft suture in peripheral nerves. *European Surgical Research* 1977; 9: 224–234.
46. Peleg, H., Rao, U.N., and Emrich, L.J. An experimental comparison of suture materials for tracheal and bronchial anastomoses. *Journal of Thoracic and Cardiovascular Surgery* 1986; 34: 384–388.
47. Minoura, N., Aiba, S., Gotoh, Y., Tsukada, M., and Imai, Y. Attachment and growth of cultured fibroblast cells on silk protein matrices. *Journal of Biomedical Materials Research* 1995; 29: 1215–1221.
48. www.eng.uab.edu/polymers/
49. Meinel, L., Hofmann, S., Karageorgiou, V., Zichner, L., Langer, R., Kaplan, D., and Vunjak-Novakovic, G. Engineering cartilage-like tissue using human mesenchymal stem cells and silk protein scaffolds. *Biotechnology and Bioengineering* 2004; 88: 379–391.
50. Altman, G.H., Horan, R.L., Lu, H., Moreau, J., Martin, I., Richmond, J.C., and Kaplan, D.L. Silk matrix for tissue engineered anterior cruciate ligaments. *Biomaterials* 2002; 23: 4131–4141.
51. Ramakrishna, S., Mayer, J., Wintermantel, E., and Leong, K.W. Biomedical applications of polymer-composite materials: a review. *Composites Science and Technology.* 2001; 61(9): 1189–1224.
52. Huda, M.S., Drzal, L.T., Mohanty, A.K., and Misra, M. Chopped glass and recycled newspaper as reinforcement fibers in injection molded poly(lactic acid) (PLA) composites: a comparative study. *Composites Science and Technology* 2006; 66: 1813–1824.
53. Shibata, M., Ozawa, K., Teramoto, N., Yosomiya, R., and Takeishi, H.. Biocomposites made from short abaca fiber and biodegradable polyesters. *Macromolecular Materials and Engineering* 2003; 288: 35–43.
54. Teramoto, N., Urata, K, Ozawa, K., and Shibata, M.. Biodegradation of aliphatic polyester composites reinforced by abaca fiber. *Polymer Degradation and Stability* 2004; 86: 401–409.
55. Lee, S.H. and Wang, S. Biodegradable polymers/bamboo fiber biocomposite with bio-based coupling agent. *Composites Part A: Applied Science and Manufacturing* 2006; 37: 80–91.

56. Shibata, M., Oyamada, S., Kobayashi, S.I., and Yaginuma, D. Mechanical properties and biodegradability of green composites based on biodegradable polyesters and lyocell fabric. *Journal of Applied Polymer Science* 2004; 92: 3857–3863.

57. Serizawa, S., Inoue, K., and Iji, M. Kenaf-fiber-reinforced poly(lactic acid) used for electronic products. *Journal of Applied Polymer Science* 2006; 100: 618–624.

58. Ohkita, T. and Lee, S.H. Thermal degradation and biodegradability of poly(lactic acid)/corn starch biocomposites. *Journal of Applied Polymer Science* 2006; 100: 3009–3017.

59. Mathew, A.P., Oksman, K., and Sain, M. Mechanical properties of biodegradable composites from poly lactic acid (PLA) and microcrystalline cellulose (MCC). *Journal of Applied Polymer Science* 2005; 97: 2014–2025.

60. Dunn, M.G., Bellincampi, L.D., Tria, A.J., and Zawadsky, J.P. Preliminary development of a collagen-PLA composite for ACL reconstruction. *Journal of Applied Polymer Science* 1997; 63(11): 1423–1428.

61. Annamaria, S., Maria, R., Tullia, M., Silvio, S., and Orio, C. The microbial degradation of silk: a laboratory investigation. *International Biodeterioration and Biodegradation* 1998; 42: 203–211.

62. Nomura, M., Iwasa, Y., and Araya, H.. Moisture absorbing and desorbing polyurethane foam and its production. Japan Patent 07-292240A, 1995.

63. Hatakeyama, H. Biodegradable sericin-containing polyurethane and its production. Japan Patent 08-012738A, 1996.

64. Ishikawa, H., Nagura, M., and Tsuchiya, Y. Fine structure and physical properties of blend film compose of silk sericin and poly(vinyl alcohol). *Sen'I Gakkaishi* 1987; 43: 283–287.

65. Nakamura, K. and Koga, Y. Sericin-containing polymeric hydrous gel and method for producing the same. Japan Patent 2001-106794A, 2001.

66. Park, S.J., Lee, K.Y., Ha, W.S., and Park, S.Y.. Structural changes and their effect on mechanical properties of silk fibroin/chitosan blends. *Journal of Applied Polymer Science* 1999; 74: 2571–2575.

67. Kweon, H., Ha, H.C., Um, I.C., and Park, Y.H. Physical properties of silk fibroin/chitosan blend films. *Journal of Applied Polymer Science* 2001; 80: 928–934.

68. Lee, K.G., Kweon, H.Y., Yeo, J.H., Woo, S.O., Lee, J.H., and Park, Y.H. Structural and physical properties of silk fibroin/alginate blend sponges. *Journal of Applied Polymer Science* 2004; 93: 2174–2179.

69. Katori, S. and Kimura, T. Injection moulding of silk fiber reinforced biodegradable composites. In Brebbia, C.A. and de Wilde, W.P. (Ed.), *High Performance Structures and Composites*. WIT Press, Boston, 2002; Section 2: pp. 97–105.

70. Satyanarayana, K.G. and Wypych, F. Characterization of natural fibers. In Fakirov, S. and Bhattacharyya, D. (Ed.), *Handbook of Engineering Biopolymers: Homopolymers, Blends, and Composites*. Hanser Gardener Publications, Inc., Cincinnati, OH, 2007; Chapter 1: pp. 3–48.

7

Flame Retardant Polymer Nanocomposites

Jihua Gou and Yong Tang

University of Central Florida

CONTENTS

7.1 Introduction

Polymeric materials are commonly used in everyday life such as in construction, electrical and electronics components, homes, and transportation. Due to their intrinsic chemical composition and molecular structures, they have poor resistance to fire and, therefore, flame retardants (FRs) are usually incorporated into them to reduce flammability. The flammability behavior of polymers is defined based on the burning rate (rate of heat release), flame spread (flame and pyrolysis), ignition characteristics (ignition time and temperature), smoke production, and toxicity [1]. The most important fire hazards can be classified as heat, smoke, and toxic gases. The heat release rate is a much more critical parameter than the ignitability and the smoke toxicity, because a high heat release rate will result in a fast ignition and flame spread. Furthermore, it controls the fire intensity and available time to escape for fire victims.

Smoke production is another important fire hazard. Dark smoke disorients people and hinders them from escaping from a burning building. In addition, firefighters have severe problems in rescuing people in low or no light conditions. The toxicity of gases is mainly controlled by the content of carbon monoxide released by fires. Carbon monoxide is responsible for over 90% of deaths due to fire. There are about 5,000 people killed by fire in Europe and more than 4,000 people in the United States every year. The direct property losses are roughly equivalent to 0.2% of the gross domestic product (GDP) with the total costs of fires around 1% of the GDP [2]. Therefore, there is an urgent need to develop FR materials to minimize these hazards. The use of FRs to reduce flammability and the production of smoke and toxic products in fire has become an important aspect of research, development and the application of new polymeric materials. FR systems are intended to inhibit or stop the combustion of polymers. The modes of action of FR systems can be divided into two categories. One is a cooling action that forms a protective barrier or consists of a diluting fuel. The other is a chemical causing a reaction in the condensed or gas phase. The FRs can interfere with the polymer combustion process at different stages such as heating, decomposition, ignition, and flame spread [1,3].

In terms of physical action, the decomposition of FR additives (i.e., $Al[OH]_3$ and $Mg[OH]_2$) induces a temperature drop by water vapor, which involves the cooling of the surrounding medium below the ignition temperature. In addition, the formation of inert gases (H_2O, CO_2, NH_3, etc.) from the decomposition of FRs will dilute the flammable gases and limit the possibility of ignition. Furthermore, FR additives promote the formation of a protective layer on the surface of a condensed phase. It is recognized that such a protective layer acts as a barrier to the flammable gas transport from the underlying polymer to the burning surface. It slows down the heat and mass transfer between the burning surface and the underlying polymer at the elevated temperature [3–5].

Flame retardancy through the chemical modification of fire process can occur in either the gaseous or condensed phase. The free-radical mechanism of the combustion process can be stopped by FR additives or their degradation products such as Cl or Br radicals. These radicals can modify the combustion reaction pathway, leading to significant decrease in the temperature and the exothermicity of reaction, and decrease in flammable volatilizations. In the condensed phase, there are two types of chemical reactions. First, the rupture of polymer chains can be accelerated by FR, which will cause pronounced polymer drips and thus moves away from the flame action zone. Second, the FR can cause the formation of a carbonized layer or a ceramic-like structure at the polymer surface. This layer acts as a physical insulating layer and protects the underlying substrate.

FRs can be classified in two categories:

1. Additive FRs. They are incorporated into the polymer matrices and do not react with polymers. These FRs are usually mineral fillers, hybrids, or organic, or inorganic compounds, such as clay, ammonium polyphosphate (APP), halogen compounds, etc. Depending on their nature, these FRs can act chemically or physically in the solid, liquid, or gas phase.
2. Reactive FRs. Unlike additive FRs, the reactive FR species are incorporated into polymers via copolymerization or other type of chemical modification (i.e., grafting). The halogen-free FRs contain elements such as P, Si, B, and N [1,3].

7.2 Laboratory Fire Testing

The flammability behavior of polymer composites can be characterized by their ignitability, flame spread rate, and heat release rate. There are small, intermediate, or bench-scale flammability test methods in both industry and academia.

7.2.1 Limited Oxygen Index (LOI)

The minimum concentration of oxygen will just support the flaming combustion in a flowing mixture of oxygen and nitrogen. The LOI test is subject to ASTM D2863 or international standard (ISO 4589). The value of the LOI is defined as the minimum oxygen concentration $[O_2]$ in the oxygen–nitrogen mixture $[O_2–N_2]$ that either maintains the flame combustion of the material for 3 min or consumes a length of 5 cm of the specimen. A specimen is positioned vertically in a transparent test column, and a mixture of oxygen and nitrogen is forced upward through the column. The specimen is ignited at the top with a burner. The LOI can be calculated as:

$$LOI = 100 \cdot [O_2]/[O_2] + [N_2]$$

7.2.2 UL 94 V

UL 94 V is used to measure the ignitability and flame spread of vertical specimens exposed to a small flame. This test is an international standard (IEC 60695-11-10) for small flames (50 W). Five different classifications are included in this test according to the criteria.

7.2.3 Cone Calorimeter Test

The cone calorimeter test has been regarded as the most significant bench-scale instrument to evaluate the fire properties of polymer composites. The principle is based on the measurement of the oxygen consumed during combustion. It has been shown that most fuels generate approximately 13.1 MJ of energy per kilogram of oxygen consumed [6]. Therefore, the heat release rate (HRR) is based on the fact that the oxygen consumed during combustion is proportional to the heat released. Data collected from this bench-scale real fire test include the peak heat release rate (PHRR), time to ignition, total heat release, and concentration of CO_2 and CO.

7.2.4 Microscale Combustion Calorimetry (MCC)

MCC is one of the most-effective bench-scale methods to investigate the combustion property of polymeric materials. This technique, now classified as ASTM D7309-07, was originally developed by the U.S. Federal Aviation Administration (FAA) [7]. The MCC uses the oxygen consumption calorimeter to measure the rate and amount of heat produced by a complete combustion of fuel gases generated during the controlled pyrolysis of a milligram-sized sample. The specific HRR (W/g) is obtained through dividing dQ/dt by an initial mass of sample. A derived quantity, the heat release capacity or HRC (J/g K), is obtained through dividing the maximum value of the specific heat release rate by the heating rate during test. The heat release capacity is a molecular level flammability parameter, which is a good predictor of flame resistance and fire behavior if only research quantities of polymers are available for testing.

7.3 FR Polymer Nanocomposites

In the past decades, many academic and industrial researchers have paid considerable attention to polymer nanocomposites. The polymer nanocomposites often exhibit remarkable property improvements when compared to both virgin polymers and conventional filled systems [3,8]. The enhanced performance includes mechanical properties, thermal properties, gas barrier performance, and flammability reduction. For FR applications, commonly used nanoparticles include POSS, nanoclay, carbon nanotubes (CNTs), and carbon nanofibers (CNFs). Compared to conventional fire retardants, they are environmentally friendly and highly efficient, while simultaneously improving mechanical, electrical, and thermal properties.

7.3.1 Layered Silicates

Polymer-layered silicate nanocomposites (PLSNs) exhibit a low flammability associated with other enhanced properties [8,9]. Typically, the PHHR is suppressed by 40% to 80% on a cone calorimeter test. However, the UL 94 and LOI results of PLSN are usually poor. The addition of nanoclay into the polymer matrix creates a protective layer during combustion. Upon heating, the viscosity of the molten polymer layered silicate nanocomposites decreases and facilitates the migration of the clay platelets to the surface. The accumulation of clay on the surface of the materials acts as a protective barrier that limits heat and mass transfer between the burning surface and underlying polymer at the elevated temperature. Tang and Lewin conducted a fundamental study of the migration mechanism [10–12]. It was reported that the clay migrated to the surface of the nanocomposite during the annealing process. The migration of clay to the surface is mainly due to the difference in surface free energy between the polymer and the polymer–clay blends, the temperature and viscosity gradients during the directional heating, and the gases and gas bubbles caused by the decomposition of the surfactant and polymer. The structure of the nanocomposites is changed to a microcomposite when the surfactant or polymer decomposes. The surfactant and polymer molecules are removed from the clay gallery, and the pristine clay or aggregates are formed. It was concluded that the migration of clay in the polymer matrix depends on (1) the annealing atmosphere, (2) the amount of exfoliated clay moieties, (3) the nature of the surfactant and polymer matrix, and (4) the rate of decomposition of the surfactant and polymeric matrix.

Extensive research work has been conducted on the flame retardancy of the nanocomposites with various kinds of polymer matrices such as polypropylene, polyethylene [13,14], poly (ethylene-co-vinyl acetate) EVA [15], polyurethane [16], polyamide-6 (PA6) [17], polystyrene [18], polycarbonate, acrylonitrile–butadiene–styrene [19–21], and epoxy resin [22]. The clay nanocomposites exhibited a lower flammability and approximate 75% reduction in HRR. However, the UL94 for polymer nanocomposites is usually no-rating. For most of nanocomposites, the carbonaceous char yield is very limited, and the total heat release is not decreased compared to pure polymer. The ignition time is either not increased or may be shorter.

7.3.2 Polyhedral Oligomeric Silsequioxanes

Polyhedral oligomeric silsesquioxane (POSS)-based nanocomposites have received considerable attention due to the unique three-dimensional structure of POSS macromonomes [23,24]. The addition of POSS into polymers brings about significant property improvements such as enhanced thermal stability, hardness and oxidative resistance, mechanical properties, and decreased flammability. Silsesquioxanes are a family of compounds with the general formula $(RSiO_{1.5})_n$, where R is a hydrogen or organic group such as alkyl, aryl, or their derivatives. The diameter of the POSS cage is 0.45 nm. However, it reaches the dimensions of 1–3 nm when the organic addition is included. POSS is a candidate material for the design of polymer nanocomposites. At ~300°C to 350°C, Si-C bond cleavages occur and a ceramic-like char is formed, which acts as an insulating barrier and protects the underlying substrate. Bourbigot and coworkers investigated a polyurethane (PU) containing POSS by mass loss calorimetry [25]. Compared with the pristine PU, a large reduction of PHRR was observed. It was pointed out that the formation of an intumescent structure made an efficient insulating layer at the surface of the substrate, limiting heat and mass transfer and then decreasing the heat release rate. Fina et al. incorporated the metal POSS into the polypropylene (PP) matrix via melt blending [26]. The addition of metal-containing POSS significantly improved the thermo-oxidative stability of PP; the maximum weight loss rate temperature was raised up to 60°C. Song et al. incorporated a trisilanophenyl POSS (TPOSS) into the polycarbonate matrix [27]. The combustion behavior of the PC–POSS composites has been evaluated by cone calorimeter. The results showed that the addition of TPOSS into the PC matrix significantly reduced the PHRR from 492 kW/m² to 267 kW/m². Wilkie et al. studied the trisilanolphenyl POSS in the PMMA matrix [28]. However, they did not observe any flame retardancy improvement as measured from cone calorimeter tests. These results indicate that the flammability of polymer–POSS nanocomposites depends on the type of polymer matrix, structure of POSS, and weight fraction and dispersion of POSS in the polymer matrix.

7.3.3 Carbon Nanotubes

Since CNTs were first synthesized in 1991 [29], they have generated a great deal of interest owing to their extraordinary properties. Kashiwagi et al. [30] investigated the mode of action of CNTs (multiwalled and single-walled) as FRs in a polymeric matrix [31–33]. The FR performance was enhanced through the formation of a relatively uniform network-structured floccule layer covering the entire sample surface without any cracks or gaps. It was found that the fire performance is sensitive to the fraction of CNTs in the matrix, the extent of dispersion, and the type of polymeric matrix. In the PMMA–SWNT nanocomposites, the introduction of 0.1% mass fraction of SWNT did not significantly decrease the HRR of PMMA. The significant reduction in HRR was achieved by 0.5 wt% CNT in the PMMA matrix. The HRR decreased approximately by 60% compared with the control sample. The extent of reduction is much larger than with clay. In the clay–polymer nanocomposites, the HRR was suppressed about 30%, even at 3 wt% loading. However, when the concentration of SWNT increased to 1 wt%, the HRR increased due to the poor dispersion of SWNT in the PMMA matrix. Dubois et al. reported that the addition of crushed MWNT into EVA exhibited the reduction in PHRR and the increase in the time to ignition [34]. They attributed it to the chemical reactivity of the radical species at the surface of crushed MWNT during combustion. The FR effectiveness of CNTs in terms of the HRR reduction is better than nanoclay particles per unit mass base at a low concentration.

7.4 FR Carbon Nanopaper and Nanocomposites

A variety of nanoparticles have been used to fabricate fire retardant nano-composites with improved mechanical and functional properties. These nanoparticles provide an alternative to conventional fire retardants to reduce the flammability of polymer resin. Compared to conventional fire retardants, they are environmentally friendly, highly efficient, and capable of imparting other properties to polymers [1,3,35]. Previous studies reported the fire retardant performance of carbon nanotubes in the polymers [32,33]. The efficiency of carbon nanotubes is very high, and a much lower loading level of nanotubes (1–2 wt%) can form a relatively uniform network without any cracks or gaps. Very few studies reported the fire retardant performance of carbon nanofibers (CNFs) due to the low efficiency in forming the nanofiber network when directly mixing them with the polymers [36]. In order to achieve an equivalent fire retardant effectiveness of carbon nanotubes, a higher loading level of CNFs is required. However, CNFs are much more cost-effective than carbon nanotubes.

Fiber-reinforced polymer matrix composites have become attractive engineering materials to replace conventional metallic materials in many important sectors of industry such as aircraft, naval construction, ships, home and office construction, and offshore structures. However, these materials are susceptible to combustion and fire damage due to their chemical structures. Although the nanoclay, carbon nanotubes, CNF, and POSS have been used to enhance the fire retardancy of polymer nanocomposites, their applications have been limited in fiber-reinforced polymer matrix composites. Very few studies have been reported on the fire retardant application of nanopartciles or additives in fiber-reinforced polymer matrix composites [37–41].

A unique approach to making FR nanopaper has been reported in References 42–45. A hybrid nanopaper consisting of CNF and POSS or cloisite Na+ clay has been fabricated. The as-prepared hybrid nanopaper was then coated on the surface of glass fiber (GF)-reinforced polymer matrix composites through the resin transfer molding (RTM) process. CNF was used in the form of a free-standing CNF sheet just like traditional fiber mats for fiber-reinforced polymer composites. The preformed CNF networks within the nanopaper played an important role in the fire retardancy of composites by changing the patterns of melt flow, heat transmission, and vapor fuel transportation. The morphology, fire performance, thermal stability of the nanocomposites materials, and FR mechanism were investigated.

(a)   (b)

FIGURE 7.1
SEM images of pristine CNF nanopaper: (a) top surface and (b) bottom surface.

7.4.1 Morphologies of Nanopapers and Nanocomposites

The pristine CNF nanopaper has a density of 0.4 g/cm^3 and a thickness of 0.25 mm. The top surface of the nanopaper shows that it mainly consists of well-dispersed nanofibers with larger length and smaller diameter. The bottom surface of the nanopaper contains shorter, smaller nanofibers and catalyst particles due to the deposition during filtration process, as shown in Figure 7.1.

A series of CNF–POSS hybrid nanopapers were manufactured, as shown in Table 7.1. The morphologies of hybrid nanopapers are shown in Figure 7.2. It can be seen that CNFs were uniformly dispersed and highly entangled with each other within the nanopaper. At the bottom surface of the CNF–POSS nanopaper, CNFs are more compact compared with those at the top surface. This was due to the deposition of POSS particles during the high pressure filtration process. The POSS particles were not observed in the nanopapers with CNF/POSS weight ratios of 9/1 and 1/1. However, the EDAX confirmed the existence of POSS particles in the nanopaper, as shown in Figure 7.3. When the POSS weight percentage increased to 70 wt%, the POSS particles

TABLE 7.1

Composition of Hybrid Nanopapers and Nanocomposites

Nanocomposite Sample ID	Contents (wt %)			Weight Ratios in the Nanopaper
	GF	Resin	Nanopaper	
CNF-POSS-1 laminates	52.73	45.65	1.62	CNF/POSS = 9/1
CNF-POSS-2 laminates	51.89	45.36	2.75	CNF/POSS = 1/1
CNF-POSS-3 laminates	51.96	43.76	4.28	CNF/POSS = 3/7
CNF-POSS-4 laminates	50.58	36.20	13.20	CNF/POSS = 1/9

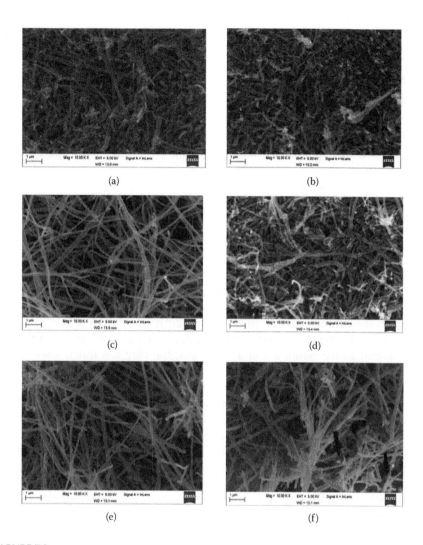

FIGURE 7.2
SEM images of hybrid nanopapers: (a) top surface of CNF-POSS-1 nanopaper, (b) bottom surface of CNF-POSS-1nanopaper, (c) top surface of CNF-POSS-2 nanopaper, (d) bottom surface of CNF-POSS-3 nanopaper, (e) top surface of CNF-POSS-3 nanopaper, (f) bottom surface of CNF-POSS-3 nanopaper, (g) top surface of CNF-POSS-4 nanopaper, (h) bottom surface of CNF-POSS-4 nanopaper.

(g)   (h)

FIGURE 7.2
(Continued)

were observed at the bottom surface of the nanopaper, and the EDAX showed that the silicon content increased dramatically (see Figure 7.3(c)). In the nanopaper with CNF/POSS weight ratio of 1/9, the POSS particles were observed even on the top surface (see Figure 7.2(g)). The large POSS aggregates were found on the bottom surface (see Figure 7.2(h)). The EDAX shows that the intensity of silicon is much higher than that of carbon (see Figure 7.3(d)).

Compared to the nanotube and nanofiber, the clay has much lower thermal conductivity and nanometer size in one dimension, which makes it a barrier to mass transmission. Cloisite Na$^+$ clays at 0.05 wt% and 0.20 wt% were added to CNF to prepare a hybrid nanopaper, labeled as CCNS5 and CCNS20, respectively (see Table 7.2). The CCNS5 has a thickness of 0.27 mm and a density of 0.411 g/cm^3; the CCNS20 has a thickness of 0.28 mm and a density of 0.502 g/cm^3. The addition of cloisite Na$^+$ clay increased the density of the nanopaper. Figure 7.4 shows the top and bottom surfaces of CCNS5 and CCNS20, respectively. In the CNF–clay nanopaper, clay particles were not observed on the top or bottom surface. It was also noted that it took much more time to filter out the CNF–clay solution compared to the CNF–POSS solution during the papermaking process. This indicates that the permeability of the CNF–clay nanopaper is much lower than the permeability of the CNF–POSS nanopaper. The cross section of CCNS20 indicates the close packing between the nanofibers and the clay particles in the nanopaper, as shown in Figure 7.4(e).

Figure 7.5 displays the morphologies of hybrid nanopapers infused with the polyester resin. It is clearly shown that CNFs were homogeneously dispersed in the nanopapers. The polyester resin completely penetrated the nanopaper throughout the thickness and fully filled the pores of the nanopaper. The POSS and clay particles cannot be identified in the SEM images, probably due to the chemical reaction during resin curing. In the hybrid POSS–CNF nanopaper, the vinyl groups on the POSS particles (initiated by the curing agent) may cross-link with the polyester resin, which increased the interaction between the resin and CNF.

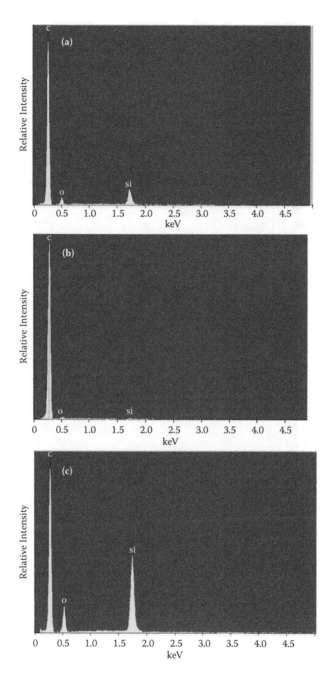

FIGURE 7.3
EDAX on the bottom surface of hybrid nanopapers: (a) CNF–POSS-1 nanopaper, (b) CNF–POSS-2 nanopaper, (c) CNF–POSS-3 nanopaper, (d) CNF–POSS-4 nanopaper, and (e) CNF–clay-1 nanopaper.

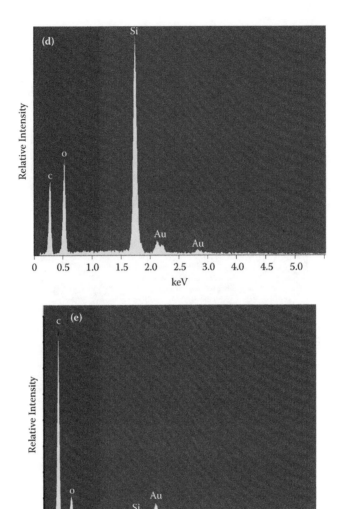

FIGURE 7.3
(Continued)

TABLE 7.2

Compositions of Different Composite Panels

	Composition (wt%)			
Sample ID	Glass Fiber	Polyester Resin	Cloisite Na+ Clay	Carbon Nanofiber
GE-CCNS20	53.22	45.68	0.20	0.88
GE-CCNS5	53.41	45.66	0.05	0.87
GE-CNS-1	54.60	44.96	0	0.84
GE-CNS-0	53.12	46.88	0	0
GE-CNS-1-20A	53.50	44.72	0.91	0.87
GE-CNS-0-20A	53.02	46.04	0.94	0

7.4.2 Thermal Stability of Nanopapers and Nanocomposites

The thermal stability of hybrid nanopapers is related to the fire performance of composites. TGA is one of the most widely used techniques for a rapid evaluation of the thermal stability of various polymer composites. The TGA results of the pristine CNF, POSS, and polyester resin are shown in Figure 7.6 and Table 7.3. The decomposition of pristine POSS occurred in the range of 210°C to 220°C with about 2 wt% residue at 600°C, as shown in Table 7.2 and Figure 7.6. The polyester resin began to decompose at about 274°C, and the residue was about 5.5 wt% at 600°C. The weight loss of CNFs at 600°C is about 1.5 wt%, which indicates that the thermal stability of CNFs was very high. The thermal properties of hybrid nanopapers are shown in Figure 7.7 and Table 7.2. The decomposition of the CNF–POSS nanopaper occurred in the range of 190°C to 206°C, as shown in Table 7.2. The residues decreased with the POSS weight fraction in the nanopaper. When the weight fraction of POSS increased from 10% (CNF–POSS-1 nanopaper) to 90% (CNF–POSS-4 nanopaper), the residue decreased approximately from 96 wt% to 15 wt%. For the nanopapers infused with the polyester resin, the decomposition temperature increased significantly, occurring in the range of 222°C to 285°C. The increase in degradation temperature is due to the higher thermal stability of the polyester resin (see Figure 7.6). During resin curing, the POSS particles could cross-link with the unsaturated polyester resin molecules. It is interesting to note that when the weight fraction of POSS in the nanopaper is more than 70 wt% (CNF–POSS-3 nanopaper with resin), the residue is more than that of the nanopaper without the resin, as shown in Table 7.2. Similar results of residue were obtained for CNF–POSS-4 nanopaper with the resin (90 wt% of POSS). This is probably caused by the increase in cross-link density between POSS particles and unsaturated polyester resin molecules. Figure 7.7(e) shows TGA results of the pristine clay and CNF–clay nanopaper. Their weight losses at 600°C are about 2.5 wt% and 4.2 wt%, respectively. This reveals that clay and the CNF–clay nanopaper are significantly more stable than POSS and the CNF–POSS nanopaper, respectively.

FIGURE 7.4
SEM images: (a) top surface of CCNS5, (b) bottom surface CCNS5, (c) top surface of CCNS 20, (d) bottom surface of CCNS20, and (e) cross section of CCNS20.

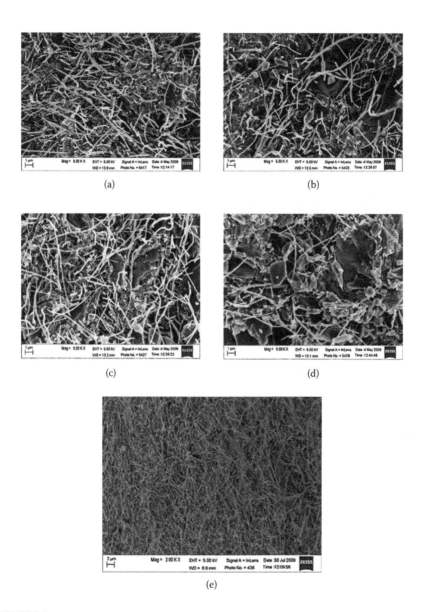

(a)

(b)

(c)

(d)

(e)

FIGURE 7.5
SEM images of hybrid nanopapers infused with the polyester resin: (a) CNF–POSS-1, (b) CNF–POSS-2 nanopaper with resin, (c) CNF–POSS-3 nanopaper with resin, (d) CNF–POSS-4 nanopaper with resin, and (e) CCNS20 nanopaper infused with the polyester resin.

TABLE 7.3

Thermal Properties of Particles, Nanopapers, and Nanocomposites from TGA

TGA Test Samples	$T_{-5wt\%}$ (°C)[a]	Char (wt%) @ 600°C
POSS	213.2	2.05
Resin	273.6	5.52
CNF	/	98.5
Clay	/	97.5
CCNS20 nanopaper	/	95.8
CNF–POSS-1 nanopaper	199.0	92.3
CNF–POSS-1 nanopaper with resin[b]	285.6	19.6
CNF–POSS-2 nanopaper	190.1	64.6
CNF–POSS-2 nanopaper with resin[b]	257.2	19.3
CNF–POSS-3 nanopaper	210.1	33.0
CNF–POSS-3 nanopaper with resin[b]	231.1	34.7
CNF–POSS-4 nanopaper	206.5	15.2
CNF–POSS-4 nanopaper with resin[b]	222.1	16.4

[a] Temperature at 5 wt% weight loss.
[b] There is no glass fiber in these samples.

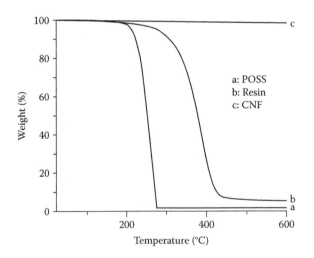

FIGURE 7.6
TGA results of POSS, polyester resin, and CNF.

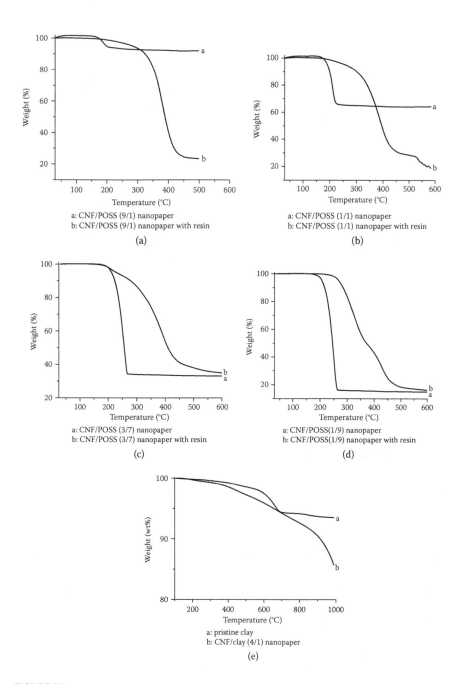

FIGURE 7.7
TGA results of testing hybrid nanopapers and clays: (a) CNF–POSS (9/1) nanopaper (with and without resin, (b) CNF–POSS (1/1) nanopaper (with and without resin), (c) CNF–POSS (3/7) nanopaper (with and without resin), (d) CNF–POSS (1/9) nanopaper (with and without resin), and (e) pristine clay and CNF-clay (4/1) nanopaper.

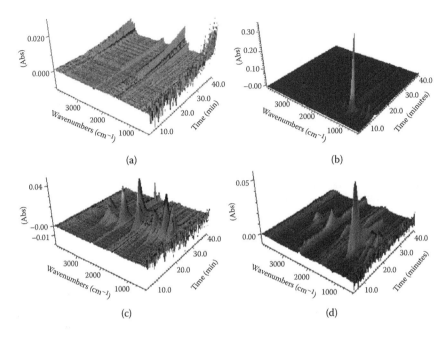

FIGURE 7.8

Three-dimensional FTIR spectra of evolved gases produced by the pyrolysis of CNF–POSS nanopaper with and without resin: (a) CNF–POSS-1 nanopaper, (b) CNF–POSS-4 nanopaper, (c) CNF–POSS-1 nanopaper with resin, and (d) CNF–POSS-4 nanopaper with resin.

7.4.3 Real-Time FTIR of Nanopapers and Nanocomposites

The changes in chemical structures of hybrid nanopapers with and without the polyester resin during thermal degradation were monitored by real-time FTIR. The TGA-FTIR technique can directly identify volatilized products to help explain thermal degradation mechanisms. The characterization of the volatilized products by the TGA-FTIR instrument was conducted in a nitrogen atmosphere. The 3D FTIR spectra of the evolved gases of hybrid nanopapers with and without the resin at different time are shown in Figure 7.8. The FTIR spectra of hybrid nanopapers without the resin (CNF–POSS-1 nanopaper and CNF–POSS-4 nanopaper) are shown in Figure 7.8(a) and (b). The peaks in the regions of around 3400–4000 cm^{-1}, 2250–2400 cm^{-1}, 2700–3000 cm^{-1}, and 1110 cm^{-1} were observed. For the nanopapers infused with the resin (Figure 7.8c,d), the peaks in the regions of around 3400–4000 cm^{-1}, 2700–3070 cm^{-1}, 2250–2400 cm^{-1}, 1400–1867cm^{-1}, and 907–1262 cm^{-1} are shown in Figures 7.8(c) and 7.6(d). The spectra fit well to the reported FTIR features of gas precuts such as H$_2$O (3400–4000 cm^{-1}), aromatic C–H (3060 cm^{-1}, stretching vibration), hydrocarbons

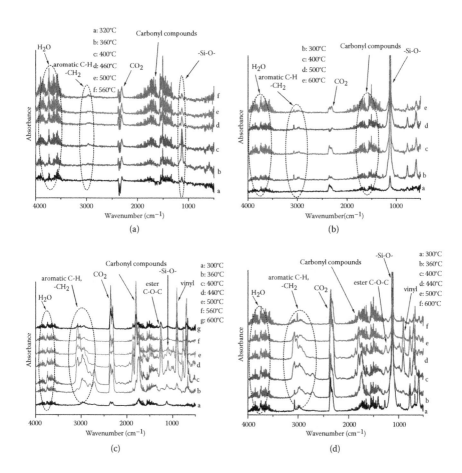

FIGURE 7.9
FTIR spectra of volatilized products at different temperature during thermal degradation: (a) CNF–POSS-1 nanopaper, (b) CNF–POSS-4 nanopaper, (c) CNF-POSS-1 nanopaper with resin, and (d) CNF–POSS-4 nanopaper with resin.

(2800–3000 cm^{-1}), CO_2 (2250–2400 cm^{-1}), carbonyl compounds (1400–1867 cm^{-1}), and ester C–O–C (1260–1000 cm^{-1}, asymmetric and symmetric stretching vibrations), and Si–O (stretching vibration at 1110 cm^{-1}) [46–48].

The characteristic spectra of the selected nanopapers (without the resin) at different temperatures (from 240°C to 600°C) are shown in Figure 7.9(a) and (b). Here, CNF–POSS-1 and POSS-4 nanopapers were selected for their vast differences in CNF–POSS ratios. Some small molecular gaseous products such as CO_2 and H_2O are easily identified by their characteristic absorbance

(CO_2 at 2357 cm^{-1}, and H_2O at 3400–4000 cm^{-1}) [46]. The alkyl and carbonyl compounds are observed at around 2980 cm^{-1} and 1740 cm^{-1} from 320°C to 560°C, respectively. The aromatic C-H is also observed at around 3060 cm^{-1}, which was probably due to the cross-linking between volatiles from the decomposition of POSS particles. Figure 7.9(b) shows that the –Si–O– peak was detected at about 200°C in the CNF–POSS-4 nanopaper, and the Si-O absorption band was detected at about 320°C in the CNF-POSS-1 nanopaper. The nanopapers are varied from the CNF-dominated structure to the POSS-dominated structure, depending on their weight ratios (CNF/POSS = 9/1 to 1/9). During the TGA test, POSS particles were relatively easy to migrate and sublimate [23], and therefore the IR detected the Si-O absorption earlier and at lower temperatures in the nanopaper with CNF/POSS = 1/9. The intensity of Si-O increased with the temperature up to 400°C, where 98% weight loss of POSS was observed as shown in Figure 7.6.

The FTIR spectra of pyrolyzed products of the nanopapers infused with the polyester resin are shown in Figure 7.9(c) and (d) again for CNF–POSS-1 and -4 nanopaper with the resin. The most important absorption bands in the IR spectrum of such nanocomposites are associated with Si-O (1110 cm^{-1}, stretching vibration), carbonyl compounds C=O (1744 cm^{-1}, stretching vibration), ester C–O–O (1262 and 1002 cm^{-1}, asymmetric and symmetric stretching vibrations), vinyl C=C (910 cm^{-1} stretching vibration), aliphatic C–H (2950–2800 cm^{-1}, asymmetric and symmetric stretching vibration), carboxylic groups (3000–2600 cm^{-1}), aromatic C–H (3069 cm^{-1}, stretching vibration), and CO_2 at 2367 cm^{-1}[46–48]. The early volatilized products at 300°C were identified as CO_2, Si-O, aliphatic, and carbonyl moieties, which could come from the decomposition of functional groups of POSS and the scission of the polyester resin. It is interesting to note that, in the CNF–POSS-4 nanopaper with the resin, the aromatic C–H exhibited a strong absorption at 3069 cm^{-1} at 300°C. However, in the CNF–POSS-1 nanopaper with the resin, this peak was not observed. The intensity of the preceding pyrolyzed products from the CNF–POSS-4 nanopaper with the resin is much greater than those from the CNF–POSS-1 nanopaper with the resin. The difference in the intensity of pyrolyzed products indicates that the formation of char materials in the CNF–POSS-4 nanopaper with the resin was earlier at a lower temperature than the CNF–POSS-1 nanopaper with the resin. Heated to 360°C, the new ester C–O–C (1263 cm^{-1}) and vinyl C=C (910 cm^{-1}) absorption band caused by the decomposition of the polyester resin were observed in both nanocomposites. This decomposition process was likely to be completed at 500°C, leading to the decrease in the intensity of C–O–C and CH_2=C- absorption bands. Based on the combined results of TGA and IR studies, it can be concluded that POSS particles decomposed early, generating volatilized aliphatic and Si-O moieties. The polyester resin molecules are then decomposed, resulting in the volatilization of aromatic and carbonyl moieties and vinyl compounds.

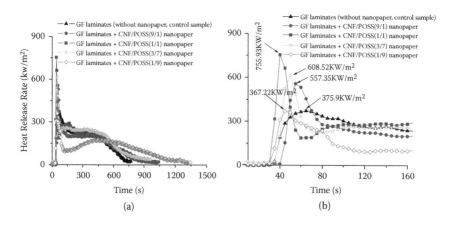

FIGURE 7.10
HRRs of composite laminates coated with hybrid nanopapers: (a) 0 to 1500 s, (b) 0 to 160 s.

7.4.4 Cone Calorimeter Test of Composite Laminates Coated with Nanopapers

The cone calorimetric results for control sample and composite laminates coated with hybrid nanopapers are shown in Figures 7.10 and 7.11. After the ignition, the HRR increases rapidly to maximum values and then decreases gradually. The PHRR of control sample (without the nanopaper on the surface) is 375.9 kW/m². For composite laminates coated with the CNF–POSS nanopaper, the PHRR increases dramatically. The PHRR reaches up to 755.93 kW/m² when the weight fraction of POSS is 50 wt% in the nanopaper. After that, the PHRR decreases with the weight fraction of POSS. The PHRR drops to 367.2 kW/m² with 90 wt% of POSS particles in the nanopaper.

Figure 7.11 shows the HRR curves of test samples. Compared to GE-CNS-0, GE-CCNS20 has a significant lower HRR, and GE-CNS-1 has a slightly increased HRR. There is a 60.5% reduction in the HRR peak for GE-CCNS20 and a 17.0% increase for GE-CNS-1. The GE-CCNS5 has a similar HRR peak with GE-CCNS20. Obviously, the existence of clay reduces the HRRs of GE-CCNS5 and GE-CCNS20. Meanwhile, an increase in HRR peak of GE-CNS-1 indicates that the nanopaper does not improve the fire performance of composites. Compared to GE-CNS-0 and GE-CNS-1, both GE-CNS-0-20A and GE-CNS-1-20A have their HRR curves slightly decreased. However, their HRRs are still much higher than those of GE-CCNS20 and GE-CCNS5, even though they have 0.20 wt% and 0.05 wt% of cloisite 20A clay compounded with the polymer resin, respectively. Therefore, the method of directly compounding clay with the resin has lower efficiency to improve the fire

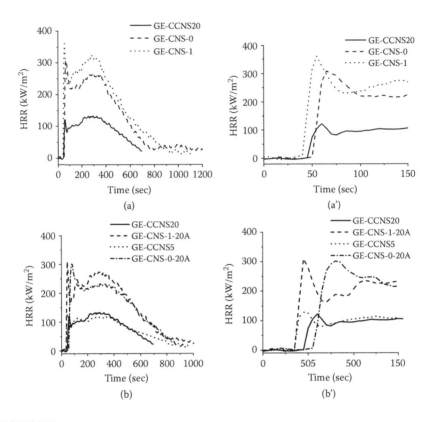

FIGURE 7.11
Heat release rate curves of: (a, a') GE-CCNS20, GE-CNS-0, and GE-CNS-1, and (b, b') GE-CCNS20, GE-CCNS5, GE-CNS-0-20A, and GE-CNS-1-20A.

performance of composites compared to the use of clay–carbon nanofiber hybrid nanopaper.

Although GE-CCNS20 and GE-CCNS5 have similar HRR curves, GE-CCNS20 has much longer ignition time than GE-CCNS5. The shorter ignition time of GE-CCNS5 is partly due to the complete penetration of CCNS5 by the polyester resin. A very thin layer of resin on the surface of GE-CCNS5 was formed owing to the penetration of the resin into the nanopaper. However, there is no obvious resin layer on the surface of GE-CCNS20 due to the lower permeability of the nanopaper with 20 wt% of cloisite Na$^+$ clay. Compared to CNF, the polyester resin has lower decomposition and ignition temperature. The excess resin on the sample surface decreases the ignition time of GE-CCNS5. Hence, the loading level of clay should be high enough to delay the ignition, which is important to prevent composites from quickly catching fire. A similar situation takes place in GE-CNS-1 and GE-CNS-1-20A. Their ignition times are obviously shorter than those of GE-CNS-0 and GE-CNS-0-20A.

The CNF–clay nanopaper has better fire retardant performance than the CNF–POSS nanopapers [42,44]. The thermal stability of clay is much higher than POSS particles and polymer resin, as shown in Table 7.2 and Figure 7.6. The flammable volatilizations come from the decomposition of the resin instead of clay. Clay has a layered structure, serving as barrier to mass and heat transfer. In the CNF–clay nanopaper, the clay layers will attenuate the concentration of the heat absorbed by CNFs from the external heat flux. The heating up to the ignition point of the polymer resin would need relatively more time. Furthermore, the CNF networks and the clay layers in the nanopaper are reinforced with each other, providing a lower permeability and therefore limiting the diffusion of O_2 to the substrate and the volatilization gases to the surface to feed the flame. However, POSS has a cage-like structure with organic groups attaching to each corner of the cage. There is no barrier effect for such a cage-like structure. In the CNF–POSS nanopaper, large aggregates of POSS particles were observed as shown in Figure 7.2(h), which could not decrease the permeability of the nanopaper.

7.4.5 Morphologies of Char Materials after Cone Calorimeter Tests

The analysis of char materials after cone calorimeter tests can provide an insight into FR mechanisms. Figure 7.12 shows the morphologies of the residues of composite laminates coated with hybrid nanopapers. The morphologies of the char vary in appearance, shape, and the extent of compaction. In Figure 7.12(a), only CNFs and a few floccules were observed, which is probably due to the pyrolysis of POSS particles and the polyester resin. In Figure 7.12(b), a number of cracks or channels were observed on the surface. The quality of the char corresponds to the fire retardant performance of composite laminates. From the cone calorimeter test, the composite sample had much higher PHRR. With the increase of weight fraction of POSS particles up to 70 wt%, more floccules can be seen in the sample, and the CNFs are much shorter. For the samples containing 90 wt% POSS particles, the char exhibits different morphology, as shown in Figure 7.12(d). There are no floccules on the surface. The entire surface is covered by a ceramic-like sintered layer, and nearly no CNFs or cracks can be seen. This char layer could serve as a barrier, decreasing the heat and mass transfer rates and the heat release rate, and slowing the combustion and degradation of the polymer resin. Figure 7.12(e) shows the morphology of the char of composite laminates coated with the CNF–clay nanopaper. In the case of CCNS5, the CNF on the surface is almost burned off. It seems that only clay particles remain on the top surface. However, the morphology changes in CCNS20. In this sample, it can be seen that the clay particles and CNF networks are reinforced with each other and formed a compact protective layer. This layer could effectively limit the transmission of the heat and the supply of the flammable fuel, resulting in a dramatic decrease in PHRR.

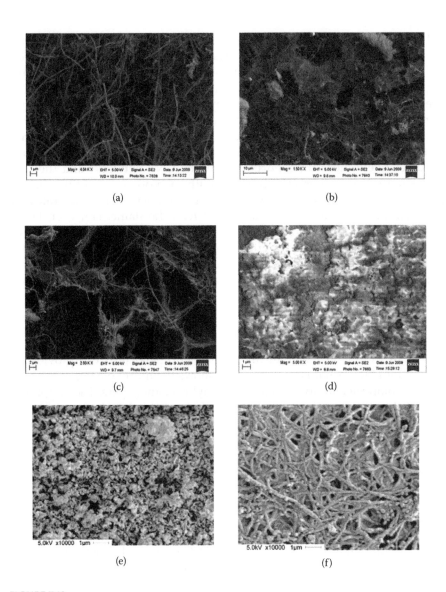

(a)

(b)

(c)

(d)

(e)

(f)

FIGURE 7.12
SEM of char materials of composite laminates coated with hybrid nanopapers after cone test:
(a) CNF–POSS-1 laminates, (b) CNF–POSS-2 laminates, (c) CNF–POSS-3 laminates, (d) CNF–
POSS-4 laminates, (e) CCNS5 laminates, and (f) CCNS20 laminates.

7.5 Conclusions

A unique concept of making hybrid POSS or clay–CNF nanopapers for fire retardant application has been explored. The hybrid nanopapers were coated to composite laminates through RTM processes. The SEM images show that CNFs were dispersed homogeneously in the hybrid nanopapers. POSS particles at a high content of 90 wt% formed aggregates. The SEM-EDAX verified the increase in Si intensity with the content of POSS particles in the nanopaper. In the CNF–clay nanopaper, the clay layers and CNFs were entangled with each other, resulting in a lower permeability of the nanopaper. The TGA test results show that the addition of POSS particles decreased the thermal stability of the nanopapers, whereas the addition of clay in the paper increased the thermal stability. The real-time TGA-FTIR indicates that the Si moiety was produced at a low temperature. The main products of the thermal decomposition of the nanopaper infused with the polyester resin are CO_2, H_2O, aromatic, vinyl, hydrocarbon moieties, etc. It was found that the heat release rates of composite laminates coated with the CNF–POSS nanopaper did not decrease as expected, even at 90 wt% POSS in the nanopaper. The addition of POSS particles to the nanopaper had an adverse effect, which may arise from the lower thermal stability of POSS particles and the migration and sublimation of POSS to the flame.

A significant decrease in PHRR was observed for composite laminates coated with CNF–clay nanopaper. The PHRR was decreased by 67% compared to that of control sample. Unlike POSS particles, the thermal stability of clay is much higher. The char materials obtained from cone calorimeter tests varied in appearance and shape, depending on the composition of composite samples. The cracks or channels were observed in the char of POSS–CNF nanopapers. The layer of clay and CNF networks formed a compact protective layer, which could effectively limit the transmission of heat and the supply of flammable fuels. Therefore, the clay is effective in decreasing the PHRRs of composite laminates coated with CNF–clay nanopaper.

References

1. Laoutid, F., Bonnaud, L., Alexandre, M., Loper-cuesta, J., and Dubois, Ph., 2009. New prospects in flame retardant polymer materials: from fundamentals to nanocomposites. *Mater. Sci. Eng.* R 63: 100–25.
2. Stevens, G. C., *Conference Flame Retardants 2000*, London, Elsevier Applied Science.
3. Bourbigot, S. and Duquesne, S., 2007. Fire retardant polymers: recent developments and opportunities. *J. Mater. Chem.* 17: 2283–2300.
4. Troitzsch, J., 1990. *International Plastics Flammability Handbook*, 2nd ed., Hanser Publishers, Munich.

5. Horrocks, A. R. and Price, D., 2001. *Fire Retardant Materials*, CRC Press, Boca Raton, FL.
6. Huggett, C., 1980. Estimation of rate of heat release by means of oxygen consumption measurements. *Fire Mater.* 4: 61–5.
7. Lyon, R. E. and Walters, R. N., 2004. Pyrolysis combustion flow calorimetry. *J. Anal. Appl. Pyrolysis* 71: 27–46.
8. Paul, D. R. and Robeson, L. M., 2008. Polymer nanotechnology: nanocomposites. *Polymer* 49: 3187–3204.
9. Paulidou, S. and Papaspyrides, C. D., 2008. A review on polymer-layered silicate nanocomposites. *Progress in Polymer Science* 33: 1119–98.
10. Lewin, M. and Tang, Y., 2008. Oxidation-migration cycle in polypropylene-based nanocomposites. *Macromolecules* 41: 13–17.
11. Tang, Y. and Lewin, M., 2007. Maleated polypropylene OMMT nanocomposites: annealing, structural changes, exfoliated and migration. *Polym Degrad Stab* 92: 53–60.
12. Tang, Y. and Lewin, M., 2008. New aspects of migration and flame retardancy in polymer nanocomposites. *Polym Degrad Stab* 93: 1986–95.
13. Tang, Y., Hu, Y., Li, B., Liu, L., Wang, Z., Chen, Z., and Fan, W., 2004. Polypropylene/montmorillonite nanocomposites and intumescent, flame-retardant montmorillonite synergism in polypropylene nanocomposites. *J. Polym. Sci. Part A: Polym. Chem.* 42: 6163–73.
14. Zhang, J. and Wilkie, C. A., 2006. Polyethylene and polypropylene nanocomposites based on polymerically-modified clay containing alkylstyrene units. *Polymer* 47: 5736–43.
15. Zanetti, M., Camino, G., Thomann, R., and Muelhaupt, R., 2001. Synthesis and thermal behaviour of layered silicate/EVA nanocomposites. *Polymer* 42: 4501– 07.
16. Xiong, J., Liu, Y., Yang, X., and Wang, X., 2004. Thermal and mechanical properties of polyurethane/montmorillonite. *Polym Degrad Stab* 86: 549–55.
17. Kashiwagi, T., Harris Jr, R. H., Zhang, X., Briber, R. M., Cipriano, B. H., Raghavan, S. R., Awad, W. H., and Shields, J. R., 2004. Flame retardant mechanism of polyamid 6-clay nanocomposites. *Polymer* 45: 881–91.
18. Zheng, X., Jiang, D. and Wilkie, C. A., 2006. Polystyrene nanocomposites based on anoligomerically-modified clay containing maleic anhydride. *Polym Degrad Stab* 91: 108–13.
19. Wang, S., Hu, Y., Wang, Z., Tang, Y., Chen, Z., and Fan, W., 2003. Synthesis and characterization of polycarbonate/ABS/montmorillonite nanocomposites. *Polym Degrad Stab* 80: 157–61.
20. Wang, S., Hu, Y., Zong, R., Tang, Y., Chen, Z., and Fan, W., 2004. Preparation and characterization of flame retardant. *Applied Clay Science* 25: 49–55.
21. Pawlowski, K. H. and Schartel, B., 2008. Flame retardancy mechanisms of aryl phosphates in combination with boehmite in bisphenol A polycarbonate/acrylonitrile-butadiene-styrene blends. *Polym Degrad Stab* 93: 657–67.
22. Morgan, A. B. and Galaska, M., 2008. Microcombustion calorimetry as a tool for screening flame retardancy in epoxy. *Polym. Adv. Technol.* 19: 530–46.
23. Tang, Y. and Lewin, M., 2009. Migration and surface modification in polypropylene (PP)/polyhedral oligomeric silsequioxane (POSS) nanocomposites. *Polym. Adv. Technol.* 20: 1–15
24. Dasari, A., Yu, Z., Mai, Y., Cai, G., and Song, H., 2009. Roles of graphite oxide, clay and POSS during the combustion of polyamide 6. *Polymer* 50: 1577–87.

25. Bourbigot, S., Turf, T., Bellayer, S., and Duquesne, S., 2009. Polyhedral oligomeric silsesquioxane as flame retardant for thermoplastic polyurethane. *Polym Degrad Stab* 94: 1230–37.

26. Fina, A., Abbenhuis, H. C. L., Tabuani, D., Frache, A., and Camino, G., 2006. Polypropylene metal functionalised POSS nanocomposites: a study by thermogravimetric analysis. *Polym Degrad Stab* 91: 1064–70.

27. Song, L., He, Q., Hu, Y., Chen, H., and Liu, L., 2008. Study on thermal degradation and combustion behaviors of PC/POSS hybrids. *Polym Degrad Stab* 93: 627–39.

28. Jash, P. and Wilkie, C. A., 2005. Effects of surfactants on the thermal and fire properties of poly (methyl methacrylate)/clay nanocomposites. *Polym Degrad Stab* 88: 401–06.

29. Iijima, S., 1991. Helical microtubules of graphitic carbon. *Nature* 354: 56–58.

30. Kashiwagi, T., Du, F., Winey, K. I., Groth, K., Shields, J., Bellayer, S., Kim, H., and Douglas, J., 2005. Flammability properties of polymer nanocomposites with single-walled carbon nanotubes: effects of nanotube dispersion and concentration. *Polymer* 46: 471–81.

31. Beyer, G., 2002. Carbon nanotubes as flame retardants for polymers. *Fire Mater.* 26: 291–93.

32. Kashiwagi, T., Grulke, E., Hilding, J., Groth, K., Harris, R., Butler, K., Shields, J., Kharchenko, S., and Douglas, J., 2004. Thermal and flammability properties of polypropylene/carbon nanotube nanocomposites. *Polymer* 45:4227–39.

33. Kashiwagi, T., Du, F. M., Douglas, J. F., Winey, K. I., Harris, R. H., and Shields, J. R., 2005. Nanoparticle networks reduce the flammability of polymer nanocomposites. Nat. Mater 4: 928–33.

34. Bredeau, S., Boggioni, L., Bertini, F., Tritto, I., Monteverde, F., Alexandred, M., and Dubois, P., 2007. Ethylene-norbornene copolymerization by carbon nanotube-supported metallocene catalysis: generation of high-performance polyolefinic nanocomposites. *Macromol. Rapid Commun.* 28: 822–27.

35. Gilman, J. W., 1999. Flammability and thermal stability studies of polymer layered-silicate (clay) nanocomposites. *Appl. Clay Sci.* 15: 31–49.

36. Zammarano, M., Kramer, R. H., Harris, R., Ohlemiller, T. J., Shields, J. R., Rahatekar, S. S., Lacerda, S., and Gilman, J. W., 2008. Flammability reduction of flexible polyurethane foams via carbon nanofiber network formation. *Polymers for Advanced Technologies* 19: 588–595.

37. Kandola, B. K., Horrocks, A. R., Myle, P., and Blair, D., 2003. Mechanical performance of heat/fire damaged novel flame retardant glass-reinforced epoxy composites. *Composites: Part A* 34: 863–873.

38. Kandola, B. K., Akonda, M. H, and Horrocks, A. R., 2005. Use of high-performance fibres and intumescents as char promoters in glass-reinforced polyester composites. *Polym. Degrad. Stab.* 88: 123–129.

39. Baljinder, K., Kandola, A., Horrocks, R., Myler, P., and Blair, D., 2003. New developments in flame retardancy of glass-reinforced epoxy composites. *J. Appl. Polym. Sci.* 88: 2511–2521.

40. Kandola, B. K., Horrocks, A. R, Myler, P., and Blair, D., 2002. The effect of intumescents on the burning behaviour of polyester-resin-containing composites. *Composites: Part A* 33: 805–817.

41. Sorathia, U., Ness, J., and Blum, M., 1999. Fire safety of composites in the US navy. *Composites: Part A* 30:707–713.

42. Zhao, Z., Gou, J., Bietto, S., Ibeh, C., and Hui, D., 2009. Fire retardancy of clay/carbon nanofiber hybrid sheet in fiber reinforced polymer composites. *Composites Science and Technology* 69: 2081–2087.
43. Gou, J., Tang, Y., Zhuge, J., Zhao, Z., Chen, R., Hui, D., and Ibeh, C., 2010. Fire performance of composite laminates embedded with multi-ply carbon nanofiber sheets. *Composites: Part B.* 41: 176–181.
44. Tang, Y., Zhuge, J., Gou, J., Chen, R., Ibeh, C., and Hu, Y., 2010. Morphology, thermal stability, and flammability of polymer matrix composites coated with hybrid nanopapers. *Polym. Adv. Technol.* In press.
45. Gou, J., Tang, Y., Liang, Fei., Zhao, Z., Firsich, D., and Fielding, J., 2010. Carbon nanofiber paper for lightning strike protection of composite materials. *Composites: Part B.* 41: 192–198.
46. Xing, W.Y., Hu, Y., Song, L., Chen, X. L, Zhang, P., and Ni, J. X., 2009. Thermal degradation and combustion of a novel UV curable coating containing phosphorus. *Polym. Degrd. Stab.* 94: 1176–82.
47. Bhandare, P. S., Lee, B. K., and Krishnan, K., 1997. Study of pyrolysis and incineration of disposable plastics using combined TG/FTIR technique. *J Thermal Anal.* 49: 361–66.
48. Balabanovich, A. I., Pospiech, D., Korwitz, A., Haubler, L., and Harnisch, C., 2009. Pyrolysis study of phosphorus-containing aliphatic aromatic polyester and its nanocomposites with layered silicates. *Polym. Degrd. Stab.* 94: 355–64.

8

Polyurethane Nanocomposite Coatings for Aeronautical Applications

Hua-Xin Peng

University of Bristol

CONTENTS

8.1 Introduction

There is evidence that lacquer technology was developed in China many thousands of years ago, and these lacquers produced very hard, durable finishes that are both beautiful and very resistant to damage by water, acid, alkali, or abrasion. According to archaeological digs, ancient applications included coffins, plates, musical instruments, and furniture. In the lacquering of the Chinese musical instrument gu-qin, the lacquer was mixed with deer horn powder or ceramic powder to give it more strength as protection against scratches. The varnish resin used was derived from the varnish tree, unlike the modern chemical resins used nowadays.

Modern paint systems have been undergoing continual change with the fast development of modern industries. After the first powered wood-and-cloth plane flight in 1903, the aerospace industry has grown to be one of the most exciting, diverse, and fast-paced fields [1]. Especially in the last few decades, the aerospace industry is moving in a major way toward advanced composite materials [2,3] (see Figure 8.1). Composite structures have found commercial applications where high performance is required since the early 1970s, as an alternative to traditional metal-based aircraft components and structures [4]. Composite materials offer a wide range of benefits over traditional materials commonly used in aircraft, including improved chemical stability and good fire resistance, as well as their obvious weight and cost savings not only during manufacture but also during the lifetime of the aircraft [5].

There are disadvantages in composite materials, however. An obvious one is their considerably higher electrical resistance compared with aluminum, with limited applications such as electromagnetic shielding, circuits, antennas, and lighting strike protection [6]. Similarly, the lower thermal conductivity of composites poses a threat to composite fuel tanks and affects de-icing. Composite materials are also less resistant to impact, which is a major issue for aircraft manufacturers [7]. Another problem regarding composites is moisture absorption and aging over time [8–10].

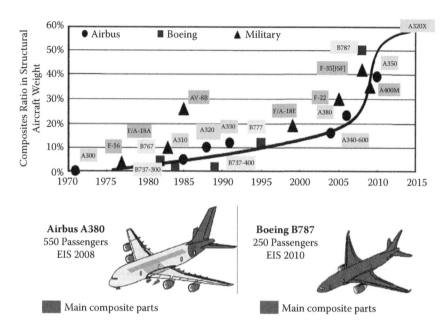

FIGURE 8.1
An estimate of the composites ratios in structural aircraft. (From Bamford, F. GKN Aerospace—A Vision of the Future. April 2007.)

Typically, the protective painting systems for aerospace include a primer and topcoat (see Figure 8.2). The primer coat is porous and brittle, with no durability but providing corrosion protection and adhesion of the coating to the pretreated substrate surface. The topcoat is supposed to be flexible, have a matte finish, and be washable and chemically resistant with optimum exterior durability [11]. Because of high-performance requirements in terms of durability, flexibility, and corrosion protection, aerospace and military camouflage coatings have traditionally used two-component polyurethane systems [12–15]. However, many of the materials that are most critical

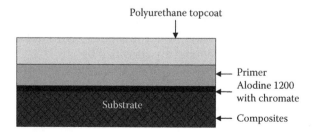

FIGURE 8.2
Typical aerospace coating system.

to the function of current aircraft coating systems are limited in that they are either not maintained for the life of the coating system (e.g., cladding), or have limited mission function [16]. Nanocomposites obtained by dispersing nano-scaled fillers of various functionalities within polymeric matrices are of particular importance because they offer a great opportunity to improve and tailor the properties of the polymer [17], which allows the adjustment of mechanical (wear, friction), chemical (corrosion, permeation, temperature insulation, biocompatibility, wettability), transport (electrical and thermal conductivity), and optical (transmission, reflection, absorption, color) properties of surfaces [18].

8.2 Background: In-Flight Environments and Damage Modes— Challenges and Opportunities in Aeronautical Engineering

Commercial and military aircraft operate in diverse worldwide environments. Composite structures for these aircrafts must be designed to withstand all these environments, including large variations in temperature and moisture, contact with aircraft liquids such as jet fuel and hydraulic fluid, and lightning strikes. The coatings that are applied to airplanes represent a highly technical challenge for paint makers as they are the most demanding of all paint systems (Figure 8.3). They need to be able to withstand temperatures ranging from −48°F in the air to 120°F on the ground, as well as strong UV exposure at 30,000 ft. In addition to coping with extreme temperature fluctuations, they need to resist cracking due to rapid changes in air pressure. They also need to resist erosion from air drag and deflect the impact of dust traveling at 500 mph, which has an effect similar to sandpaper. On the ground, aircraft coatings also have to withstand contact with aggressive fluids such as fuel, de-icing fluids, and hydraulic liquids. Additionally, the appearance of airplanes is important. Attractive colors and unique paint schemes are desirable for identifying and distinguishing one airline from another [19,20].

Thus, the design of modern aircraft paint to withstand these demands is reflected in its detailed specification. This includes requirements for temperature, UV, flexibility, adhesion, water resistance, and fluid and corrosion resistance, which can be systematically classified by

- Mechanical properties: increased modulus, strength, impact toughness
- Barrier properties: reduced moisture absorption, increased chemical resistance, reduced gaseous diffusion
- Fire retardant properties: reduced burn-through, enhanced FST properties, increased charring

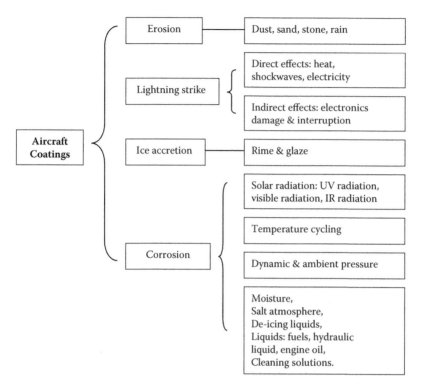

FIGURE 8.3
Classification of environments encountered by aircraft coatings.

- Thermal properties: increased HDT and stability, enhanced conductivity, controlled coefficient of thermal expansion (CTE)
- Tribological properties: increased surface hardness, reduced wear rate, scratch resistance
- Electrical properties: conductivity, electrostatics, EMI shielding, EMH protection

The necessity to meet the requirements of such specifications has a strong influence on the paint formulation [21].

8.2.1 Interaction between Lightning and Aircraft

The effects of lightning strikes on aircraft can be divided into two types, direct and indirect effects. The direct effects refer to causes of damage, including heat, acoustic shockwave, and electricity. These can affect different components or areas of the aircraft with varying degrees of severity. The figures associated with a typical lightning strike are quoted by Reference 22 as ~30,000°C temperature generation, ~500 psi impact force from the acoustic

TABLE 8.1

Typical Lightning Strike Levels and Airframe Requirements

Threat	Criteria	Requirement
High-energy strike	Rare lightning strike 50–200 kA	Striking level in accordance with zoning diagram
		Continued safe flight (70% DLL)
		Ready detectable damage
Intermediate-energy strike	Medium lightning strike 30–50 kA	Repair needed (100% DLL) Visible damage
Low-energy strike	Nominal lightning strike 10–30 kA	No repair needed (150% DLL)
		No or barely visible damage

DLL = design limit load.
Source: Feraboli, P. and M. Miller, *Composites Part A,* 2009. 40 (6–7): p. 954–967.

shock, and ~200,000 amp of electric current [23] (see Table 8.1). The indirect effects refer to damage or interruption of electrical equipment onboard the aircraft by induced voltages. Unlike the direct effects, indirect effects can occur even if the lightning strike doesn't make contact with the aircraft.

Aircraft can be divided into zones that define their susceptibility to lightning arc attachment. Figure 8.4 highlights the distribution of these zones across the aircraft [24]. There are three main zones, which are then subdivided into subzones. Typically the areas illustrated as zone 1 are the most likely locations for initial lightning strike attachments. Zone 2 locations are susceptible to subsequent swept stroke attachments after the initial attachment, and, finally, zone 3 locations are where arc attachments are unlikely. The zones are summarized as follows:

Zone 1—Initial attachment point with probability of flash hang-on

Zone 2—High probability of swept stroke reattachment from zone 1

Zone 3—Low probability of lightning arc attachment

8.2.1.1 Damage Mechanism

As already mentioned, the direct effects of a lightning strike involve causes of damage, including the heat generated by the resistance of the paint. This resistive or Joule heating originally caused most damage. More recently, however, with the achievement of reduced dwell times in lighting strike testing, the question has been raised as to whether the acoustic shockwave that occurs in the first few microseconds of a lightning strike causes more damage than originally suspected. Lastly, the electrical charge that has already been found responsible for resistive heating may also causes a breakdown of the matrix because of the dielectric properties of the paint.

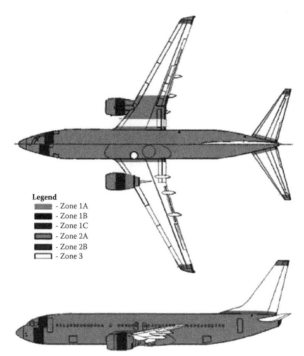

Zone 1 Initial attachment point with probability of flash hang-on,
Zone 2 High probability of swept stroke reattachment from zone 1.
Zone 3 Low probability of lighting arc attachment.

FIGURE 8.4
Lightning zone diagram for a typical large commercial transport per SAE 5414. (From Feraboli, P. and M. Miller, *Composites Part A*, 2009. 40(6–7): p. 954–967.)

Puncture is more likely to occur with a composite material than with a homogeneous plastic such as a polycarbonate because composites have microscopic holes (porosity) and material interfaces through which or along which an electric discharge can propagate, as shown in Figure 8.5, where projected damage area was obtained via C-scan for filled-hole specimens [24]. The field required for punching a given thickness of glass fiber or aramid fiber composite is, in fact, only slightly greater than that required to ionize a similar thickness of air. A measure of the ability of a nonconductive material to resist puncture is its dielectric strength. Homogeneous materials, such as acrylic and polycarbonate sheets, have very high dielectric strengths and are more resistant to puncture.

8.2.1.2 Protection Methodology

There are two basic ways of providing protection to nonconductive composites. One method employs diverter strips or bars on the exterior surface to serve as preferred streamer initiation points and to intercept lightning flashes,

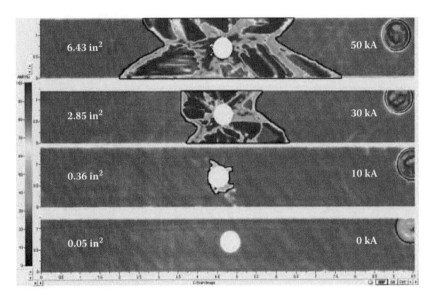

FIGURE 8.5
Projected damage area obtained via C-scan for filled-hole specimens. (From Feraboli, P. and M. Miller, *Composites Part A*, 2009. 40(6–7): p. 954–967.)

while allowing the skin to be transparent to electromagnetic waves. This is the approach used for protection of radomes and some antenna fairings. The other method is to apply an electrically conductive material over the exterior of the structure. This latter method provides the most effective lightning protection and should be employed whenever possible. It also provides improved protection of enclosed systems against the indirect effects of lightning.

Where radio frequency transparency is not required, or conductive structural materials are utilized, a conductive coating can be applied to the exterior surfaces of composites to prevent electric field penetration and puncture, and to conduct lightning currents. Protective materials include arc- or flame-sprayed metals, woven wire fabrics, expanded metal foils, aluminized fiberglass, nickel-plated aramid fiber, and metal-loaded paints.

The initial lightning strike must be dispersed quickly around the airframe to prevent concentrated damage. Also, the airplane's electronic flight instruments must be shielded from disruption by the intense electromagnetic field. To accomplish this, a thin metal mesh or foil is embedded in the outer layers of the composite fuselage and wing [25]. A thin metal mesh embedded in the outer layer of the composites also shields the electrical systems [26].

But the metal mesh or foil increases the weight of the plane. If highly conductive composites were available to replace metal meshes or foils, the weight would be sharply reduced. Carbon nanotubes hold the potential for conducting composite structures, with good surface finish, to deliver an overall weight savings. Also, conducting resin systems have been investigated for lightning strike protection [27]. Ben Wang's group [27] developed a unique technical approach to incorporate SWNT buckypaper materials into conventional fiber-reinforced and foam composite structures for improving EMI and lightning strike protection. The results show that foam structures with buckypaper of 700 mg surface can achieve as much as 26 dB of EMI shielding over the test range from 455 to 500 MHz, compared with a control panel made from a pure foam structure. Also, the results showed that random buckypaper samples exhibited better EMI shielding properties. For lightning strike resistance, no visible improvement or further improvement in electrical conductivity of buckypaper composites is needed in order to utilize SWNTs for the purpose of improving the EMI and lightning strike resistance properties of composite structures.

8.2.2 Rain, Sand, and Ice Particle Erosion

An aircraft is most likely to encounter dust and sand during take-off and landing, and rain during ascent and descent, as cruising altitudes are generally above cloud levels. However, military aircraft and missiles may experience all types of conditions [28,29].

Figure 8.6 shows the synergistic effects of liquid and sand erosion on a laboratory scale. When sand particles impact on a preexisting crack caused by liquid impact, the damage site is enhanced, since the crack is easily opened further by sand particle impact. These cracks are further exploited after prolonged exposure. If not caught in a timely manner, the sand may erode into the fiberglass blade skin, degrading its structural integrity. Additionally, since an expanded copper mesh is often embedded on the surface of the composite skin for lightning strike protection, the erosion damage may locally degrade the protection of the blade. Since lightning strikes tend to be concentrated at the outboard section of the blade, erosion damage may degrade lightning strike protection in an area where it is most necessary [30]. Current blade protection against erosion is either a metallic shield, usually nickel or titanium, or a polymeric coating or tape, which is usually either pure polyurethane (PU) or a blend of polyurethane with other polymers such as polyethylene [31].

Recent efforts have been concentrating on the development of new advanced polymeric materials as they already offer greater protection than metallic systems, do not alter blade dynamics in the same way that a metallic or ceramic system would, allow metallic elements to be eliminated from blades, and have the potential for much easier and lower cost replacement. A particular area that attracts much interest in research is the development of nanocomposite coatings, to be outlined in later sections.

8.2.2.1 Classification of Erosion Behavior

The erosion classification of rubber-like elastomers is complex and shows brittle, ductile, or rubber-like erosion behavior, depending on temperature

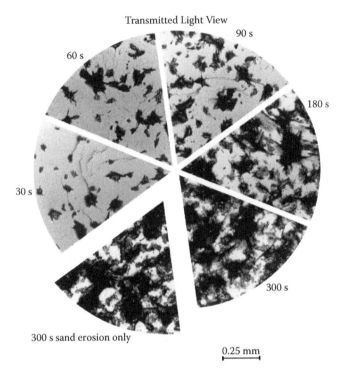

Transmitted Light View

FIGURE 8.6

FLIR ZnS previously liquid impacted by 0.8 mm jets at 185 m s^{-1} (100 impact random array over 14 mm diameter circle) followed by sand erosion at 0.15 kg m^{-2} s^{-1} at 20 m s^{-1} by C25/52 sand. (From Jilbert, G.H. and J.E. Field, *Wear*, 2000. 243(1–2): p. 6–17.)

and deformation time [31,32]. They react differently from brittle materials (ceramics and brittle polymers) [33,34] or ductile materials (metals and ductile plastics) [35]. Single impact seems to cause no visible damage at all, which is in agreement with the incubation time observed in the erosion process before a steady state is obtained [36]. During this incubation time, the substrate may even increase a little in weight as a result of particles embedded in the substrate surface. In the case of polyurethane elastomers, the incubation time is found to decrease with increasing brittleness and decreasing angle of attack [37]. At oblique impact, maximum erosion is found at glancing impact, as with ductile materials. The erosion rate of PUs, for example, is found to differ by an order of magnitude between normal impact and 30° angle of incidence [38]. The difficulty of erosion classification is fundamental to polymeric materials. Depending on the temperature and rate of deformation, the same material may behave in a dominantly ductile, rubber-like, or brittle fashion.

A great difference in the classification of various materials in respect to their erosive wear exists when the variation in impact angle and time is

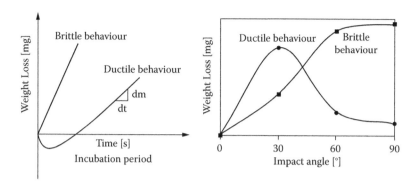

FIGURE 8.7
Schematic depiction of typical erosion behaviors. (From Barkoula, N.M. and J. Karger-Kocsis, *Journal of Materials Science*, 2002. 37(18): p. 3807–3820.)

considered. Figure 8.7 shows typical erosion diagrams as a function of impact time and angle, respectively. Taking impact time into account, an incubation period is involved in ductile erosion with the weight increases first followed by settling down to a steady state, as shown in Figure 8.7 (left). Normal impacts (the impact angle $\alpha \approx 90°$) are due to the initial embedding of erosion sands in the target material surface. Figure 8.7 (right) demonstrates that the maximum material removal occurs at low impingement angles in ductile behavior, whereas at high impingement angles brittle erosion dominates. The differences in the erosion behavior can be traced to material removal mechanisms that can range from tearing and fatigue for rubbers, through cutting and chip formation for ductile metals and polymers, to crack formation and brittle fracture for ceramics, glasses, and brittle polymers [39].

8.2.2.2 Sand Abrasion Mechanism

The abrasive action of sand and dust is probably one of the simpler mechanisms to understand and explain. At low speeds, tangential abrasion is generally performed using a Taber abraser which is a rotary platform; other tests have been developed using sand fed onto rubber discs. At the high speeds encountered by propellers and rotor blades, we see damage also occurring at perpendicular impact. Perpendicular impact damage is generally in the form of micro-pitting of the surface, and embedding of sand into the pits is not uncommon on soft polymeric coatings. At oblique angles (typically tested at 30°), we see a combination of scrubbing abrasion and micro-cutting. This scrubbing action is generally resisted very well by resilient materials such as polyurethane, as is sand abrasion on the whole; however, when it does occur along with micro-cutting we see an effect called "comet tailing" where material is plastically deformed by micro-fatigue, forming ridges which are then more susceptible to removal due to the exposure of ridges and a structure

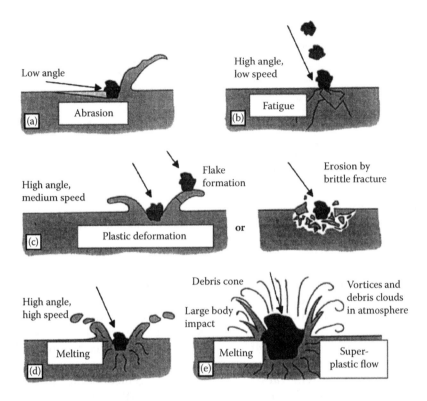

FIGURE 8.8
Possible mechanisms of solid particle erosion: (a) abrasion at low impact angles, (b) surface fatigue at low speed, high impingement angle, (c) brittle fracture or multiple plastic deformation at medium speed, large impingement angle, (d) surface melting at high impact speeds, (e) macroscopic erosion with secondary effects. (From Barkoula, N.M. and J. Karger-Kocsis, *Journal of Materials Science*, 2002. 37(18): p. 3807–3820; G.W. Stachowiak, A.W.B., *Engineering Tribology*, Amsterdam; New York: Elsevier, 1993 (Tribology series; 24).)

that is weakened by fatigue cracks [41]. Mechanisms of erosive wear were summarized schematically by Stachowiak and Batchelor [42], as shown in Figure 8.8. Details of the solid particle erosion behavior of polymers and polymeric composites were reviewed by Barkoula [40].

8.2.2.3 Rain Erosion Mechanism

Liquid impact can damage materials since, during the very early stages of impact, compressible effects occur in the liquid. A basic theory for understanding the major features of high-speed liquid impact (impact speed > 50 ms^{-1}) was the "water-hammer" pressure first analyzed by Joukowski in 1898 [43,44]. These high pressures are responsible for most of the damage resulting from liquid impact and are maintained while the edge of the contact area between the impacting liquid and the solid moves supersonically at the shock speed in the liquid. However, the water-hammer pressure idea remained unchanged until the mid-1960s, when aircraft erosion became a big issue [45]. By using high-speed photography and clever experimental design, Bowden and Field were able to make visible a number of features associated with the initial stages of impact [46]. Thereafter, 30 years of research led to a reasonable understanding of the basic mechanics of the theory of guided acoustic waves [47]. One of the largest databases available for the erosion resistance of composite materials is based on rain erosion testing in the aerospace industry, performed by Hammond [48]. Existing polymeric coatings are elastomers and show rubber-like behavior under liquid impact. One critical aspect that seems to be a common cause for the failure of existing polymeric coatings is across-surface shearing as the water droplet is deflected outward across the surface following normal impact, causing a radial pattern of deformation. Another aspect is penetration of water into tiny surface defects. Of course, surface shearing could eventually cause surface defects due to tearing and fatigue cracking which can then be penetrated by the water, causing a rapid failure of the coating. At high speeds, water droplets have been shown to behave like hammers, which also opens the possibility of impact induced de-bonding of the coating from the substrate, increasing the likelihood of tearing and fatigue damage. The edges of polymeric tapes are also very vulnerable to water impact, and tests have revealed de-bonding of tapes due to water droplets striking the edges of tapes, causing them to lift (see Figure 8.9).

8.2.2.4 Protection Methodology

Prior to the 1970s, metal erosion strips were applied to protect blades; they performed excellently against rain erosion but suffered severely from sand,

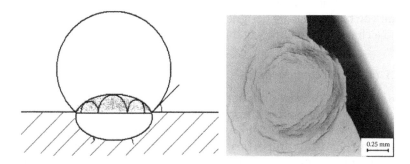

FIGURE 8.9
(Left) Initial stage of impact between a water drop and a solid target with the contact edge moving faster than the shock velocity in the liquid (i.e., supersonically). The shock envelope is made up of many wavelets, which can be found from a Huygens-type construction. The liquid behind the envelope is compressed and the target beneath this area is subjected to high pressure. (Right) Broken edge of ZnS impacted 100 times in a random array over a 14 mm diameter circle by 0.8 mm jets after the hydraulic bursting disc test. (From Jilbert, G.H. and J.E. Field, *Wear*, 2000. 243(1–2): p. 6–17.)

because titanium created a corona at night from the oxidation of particles chipped off by sand. Easy filled repair and no-corona Task L-100 paint and L-101 polyurethane paint were tested and were finally approved in 1991 to replace unstable estane and P0655 polyurethane elastomers, which were prone to hydrolysis. Hydrolysis in this case is a mechanism whereby an ester group in the primary backbone chain can be split into its two original organic acid and alcohol reactant groups [49]. The result is a broken bond in the molecular backbone, similar to a broken or missing link in a chain. In 1990, 3M 8663 and 8545 with a short shelf life and high recurring maintenance were developed; 3M 8663 was used extensively and proved to be efficient. Bayer 535330A has also been widely used in erosion protection. Since 1990, there have been two solutions for the harsh sandy environment common to rotor blades: existing metallic shields and resilient polyurethane coatings or tapes. Metallic shields are generally built into the structure of the blade to provide a near-seamless join and are usually produced from titanium or nickel. Nickel has been found to resist sand erosion more effectively than titanium. But compared with polymeric tapes, both materials are heavy. Commercial protective polyurethane topcoat is applied in an accurately controlled spray process to both civil and military aircraft. The spray ensures a controlled thin thickness coat can be applied, which allows the solvent to evaporate to produce a high quality coating. As a typical protection material for rotor blades, 3M polyurethane protective tape is made from an abrasion-resistant polyurethane elastomer that resists puncture, tearing, abrasion, and erosion [50]. It is formulated for resistance to ultraviolet light. When used in conjunction with an applied adhesive, 3M black tape provides protection of the surface from corrosion, abrasion, and minor impact damage. Tapes can easily be replaced as they are supplied in self-adhesive form. Hard ceramic materials are deemed unsuitable due to the processing conditions that could damage composite blades and to a lack of flexibility on blades that are designed to flex, causing both cracking of the coating and a change in dynamics of the blade.

8.2.2.5 Erosion-Testing Methods

The aim of solid particle impact testing is to investigate the resistance to erosion by solid particle impact of materials with a wide range of mechanical properties, and to explore the correlation between erosion rate and mechanical properties. Erosion testing has been carried out with different particle sizes, impact velocities, and impact angles in sand blast-type erosion test rigs [36,51,52]. There are four main types of erosion rigs: the sand or gas-blast rig, recirculating liquid slurry loop, centrifugal accelerator, and whirling-arm rig [53]. The whirling-arm rig has been used to simulate erosive wear conditions on a number of components, for example, helicopter rotor and gas turbine compressor blades, which are vulnerable to erosive wear from sand and dust particles, especially in desert terrain [54,55]. The main specification is to have a double-arm rotating blade at a tip speed near sonic velocity (up to 550 m/s) and to simulate rainfall of 1 to 2 in./h with fine control. A schematic diagram of a rain erosion testing rig is given in Figure 8.10, which consists of the following subassemblies:

1. High speed motor and control: ~45 kW power, up to ~5000 rpm.
2. Container (concrete wall): ~3 m diameter and ~2 m high; the wall consists of ~5 in. thick oak timbers, 1~2 in. thick steel, and ~10 in. of double-reinforced concrete
3. Rotating arm and stands: The rotating arm apparatus consists of an ~2 m diameter, double-arm blade with sample holders; tip velocities between 0 and 600 mph.
4. Diagnostic and monitoring system: This includes digital recording system and synchronized strobe light system. Closed-circuit television cameras or displays provide a means to observe and record testing.
5. Feeding system and control (to simulate rainfall): A circular, overhead manifold with capillary tubes/needles produces artificial sand/rainfall; a central tank supplies temperature-controlled water/sand to manifold. Drop size and drop rate are controlled by manipulating the water temperature, capillary tube diameter, and storage tank pressure.
6. Control panel/room: Monitors and controls the tests remotely, with sound proofing (not shown in Figure 8.10).

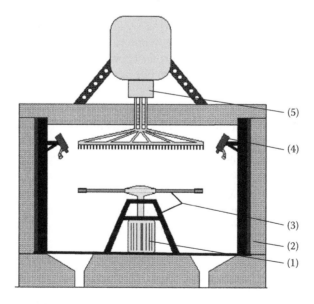

FIGURE 8.10
Schematic illustration of the rotating-arm testing rig consisting of the subassemblies.

8.2.3 Ice Accretion

Ice accretion on aircraft surfaces is a serious hazard. Ice distorts the flow of air over the wing, diminishing the maximum lift, reducing the angle of attack for maximum lift, adversely affecting the aircraft handling qualities, and significantly increasing drag. Wind tunnel and flight tests have shown that frost, snow, and ice accumulations (on the leading edge or upper surface of the wing) no thicker or rougher than a piece of coarse sandpaper can reduce lift by 30% and increase drag by up to 40%. Larger accretions can reduce lift even further and increase drag by 80% or more. A NASA study [56] revealed close to 50% of the total drag associated with an ice encounter remained after all the protected surfaces were cleared. Also icing causes damage to external equipment such as antennae, can clog inlets, and causes impact damage to fuselage and engines [57].

8.2.3.1 Types of Ice

Ice forms on aircraft surfaces at 0° Celsius (0°C) or colder when liquid water is present. The following list illustrates the icing risk in terms of cloud type and ambient temperature. Structural ice adheres to the external surfaces of aircraft. It is described as rime, clear or glaze, or mixed [58]:

1. Rime ice has a rough, milky-white appearance resulting from air trapped when it strikes the leading edge of an airfoil and freezes. It typically occurs with temperatures between –15°C and –20°C. It is less dense, and can be easier to remove than clear ice.
2. Clear or glaze ice typically forms when temperatures are around –10°C to 2°C, with large water droplets, freezing drizzle, or freezing rain. It is the most dangerous type of structural ice not only because it is hard to see, but also because it can change the shape of the airfoil.
3. Mixed ice is a mixture of clear ice (from large drops) and rime (from small droplets) that has the worst characteristics of both, and can form rapidly when ice particles become embedded in clear ice and build a very rough accumulation. Mixed ice is most likely to form at temperatures between –15°C and –10°C.
4. Hoar frost occurs when moist air comes in contact with a surface at subzero temperatures. The water vapor, rather than condensing to form liquid water, changes directly to ice and is deposited in the form of frost.

8.2.3.2 Current Icing Solutions

Extensive research has been undertaken to identify and prevent icing, and the methods can be classified as de-icing and anti-icing. The former removes

TABLE 8.2

Comparison of Anti-icing and De-icing

	Anti–icing	De-icing
Features	Preemptive, turned on before the flight enters icing conditions	Reactive, used after significant ice build up
Methods	Thermal heat, prop heat, pitot heat, fuel vent heat, windshield heat, and fluid surface de-icers	Surface de-ice equipment such as boots, weeping wing systems and heated wings

the ice from the surface after its formation, while the latter prevents the initiation of icing [57] (see Table 8.2).

To prevent ice build-up on rotor blades under icing conditions, equipment for anti-icing consists of an electrical matrix which covers 20% of the leading edge chordwise, from the tip along the length of the blade. Heat is phased into this matrix in different sectors, timed to coincide with the natural shedding cycle, that is, when sufficient ice has built up. The power supply for the matrix equipment is a drain on the electrical resources and since the only satisfactory solution would be to heat the whole blade, a generator large enough to do this would impose weight problems. A CNT paper resist heater has been fabricated in TEG, Germany [59]. The added value is the combination of higher flexibility of design compared to a metal heater, and high uniformity of temperature. Also, Han et al. reported a rapid thermal response and stable reversibility of transparent SWNT films made by a vacuum filtering method [60].

Also, the addition of a thermally insulating coating to the blade surface raises several concerns. The primary concern is that the coating might hinder heat transfer to the surface, blocking the heating needed to drive ice melting and shedding. A secondary concern is that the insulating coating might retain the heat in the blade structure, potentially causing thermal degradation of the composite structure [30]. Thus, effort to enhance the thermal conductivity of coatings to enhance their de-icing performance may be fruitful [61]. Current research is heading toward nanocomposite coatings, which create very high contact angles with water and where the reinforced polymer acts to absorb and dissipate the higher impact energy caused by repeated impact of particulate matter. A mixture of Rain-X and MP55 PTFE (powder Teflon) has also been claimed to be an outstanding coating to reduce ice adhesion to the surface of the space shuttle [62]. An anti-icing coating has also been developed by Cape Cod Research, Inc. for marine applications and is called PCM-based ice-phobic coating [63].

8.2.4 Environmental Corrosion

Corrosion damage to aircraft fuselages is an example of atmospheric corrosion caused by ultraviolet radiation, heat, and moisture, as well as salt and gas concentrations. Corrosion is manifested in many different forms. Concentration cell or crevice corrosion is the most common type found on airplanes, occurring whenever water is trapped between two surfaces, such as under loose paint, within a delaminated bond line or in an unsealed joint. It can quickly develop into pitting or exfoliation corrosion, depending on the alloy, form, and temper of the material under attack [64]. The Aloha Airlines incident in 1988 highlighted crevice corrosion damage as a major safety concern [65].

Among criteria for material properties, fatigue and corrosion are the primary failure modes of aircraft structural components and are therefore the most significant criteria in materials selection. To meet these key requirements, normally coatings are used to provide corrosion properties while the substrate gives fatigue resistance [15,66]. For naval aircraft, sea vapor poses a threat to the substrate. The introduction of de-icing chemicals brings corrosion to the aircraft surface [67]. Also composites eliminate a lot of corrosion problems, but introduce different galvanic corrosion: conducting carbon fibers can become exposed (anodic), and aluminum airframes and fasteners can become cathodic [68]. Coating systems exist to prevent galvanic corrosion at the composite/fastener/structural frame interface and provide sealing fastener coatings, coatings for interface regions, conductive gap fillers, and repair methods. In this case, the topcoat provides water resistance and barrier properties. The primer provides corrosion protection when the coating system is damaged as well as adhesion to the metal substrate.

There are three basic protective mechanisms of anticorrosion coatings: barrier protection, passivation of substrate surface (inhibitive effect), and sacrificial protection (galvanic effect) [69]. Barrier protection is obtained by impeding the transport of aggressive species into the surface of the substrate by application of a coating system with low permeability for liquids, gases, and ions. Passivation of the substrate surface can be obtained by a chemical conversion layer or by addition of inhibitive pigments to the coating. Metallic, organic, and inorganic coatings have all been widely applied for protection of metals against corrosion by means of sacrificial protection; that is, protection is obtained by sacrificial corrosion of an electrochemically more active metal, which is in electrical contact with the substrate. Recently, the introduction of acetate- and formate-based de-icing chemicals led to the discovery that corrosion damage on aircraft has become significantly more frequent [70].

In summary, the engineering challenges of lightning strike hazards, icing, fatigue, erosion, and corrosion environments warrant the need to develop novel coatings systems with multifunctional compensating properties. Nanotechnology-enabled composite systems offer optimal opportunities to achieve these requirements.

8.3 Coating Material Systems

A coating system usually consists of multiple layers of different coatings with different properties and purposes. Depending on the required properties of the coating system, the individual coats can be metallic, inorganic, or organic [71]. As described above, the topcoat is exposed to the external environment and must provide the surface with the required color and gloss. In addition to adequate resistance to alternating weathering conditions and impacts from objects, the topcoat should also have a high resistance to UV. Nanocomposites obtained by dispersing nano-scaled fillers of various functionalities within polymeric matrices are of particular importance because they offer a great opportunity to improve and tailor the multifunctional properties of the polymer.

8.3.1 Polyurethane Matrix

Film formation (i.e., the transition from a liquid product to a solid coating) can occur in three ways: either by evaporation of the solvent, by chemical reaction, or by a combination of both [72]. Chemically curing binders for coatings curing at room temperature may be divided into three subgroups, depending on the type of chemical process that forms the basis for the formation of a solid coating (Figure 8.11). Oxidatively cured coatings absorb and react with oxygen from the air in the presence of a catalyst. Similarly, moisture-curing coatings, such as zinc silicates, react with moisture from air during the curing process. Two-component systems rely on a reaction between a binder and curing agent (often in the presence of catalysts such as various types of solvents) that must be supplied in separate containers by the coating manufacturer.

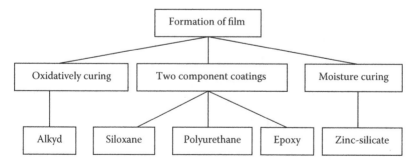

FIGURE 8.11
Classification of binders according to curing mechanisms.

Polyurethane (PU) elastomers are largely applied to industry and consumer products, particularly in the fields of heavy pressure, load, impact, and wear because they possess excellent comprehensive properties such as high wear, oil and corrosion resistance, high elasticity and damping, good adhesion to other materials and so on [73–75]. Literally thousands of different polyurethanes can be made from the array of commercially available isocyanates and active hydrogen compounds, and can mainly be divided into two-component and one-component systems. Two-component systems can take two forms: one shot systems and a prepolymer route. In one shot systems [69], the polyol is first blended with catalysts and flame retardants, plasticizers, fillers, etc., as required to form the resin component, and then reacted with the iso component, which is made up of only diisocyanate. In the prepolymer route, the first step is to produce a prepolymer by reacting one equivalent of polyol with up to two equivalents of isocyanate. Polyurethanes are then produced by the reaction of prepolymers with a chain extender (also known as a curative), such as a short-chain polyol or diamine. Typically, additives such as catalysts, flame retardants, plasticizers, fillers, etc., are blended into the prepolymer prior to chain extension.

8.3.2 Nanoscale Fillers

There exists a wide variety of nano-scale fillers used in modern polymer composite materials, among which are very basic and common fillers like titania, alumina, and silica, clays [76–80], and then carbon nanotubes (CNTs) in single and multiwalled configurations [81–85], as seen in Figure 8.12 [86]. Here, we only discuss three typical types of geometry: clay, CNTs, diamond, and alumina, with particular focus on aeronautical coating applications. This is a convenient way of discussing polymer nanocomposites, because the processing methods used and the properties achieved depend strongly on the geometry of the fillers.

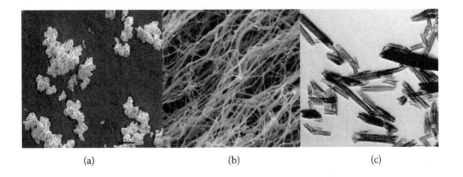

(a)   (b)   (c)

FIGURE 8.12
Typical morphologies of (a) particles, (b) tubes or rods–carbon nanotubes (CNTs), and (c) layered montmorillonite. (From Fornes, T.D. et al. *Polymer*, 2001. 42(25): p. 09929–09940.)

8.3.2.1 Nanoclay

The era of polymer nanocomposites received an impetus from Toyota in 1987. Its nylon-6–organophilic montmorillonite (MMT) clay nanocomposite showed dramatic improvements in mechanical and physical properties and heat distortion temperature at very low content of layered silicate [87]. The excellent properties lie in the unique structure of MMT, where the platelet thickness is only 1 nm, while its dimension in length and width can be around hundreds of nanometers. The unique structure makes nanoclay a very promising filler for coating applications, which require higher water barrier properties as well as robust mechanical properties, thermal stability, and flame retardance [88–90]. The effect of lamellar MMT on the permeability of protective coatings is illustrated in Figure 8.13. In coatings with spherical fillers, the aggressive species may migrate almost straight through the coating, whereas coatings formulated with lamellar nanoclay impede the transport of aggressive species by providing a tortuous diffusion path [91]. Three major processing methods have been developed to produce polymer–layered silicate nanocomposites: in

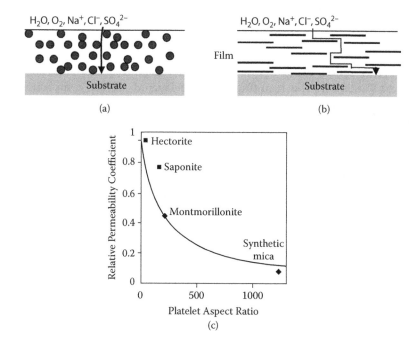

FIGURE 8.13
Idealized sketch of the effect of (a) spherical filler and (b) lamellar fillers in coatings. (c) The reduction of the relative permeability coefficient is dependent on the clay platelet aspect ratio in the polyimide–clay hybrid system with water vapor as permeate. Each hybrid contains 2.0 wt% clay. The aspect ratios for hectorite, saponite, montmorillonite, and synthetic mica are 46, 165, 218, and 1230, respectively. (From LeBaron et al. *Applied Clay Science*, 1999. 15(1–2): p. 11–29; Yano et al. *Journal of Polymer Science Part A: Polymer Chemistry*, 1997. 35(11): p. 2289–2294.)

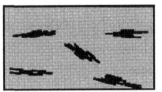

A: Conventional composite

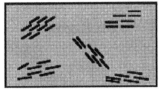

B: Intercalated nanocomposite

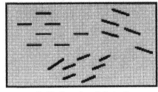

C: Ordered exfoliated
nanocomposite

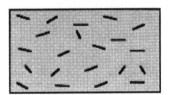

D: Disordered exfoliated
nanocomposite

FIGURE 8.14

Schematic illustrations of (A) conventional; (B) intercalated; (C) ordered exfoliated; and (D) disordered exfoliated polymer–clay nanocomposite. (From Yano et al. *Journal of Polymer Science Part A: Polymer Chemistry*, 1997. 35(11): p. 2289–2294.)

situ polymerization, the solution-induced intercalation method, and the melt processing method. The resulting polymer–layered silicate composites are ideally divided into four general types: conventional composites, intercalated nanocomposites, and ordered exfoliated and exfoliated nanocomposites, as shown in Figure 8.14 [92].

8.3.2.2 Carbon Nanotubes

The outstanding mechanical, thermal, and electrical properties of carbon nanotubes (CNTs), and their huge aspect ratio, motivated sustained research into their physical properties and potential applications, and a vast literature exists on this topic. Here, only a brief mention is made of the use of CNTs in coating applications and the research carried out by the author's research group. As shown in Figure 8.15, single-walled carbon nanotubes have outstanding mechanical properties, excellent thermal conductivity, and exceptional electrical conductivity, which demonstrates the revolution achieved by replacing micron-sized carbon fibers with carbon nanotubes and reflects the reduction in physical properties as the size of the carbon structure is increased [93].

8.3.2.2.1 Treatment of Carbon Nanotubes The nanotubes are dispersed either by direct mixing of CNTs into epoxy using a disperser or by a sonication method. However, to achieve good dispersion in a polymer matrix and strong interface adhesion between the surrounding polymer chains,

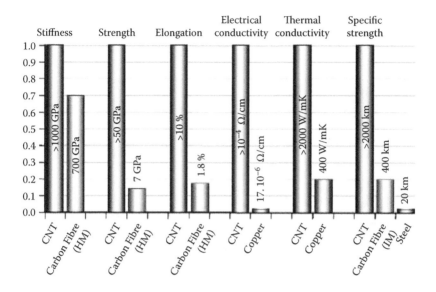

FIGURE 8.15
Comparison of the performance of carbon nanotubes against conventional materials. (From Daniel, L. and M. Chipara. *Carbon Nanotubes and Their Polymer-Based Composites in Space Environment*. September 14–17, 2009.)

carbon nanotubes have to be chemically functionalized (including copolymerization). Defects in nanotubes are important in the covalent chemistry of the tubes, because they can serve as anchor groups for further functionalization or for the covalent attachment of other chemical moieties. Therefore, defects can be utilized as a starting point for the development of the covalent chemistry of nanotubes. An analysis demonstrates that about 5% of carbon atoms in a SWNT are localized at defects [94]. The defects in a tube can be divided into four categories; one growth defect encountered frequently is the Stone-Wales defect, comprised of two pairs of five-member and seven-member rings, referred to as a 7-5-5-7 defect [95]. A Stone-Wales defect leads to local deformation of the nanotube sidewall, thereby introducing extra curvature. The strongest curvature occurs at the interface between the two five-member rings; as a result, addition reactions are most favored at the carbon–carbon double bonds in this region [96]. Removal of the nanotube caps by strong acids—for example, HNO_3—can result in decoration of the ends with carboxylate groups [94,97–99]. Also, the carboxylate groups can be bonded on defects along the nanotube sidewalls. Typically, around 1%–3% of the carbon atoms of a SWNT are functionalized as carboxylate groups after nitric acid treatment [100].

In addition to defect functionalization, other approaches have been developed to functionalize the CNTs in both molecular and supramolecular

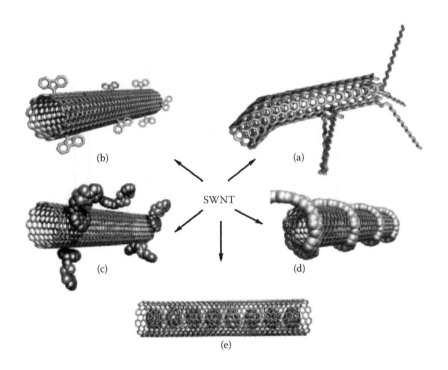

FIGURE 8.16
Functionalization possibilities for SWNTs: (a) defect-group functionalization, (b) covalent side-wall functionalization, (c) noncovalent exohedral functionalization with surfactant, (d) nonco-valent exohedral functionalization with polymers, and (e) endohedral functionalization with, for example, C60. For methods (B–E), the tubes are drawn in an idealized way, but defects are found in real situations. (From Andreas, H., *Angewandte Chemie International Edition*, 2002. 41(11): p. 1853–1859.)

chemistry. As shown in Figure 8.16, these approaches include defect func-tionalization, covalent functionalization of sidewalls, noncovalent exohedral functionalization, for example, formation of superamolecular adducts with surfactants or polymers, and endohedral functionalization [100–102].

The covalent functionalization of a nanotube with polymer is considered to be an effective way of improving compatibility and achieving a homogeneous dispersion by direct chemical linkage between the nanotubes and PU [103]. This includes "grafting onto" (directly reacting existing polymers containing terminal functional groups with the anterior functional groups on carbon nanotubes) [104], "grafting from" (growing polymers from carbon nanotube surfaces by in situ polymerization) [105] and in situ polycondensation [106]. Typically, for a PU matrix, the as-treated carboxylic carbon nanotubes are refluxed with thionyl chloride ($SOCl_2$). After the residual thionyl chloride is removed by vacuum evaporation, the acyl chloride MWNTs (MWNT-COCl) are washed repeatedly with anhydrous THF and dried under vacuum. An amine, for example, *para*-phenylenediamine (PPD), together with pyridine in DMF is added to the flask that contains MWNT-COCl, and the reaction

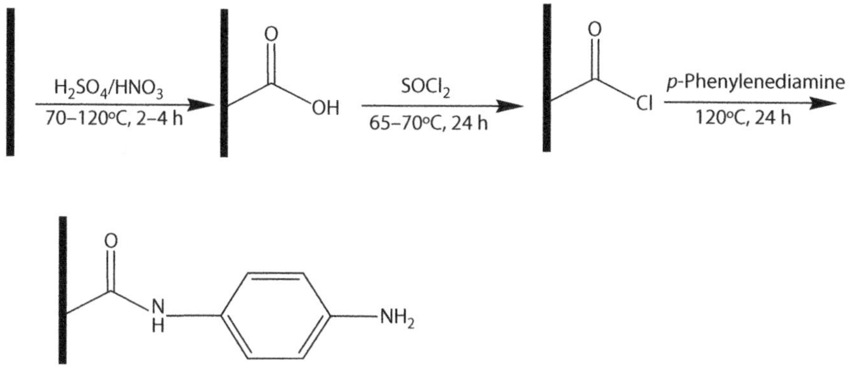

SCHEME 8.1
Reaction scheme for the synthesis and preparation of PPD-grafted MWNTs. (From Zhao, W. et al. *Macromolecular Materials and Engineering,* 2010. DOI:10.1002/mame 201000080.)

product is stirred under a pure N_2 atmosphere. After the reaction is finished, the solvent is completely removed by vacuum evaporation. The nanotubes are washed with ethanol until the filtrate shows the absence of amine by the titration method [107], and then are dried in a vacuum oven (Scheme 8.1).

8.3.2.2.2 Metal–Nanotube Nanohybrids The interactions between nanotubes and various metals are important in the formation of low-resistance ohmic contacts to nanotubes and other issues such as forming metal or superconducting nanowires on nanotube templates [108–110]. Generally, there are three methods to deposit metal nanoparticles onto a nanotube surface, including electrochemical, chemical, and physical methods, and each offers varying degrees of control of particle size and distribution along the nanotubes [111] (see Figure 8.17).

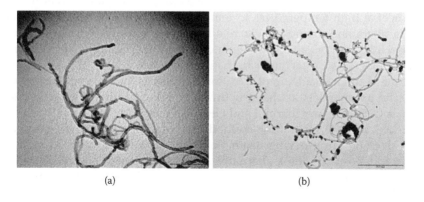

(a)   (b)

FIGURE 8.17
SEM images of (a) pristine MWNTs and (b) Ag-doped MWNTs after immersion in an aqueous solution of $AgNO_3$.

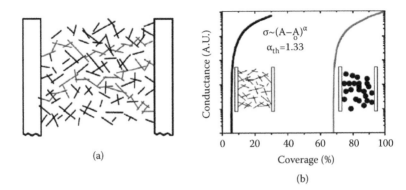

FIGURE 8.18
Nanowire network (nanonet) architecture: (a) network above percolation threshold, with pale lines indicating conducting pathways and (b) network conductance versus coverage for a random network of wires and dots. (From Gruner, G. *Journal of Materials Chemistry*, 2006. 16(35): p. 3533–3539.)

8.3.2.2.3 CNT Nanonets, Films, and Sheets   To utilize these nanostructured materials in engineering applications, it is crucial to develop processing techniques that are both scalable for producing macroscopic structures and capable of efficiently utilizing nanoscale reinforcement in the as-manufactured composite. Some promising techniques for processing precursors for macroscopic composites are briefly outlined in the following paragraphs.

Two-dimensional carbon nanotube macrostructures offer an opportunity to use carbon nanotubes with random networks of many carbon nanotubes, called *nanonets*, which enable numerous basic electronic functions at low cost [112]. Nanotube sheets (called "nanotube paper" or "buckypaper") are normally obtained by filtering well dispersed SWNT suspensions, peeling the resulting sheet from the filter after washing and drying, and annealing the sheet at high temperatures to remove impurities [113]. So far, nanotube sheets, fibers, and composites can retain the properties of the individual nanotubes. Buckypaper can be impregnated with resin, including epoxy, bismaleimide, or cyanate ester, and can then be included as a ply or multiple plies within a laminate made up of traditional prepregs. In the resulting molded part, the prepreg acts as the structural component, while the buckypaper with its CNTs imparts thermal and electrical properties [114].

Various room temperature methods for making transparent nanotube films include drop-drying from solvent, airbrushing, and Langmuir–Blodgett deposition. These alternatives, however, present severe limitations in terms of film quality or production efficiency. Nanotube films made by the filtration method are homogeneous and controllable, and surfactants are easily removed. The architecture of a network of nanotubes is illustrated in Figure 8.18. With components that are conductors or semiconductors, a two-dimensional (2D) nanowire network (nanonet) is a conducting medium, with several attractive attributes, such as electrical conductance, optical transparency, flexibility, and fault tolerance [115].

8.3.2.3 Alumina and Silica

Alumina and silica have been benchmark micron and nanoscale fillers in scratch- and abrasion-resistant nanocomposite polymers, as they are easy to obtain and the surfaces of both silica and alumina feature hydroxyl groups (OH) [116] that can be exploited with silane chemicals to add a reactive group to produce a good chemical interface with the resin [117–119]. According to Zhou et al. [85], the erosion wear mechanism of PUR composites reinforced by $Al2O_3$ particles can be described as follows: the matrix is cut apart and torn by eroding particles at sharp angles and high velocity, but the $Al2O_3$ particles are worn more slowly than the matrix due to their high hardness, so they protrude progressively more above the worn surface and protect the matrix underneath in the impact direction; meanwhile, cracks begin to initiate at the interface between the $Al2O_3$ particles and the matrix, further propagate, and finally lead to removal of the $Al2O_3$ particles. A schematic diagram of the entire erosion process is shown in Figure 8.19.

Even without surface treatment, silica and alumina are still demonstrated to improve sharply the hardness and elastic modulus of poly(-ether-ketone) (PEEK) matrices [120]. Figure 8.20 shows that for 10 wt% loading of fillers,

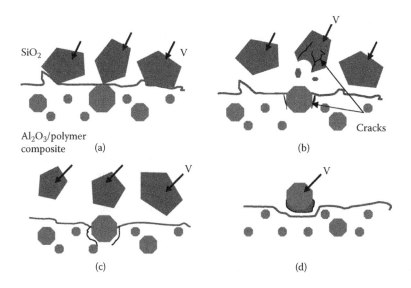

FIGURE 8.19
Schematic diagram of the erosion wear mechanism of composites: (a) beginning of erosion wear; (b) matrix has been ploughed, $Al2O_3$ particles protrude, cracks born at the interface; (c) interfacial cracks become extended, and (d) $Al2O_3$ particles are de-bonded and eroded away. (From Zhou, R. et al. *Wear*, 2005. 259(1–6F): p. 676–683.)

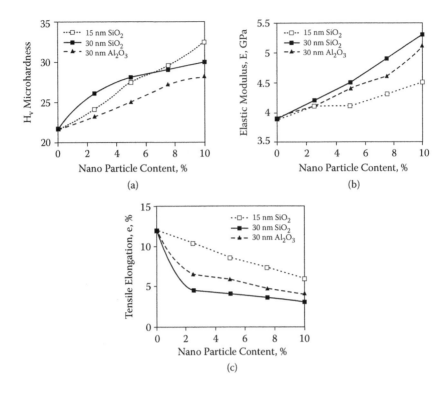

FIGURE 8.20
Variation of microhardness of nanocomposites as a function of the nanoparticle content in wt%. Variation of (a) Young's modulus E, (b) ultimate tensile stress (UTS), and (c) tensile failure elongation (e) of nanocomposites as a function of particle content in wt%. (From Kuo, M.C. et al. *Materials Chemistry and Physics*, 2005. 90(1): p. 185–195.)

the microhardness is seen to increase steadily from 21.7 Hv to > 30 Hv for silica (15 nm), to 30 Hv for silica (30 nm) and 28 Hv for alumina (30 nm), respectively. The elastic modulus also increases steadily from 3.9 GPa to > 5 Gpa for silica (30 nm) and alumina (30 nm) and 4.5 GPa for silica (15 nm). Unavoidably, however, the tensile elongation decreases drastically from 12% for pure PEEK to just around 5% for alumina (30 nm) and silica (30 nm) at only 2 wt% loading. There was then a steady further decline in elongation down to ~4% for 10 wt% loading. At the same 10 wt% loading, the smaller silica particles (15 nm) had a more than linear drop from 12% PEEK to approximately 6%, which is still a dramatic reduction.

For polyurethane (PU)/alumina composites treated with silane coupling agents A-187 and A-1100, the results of erosion tests are shown in Figure 8.21. With increasing content of Al2O$_3$ particles, the wear resistances of both composites rose gradually, reaching a maximum and then declined gradually. The peak wear resistance for the A-187-treated composites is 1.14 times that of pure PU when the Al2O$_3$ particle content reaches 21 wt%. For the A-1100

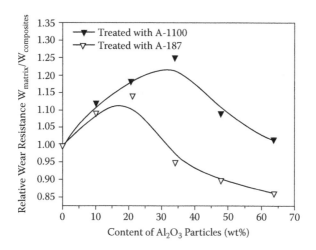

FIGURE 8.21
Content of Al2O₃ particles versus wear resistance in composites. (From Zhou, R. et al. *Wear*, 2005. 259(1–6F): p. 676–683.)

treated composites, the peak wear resistance is 1.25 times that of pure PU when the Al_2O_3 particle content reaches 34 wt%. Beside these results, there are several other good examples of interface works with alumina and polymers used in coatings. One particular work with acrylic–melamine automotive topcoats [121] examines the difference between various hydroxyl and amine functional group additions acting as chain extenders and further improving the mechanical performance. Figure 8.22 shows that the ultimate tensile strength decreases with increased loading of an alumina that had been given an apolar surface treatment (alkylbenzene sulfonic acid and toluene sulfonic acid). However, the introduction of polar and reactive tails on the alumina surface improves the mechanical performance up to levels of approximately 10% before falling off; the hydroxyl functional group (Lewis acid) gives better results than the amine group (Lewis basic) because of the acid–base interaction with acrylic-melamine polymer.

8.3.2.4 Nanodiamond

The use of chemical vapor deposition (CVD) diamond as a coating and diamond powder as a polishing and cutting medium is common; however, their use as a reinforcement is relatively new [122]. Nanodiamond (ND), also known as ultra-dispersed diamond [123], combines the unique properties of a diamond core such as superior hardness and thermal conductivity with a large readily modifiable surface, imparting enhanced mechanical strength, wear resistance, and thermal stability to composite materials even when pristine ND powders with a size of 2–10 nm are physically dispersed in the polymer matrix as reinforcement [123–125]. ND can be obtained in large quantities by detonation

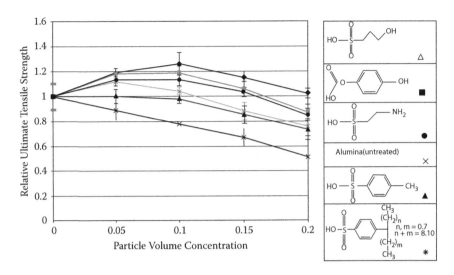

FIGURE 8.22

Relative ultimate tensile strength of composites of alumina particles with different surface treatments and with varying particle volume concentration. (From Hosseinpour, D. et al. *Progress in Organic Coatings*, 2005. 54(3): p. 182–187.)

synthesis in a process that is relatively inexpensive and has broad applicability [126]. Evidence of functional groups such as hydroxyl (OH), amine (NH$_2$), and carboxyl (COOH) on the surface of detonation-produced diamond offer promise [127] for untreated ND, and the hydrogen functional groups on diamond can be utilized for bio-sensing applications [128]. Purification can be regarded as the first chemical treatment, where ND is heated with concentrated oxidizing acids or other compounds like KNO$_3$ to remove graphitic, amorphous sp^2 carbon materials, and metallic impurities [129]. Oxidation of the ND surface can be achieved by air oxidation or treatment with oxidizing mineral acids. Other reactions for the primary functionalization of ND include reaction with gaseous fluorine at elevated temperatures [130] and reaction with ammonia [131], chlorine [132], and hydrogen [133], yielding the functionalized diamond materials (see Figure 8.23). These materials can then be modified in a secondary step with more complex moieties [126]. Recently, Krueger reported a wet chemical method for hydroxylation of the ND surface, which allows for the subsequent grafting of different trialkoxysilanes and further functionalization by covalent bonding [134]. On the other hand, covalent grafting of alkyl and aryl moieties by radical reaction was reported by Nakamura and Tsubota [135,136].

The use of chlorine in this way was also demonstrated by Ando et al. by using the newly chlorinated surface as an intermediary for further functionalization to add both hydroxyl and amine groups which, as previously

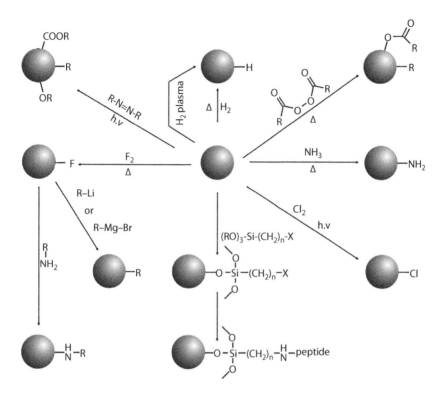

FIGURE 8.23
Surface functionalization of nanodiamond yields various covalently modified derivatives with different terminal groups for further surface modification. (From Krueger, A. *Journal of Materials Chemistry*, 2008. 18(13): p. 1485–1492.)

discussed, are suitable for use with polyurethane [137,138]. The first example to integrate ND into polymer matrix has been reported by atom transfer radical polymerization (ATRP) using a "grafting-from" approach [139], which opened up a wide variety of ND–polymer brush materials with controlled dispersibility and functional group reactivity. It is evident that both the hardness and modulus of the polymer matrix (PVA) increase significantly upon the addition of ND [140]. The average value of hardness increases by ~80%; the modulus almost doubles with the addition of 0.6 wt% ND (see Figure 8.24), which suggests that excellent adhesion between the matrix and the functionalized ND particles is the main reason for this marked improvement in mechanical performance.

8.3.3 Techniques for the Production of PU Nanocomposite Coatings

For manufacturing nanocomposites, the crucial step is achieving a well dispersed, largely homogenized mixture at the start. No matter how capable the

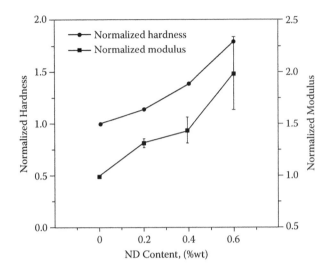

FIGURE 8.24
Variation of normalized hardness and modulus plotted as a function of ND content. (From Maitra, U. et al. *Solid State Communications*, 2009. 149(39–40): p. 1693–1697.)

manufacturing technique, without this the quality of the resulting sample is likely to be poor. Major preparation methods for polymer nanocomposites include direct blending (including melt blending, solution blending, and emulsion blending), sol–gel, and in situ polymerizations [141,142]. While the majority of these techniques are not suitable for industrial application, the variations in resulting samples produced will help to build a better understanding of how the emergent properties change as a result of structural features. Films and coatings made of randomly distributed CNTs have been shown to possess excellent performance with low sheet resistance and high optical transparency as well as robust mechanical flexibility and thermal stability.

Cast molding is one of the simplest manufacturing processes for nanocomposite materials, whereby a well dispersed suspension is cast in a mould and left to cure and post cure to produce a film (Scheme 8.2). This is an effective method for producing tapes quickly. The disadvantages of this method, however, include nonuniform thickness and unsuitability for production scale-up.

Spray painting is the current method used for aircraft, and aircraft paints include additives that help to improve the viscosity, opacity, and color for spraying. By incorporating nanoparticles into the polymer solution, the viscosity of the solution will increase quickly, making the solution harder to spray. Depending on the weight percentage of nanoparticles added, there may not be adequate wetting of the nanoparticles or other additives by the

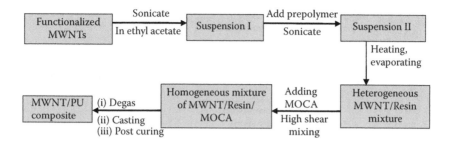

SCHEME 8.2
Schematic of the preparation of PU–MWNT nanocomposites. (From Zhao, W. et al. *Macromolecular Materials and Engineering*, 2010. DOI:10.1002/mame 201000080.)

resin, which can contribute to premature paint failure. These are all considerations that must be taken into account when attempting to disperse these particles into a paint matrix [143].

Electrospinning or electrostatic spin coating [144] is a method where a polymer solution or melt is electrically charged and fired as a jet or spray, through a capillary tube or Taylor cone. One of the main advantages of this method is continuous deposition of uniformly sized, homogeneous nanofibers. Nanofibers offer good mechanical properties, including improved surface area-to-volume ratio, increased flexibility and good tensile strength. This method is a possible solution for incorporating carbon nanotubes into nanofibers to improve the electrical properties of the fibers. In theory this could improve the dispersion and alignment of carbon nanotubes. There are inherent disadvantages, and the main issue revolves around how the resulting nanofibers should be used and encapsulated into a paint system. Other issues concern the process of nanofiber production. Typically, a large amount of solvent is required to obtain the correct viscosity for spraying. This can be expensive and potentially harmful if the solvent is toxic. While this method offers good dispersal in the nanofibers, it still requires the carbon nanotubes to be sufficiently dispersed in the original polymer solution.

8.3.4 The Need of a Roadmap

A clear roadmap for turning fundamental laboratory research in polymer nanocomposite and processes into successful innovations for coating applications is needed, such as the one illustrated in Figure 8.25. The development of nanoparticle-filled polymer systems is the core of the roadmap with the development of technology to tailor both nanoparticle–polymer interfaces and nanoparticle dispersions. The roadmap should be focused on existing polymeric systems that have been utilized by industry, but incorporating

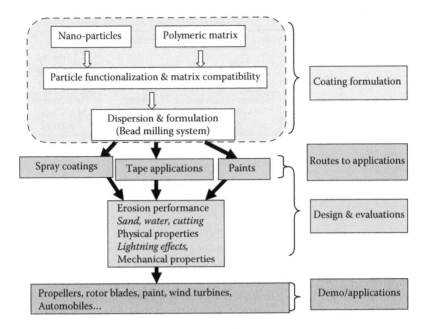

FIGURE 8.25
A proposed nanocomposite coating technology roadmap.

diamond, alumina, silicon carbide, CNT, clay, and metallic nanoparticles (e.g., Cu, Ag).

8.3.4.1 Stage 1: Formulation of Coatings

Viable nanoparticle filled polymer systems should be formulated with tailored viscosity and particle distribution to enable the materials to be sprayed for coating and to be moulded or extruded for tape application as well as painting application.

8.3.4.2 Stage 2: Routes to Application

This step investigates materials suitable for spraying and tape-making processes, as well as the painting process.

8.3.4.3 Stage 3: Evaluation of Coatings

The performance of the coatings and tapes must be assessed against sand erosion, rain erosion, and sharp object cuts. The paints need to be assessed against lightning strikes. A suitable water erosion test rig is desperately needed.

8.3.4.4 Stage 4: Demonstration Tests on Articles

The industrial sectors have the responsibility for providing guidance and assessment of the technology for future implementation. The required series of short industrial applications include:

1. Erosion trial test on tapes and paints in rotorcraft environment
2. Erosion trial test on coatings in transport aircraft environment
3. Lightning strike test

The various particles need to be selected to target different applications. For example, while Al_2O_3 and SiC are primarily for sand erosion and sharp-object cutting resistance, diamond particles are chosen due to their high thermal conductivity and high electrical resistivity, which can be potentially used in coating layers that will aid heat conduction for de-icing. Metallic nanoparticles can be used for concocting paint formulations. It needs to be pointed out that a radical improvement in performance does not have to involve a significant increase in cost, for example, diamond particles start at £4 per gram, which equates to £1 to 2 per square meter of coating.

8.4 Recent Advances in Polymer Nanocomposite Coatings for Aerospace Composites

8.4.1 Lightning Strike Properties of Composite Structures

Since many aerospace applications require transparent, durable coatings that will be exposed to charged environments during flight, electrostatic dissipation (ESD) by intrinsic electrical conductivity is necessary. The potential of nanotubes as conducting fillers in multifunctional polymer composites has been successfully realized with a very small loading of 0.1 wt% or less. A variety of applications are being pursued using these conductive composites: electrostatic dissipation, electrostatic painting, electromagnetic interference (EMI) shielding, printable circuit wiring, and transparent conductive coating. The DC electrical resistivity of SWNT–PU composite films was measured by the four-probe method and the results are shown in Figure 8.26. Compared with pure PU film, the electrical resistivity decreased from 10^{13} Ω cm (pure PU) [145] to 10^8 Ω cm (0.5 wt% MWNTs) and 10^6 Ω cm (5.0 wt% MWNTs). Figure 8.26 also shows the required resistivity levels necessary for electrical applications such as electrostatic dissipation, and a 0.5 wt% MWNT loading would be sufficient for ESD applications [15,146–148]. Since such a very low MWNT loading will impart the composite with the required level of conductivity, the base polymer's other preferred physical properties and processability would not be compromised [147].

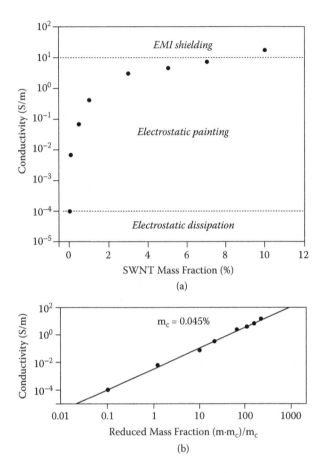

FIGURE 8.26

(a) Electrical conductivity of SWNT–polycarbonate nanocomposites as a function of nanotube loading, showing typical percolation behavior. Dashed lines represent the lower limits of electrical conductivity required for the specified applications. (b) Electrical conductivity as a function of reduced mass fraction of nanotubes, showing a threshold of 0.11 wt %. (From Ramasubramaniam, R. et al. *Applied Physics Letters*, 2003. 83: p. 2928.)

The new generation of civil aircraft depends heavily on electronic systems to implement safety-critical functions. Because these aircraft may be exposed to high intensity radiated fields (HIRFs) created by radio frequency (RF) emitters based on the ground, in the air, and at sea, civil aviation authorities have become increasingly concerned about the potential for electromagnetic interference (EMI) to these critical civilian aircraft electronic systems. Military aircraft and weapon systems must operate compatibly within an electromagnetic environment that can be even more severe than the civil HIRT environment. For example, aircraft that must take off from and land on naval ships can be exposed to a very intense electromagnetic environment

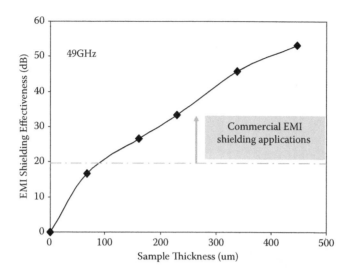

FIGURE 8.27
The relationship between thickness of carbon nanotube sheet and EMI shielding effectiveness. (From Zhao, W. et al. *International Journal of Smart and Nano Materials*, 2010, in press.)

with electric field strengths exceeding 1000 V/m [149]. There is a linear relationship between sample thickness and EMI SE value (see Figure 8.27); the EMI SE increases from 18.7 dB to 45.5 dB as the thickness of nanotube sheets increases from 68 to 339 µm. With one layer of CNT sheet of thickness 68 µm, the effectiveness of shielding increases dramatically and approaches the target value (20.3 dB) of EMI shielding materials needed for commercial applications [150].

Gou [151] developed specialty paper made of carbon nanofibers and nickel nanostrands as a surface layer on composite panels and explored the potential for replacing existing lightning strike protection materials. The lightning strike tests conducted on these composite panels showed that lightning strike tolerance is correlated with the surface conductivity of composite panels. Figure 8.28 shows the surface damage on these composite panels after a lightning strike. Clearly, the CP-CNFP-1 had the largest damaged area, where ~5.9% of the area of the paper was damaged. The carbon fibers underneath the paper were obviously damaged. However, only ~3.3% and ~1% of the area of the paper were damaged for CP-CNFP-2 and CP-CNFP-3, respectively.

8.4.2 Erosion-Resistance Coatings

Some tests have been performed to assess the effect of nanoparticles on the polymer matrix at an early stage, check equipment suitability, and provide damaged specimens to inspect and help with the determination of damage

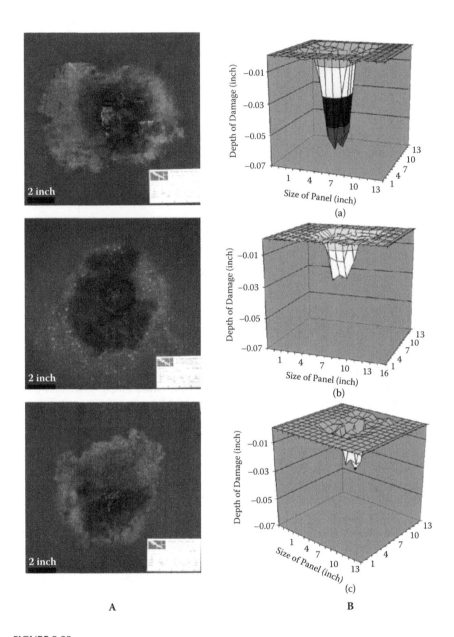

FIGURE 8.28
(A) Surface damage of composite panels and (B) damage area and thickness of composite panels after lightning strike test: (a) CP-CNFP-1 (σ = 222 S/m); (b) CP-CNFP-2 (σ = 31000 S/m); and (c) CP-CNFP-3 (σ = 34100 S/m). (From Gou, J. et al. *Composites Part B: Engineering.* 41(2): p. 192–198.)

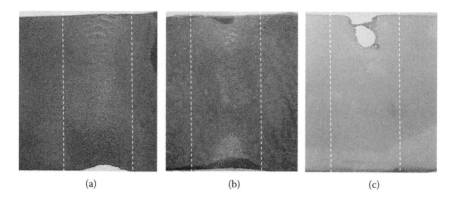

(a)   (b)   (c)

FIGURE 8.29
Surface damage pattern of (a) 3M 8663 tape, (b) 1% diamond nanocomposite, and (c) commercial defense polyurethane, under close range (25 mm) oscillation sand blasting test repeated for a total of 200 s. Dotted lines roughly indicate the range.

mechanisms. Samples of 1 wt% nanodiamond-reinforced polyurethane have been fabricated [155]. A basic comparison study was used to evaluate the rates and types of erosion that can be obtained with a sand blaster. Pieces of 3M 8663 tape measuring 30 × 30, with 1% diamond nanocomposite and commercial defense polyurethane were attached to a section of rigid substrate with curvature to test for erosion at oblique angles.

To acelerate the damage, the nozzle was brought to a distance of 25 mm and the specimens were subjected to twelve 10-s periods in cycles. Figure 8.29(a) shows the 3M tape suffered substantial thinning across the damage area, with complete loss at edges and obvious signs of plastic deformation with comet tailing. The nanocomposite also suffered substantial thinning but only at the edges, that is, with more oblique impact angles and the center still appearing as it had after the early loss of initial materials (see Figure 8.29(b)). The commercial defense PU failed along the top edge; however, material loss was not as great in the center section as in the 3M tape (Figure 8.29(c)). Overall, this was an encouraging result for the nanocomposites as potential materials protecting against hard-particle erosion.

8.4.3 Anti-Icing and De-Icing Coatings

A conventional de-icing heater uses metal foil as the heating element. One of the disadvantages is that there is not a good, durable bond with the protecting layer. The other disadvantage is that heating may not be uniform, resulting in cold spots which retain ice on the airplane, and hot spots which cause the ice to melt and refreeze at different sites. The de-icer also needs to be applied to an irregularly shaped surface in order to heat the surface uniformly. The metal foil heating element in a conventional de-icer can either be flexible but fragile, or strong but inflexible.

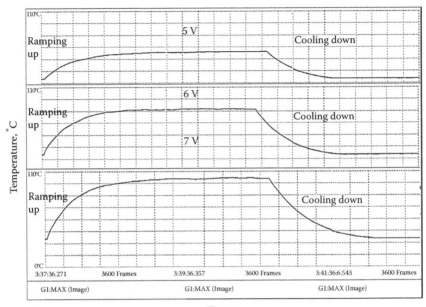

FIGURE 8.30
Thermal cycle with different input voltages: stage I, ramp-up; stage II, stabilization; stage III, cool-down.

In this context, a composite heating element has been developed as described here that is both strong and, before curing, flexible. Therefore, it can follow an irregular shape on large surfaces and provide uniform heating on such surfaces. Due to the high electrical conductivity of carbon nanotube sheets, heating is achieved in the thin layer of the heating element (0.038 and 0.045 mm thick in the model heater) without using high voltage. Since the voltage drop across the heater is not high, the thin protection layer can be used without dielectric breakdown. Since the heater can be thin, it is applied directly to the region very close to the surface to be heated. Therefore, the amount of heat accumulated in the composite heater is minimized and the rate of heating can be high despite the low transverse thermal conductivity. Also, the electricity can penetrate through the heating element in the transverse direction despite the low transverse electrical conductivity.

The heater was used in a room environment at different input voltages. The temperature of the heater was monitored by a noncontact IR thermometer. The temperature as a function of heating time was recorded (Figure 8.30). The ramp-up, steady-state, and cool-down temperatures were recorded. If I is the current conducted through the heater, E is the voltage dropped across the heater, P is the power per unit area of the heater, ρ, L, W, and t are the

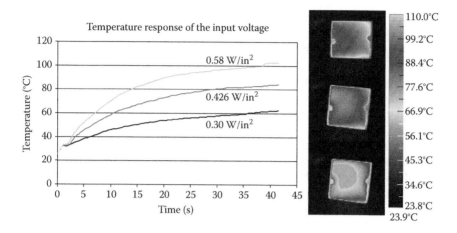

FIGURE 8.31
Brightness of an individual CNT-based specimen at different heating powers. The brightness of the sample increased exponentially with heating power. The maximum brightness was 110°C, at about 0.78 W.

resistivity, length, width, and thickness of the heater, then the equation relating the required power to heater resistivity and dimensions is

$$P_{LW} = IE = I^2 \rho L / Wt \tag{8.1}$$

Knowing that the heater in Kevlar was 0.038 mm thick and 2.10 cm wide, the distance between the foil in the middle section of the heater was 2.05 cm, and the voltmeter reading was 5 when the resistance was 62.7 Ω; the conductivity of this particular composite was calculated to be 430 S/m. If a 45.7 µm thick heating element with a conductivity of 616.52 S/m is fabricated with dimensions of 2.46 × 2.07 cm, with 0.3 W/in² power density at an input voltage of 5 V, according to the above equation, it conducts a current, and voltage drop across the de-icer, as shown in Figure 8.31.

The heating capability of the MWNT films was verified in terms of response time and input power. The MWNT sheets showed a rapid thermal response and stable reversibility when heating the target substrate. Further tests are under development to demonstrate the potential of using MWNT sheets for de-icing applications.

8.4.4 Other Related Characteristic Properties

Various mechanical testing procedures are needed to help determine the effects of particles, their quantities, sizes, and any interface treatment on the final product compared to pure polyurethane. Certain tests are also useful to help elucidate the erosion mechanisms and determine suitable steps to improve performance, which may later involve using a more compliant polymer if increased hardness is detrimental to performance. Strain to failure and elastic modulus are both properties that require evaluation to characterize the final materials, but they are also useful ways to gauge the quality of a material prior to submitting samples and should be classed as benchmark tests.

The base polymer material has a shore hardness of 90 A. Zoran et al. [152] performed shore hardness testing of a more compliant polyurethane with nano-sized and micro-scale silica and observed a 50% increase of shore hardness with nano-sized silica of up to 50% weight loading, but the micron-sized silica rose slightly to 30% weight loadings and then dropped sharply to almost half that of the pure resin at 50% weight. This corresponds to similar results mentioned earlier, with micro hardness testing of nano-sized silica and alumina PEEK [120].

Similar studies with clay in EPDM [78] saw an increase in hardness with nano-sized clay but an actual decrease of micron-sized clay platelets, but other studies have demonstrated no increase in hardness when modified clays have been added to polyurethane [153] despite an increase in breaking strain of 150%.

A hydrophobic nature of materials may be important both in resistance to climatic moisture and in prevention of ice build-up on blades; the contact angle measurement is a useful method for testing this. Chen et al. [154] looked into the effect of filler loading and surface finish of 12 nm silica on polyurethane using this method along with TEM, AFM, and peel testing. The substrate material that the polyurethane nanocomposite was prepared on had an effect on the density of silica at the surface with the silica seeming to prefer the glass substrate compared to the plastic and tin which resulted in less silica at the surface. They found that a higher loading of silica at the surface resulted in higher surface energy and thus lower bonding, and increased hydrophobic behavior to the water–glycol mixture that was chosen as it would not react with the polyurethane matrix. They also found that higher loadings had smoother surfaces when examined, which they felt was due to the fillers filling gaps; however, this could also be due to higher levels of branching from particles associated with the in situ polymerization method.

8.5 Concluding Remarks

Lightning strike, erosion, ice accretion, and environmental corrosion represent significant safety issues for aeronautical operation. The significance of these concerns is likely to increase as more and more composites are used in modern aircraft. It is important to understand the potentially damaging materials and situations that a coating system undergoes in order to identify those properties that are key to total system performance. New nanocomposites combined with nanoclay, carbon nanotube, nanodiamond, alumina, etc., have shed light on the multidisciplinary nature of protective coatings where major mechanical, physical, and chemical issues need to be addressed together. The developing ability to make uniform nanoscale components and to integrate them in complex macromolecular structures will allow technologists to have a new level of control on the physical and chemical properties of components of macroscopic materials.

Novel coatings have also been widely developed for advanced polymer matrix systems. Notably, studies on self-healing polymers have demonstrated the repair of bulk mechanical damage as well as dramatic increases in fatigue life [155]. Also, the technology of self-cleaning coatings has developed rapidly over the years [156] and self-cleaning paint is currently available in Europe [157]. Another development direction concerns volatile organic compound (VOC) emissions, because until now aeronautical coatings still have had relatively high solvent content. Fast reacting coatings requiring dual sprays show promise for very low or even zero VOC. Recently, a second generation paint removable systems devised for the A380 superjumbo by Akzo Nobel Aerospace Coatings (ANAC) enabled the topcoat to be quickly removed during maintenance, cutting the required time by up to 40% by avoiding the need to strip to bare metal [158,159].

Acknowledgments

This chapter contains research results and findings from some recent work carried out within the Centre for Nanoscience and Quantum Information (NSQI), University of Bristol, by the author's research group including Ph.D. student Mr. A.P. Limmack who helped generate Figures 8.10 and 8.29. Particular thanks to Miss W.M. Zhao, who recently passed her Ph.D. viva, for her dedicated effort in collating literature and related information. To the best of our knowledge, the information contained here is relevant and accurate. The financial support from the Engineering and Physical Science Research Council (EPSRC) of the United Kingdom is acknowledged. A thank you also goes to those who kindly authorized permission to use or reproduce figures and drawings; corresponding references are duly indicated.

References

1. Herberts, K. *Oriental Lacquer: Art and Technique*. Thames and Hudson, London, 1962: p. 513.
2. Bansemir, H. and O. Haider. Fiber composite structures for space applications—recent and future developments. *Cryogenics*, 1998. 38(51–59).
3. Bamford, F. GKN Aerospace—A Vision of the Future. http://www.aircraftbuilders.com/UserFiles/File/GKN_Vision_of_the_Future_ABC.pdf, April 2007.
4. Varvani-Farahani, A. Composite materials: characterization, fabrication and application-research challenges and directions. *Applied Composite Materials*, November 5, 2009.

5. Mallick, P.K. *Fiber-Reinforced Composites: Materials, Manufacturing, and Design.* CRC Press, Boca Raton, FL, 1993.
6. Lightning strike protection requirements for service aircraft. Interim defence standard 59-113(1), Defence Procurement Agency, Glasgow, UK, March 3, 2006.
7. Henkhaus, K. Overview of research on composite material impact behaviour. *16th ASCE Engineering Mechanics Conference, 2003.* University of Washington, Seattle.
8. Wall, R., Airbus A350 Completes Milestone 5 Review. *Aviation Week,* January 2, 2009.
9. Mahieux, C. *Environmental Degradation in Industrial Composites.* Elsevier, 2006.
10. Martin, R., Ed., *Ageing of Composites.* Woodhead Publishing, Cambridge, 2008.
11. Chattopadhyay, A.K. and M.R. Zentner, Aerospace and Aircraft Coatings. Federation Series on Coatings Technology, Federation of Societies for Paint Technology, Philadelphia, PA, 1990.
12. Sutter, J.K. et al. *Erosion Resistant Coatings for Polymer Matrix Composites in Propulsion Applications.* NASA/TM-2003-212201.
13. Chen, A.T. and R.T. Wojcik, Polyurethane coatings for metal and plastic substrates. *Metal Finishing,* 2000. 98(6): p. 143–154.
14. Aircraft Surface Coatings. NASA contractor Report 165928, June 1982. Contract NAS1-15325, Task 4.4.
15. Bierwagen, G.P. and D.E. Tallman, Choice and measurement of crucial aircraft coatings system properties. *Progress in Organic Coatings,* 2001. 41(4): p. 201–216.
16. Taylor, S.R. et al., The Development of an Environmentally Compliant, Multi-Functional Aerospace Coating Using Molecular-and Nano-Engineering Methods. 2006, Storming Media.
17. Sanchez, C. et al., Applications of hybrid organic–inorganic nanocomposites. *Journal of Materials Chemistry,* 2005. 15(35–36): p. 3559–3592.
18. Baer, D.R., P.E. Burrows, and A.A. El-Azab, Enhancing coating functionality using nanoscience and nanotechnology. *Progress in Organic Coatings,* 2003. 47(3–4): p. 342–356.
19. Wright, T., Aerospace coatings: the U.S. and European airline industries are flying through rough turbulence, coating world. http://www.coatingsworld.com/articles/2008/08/aerospace-coatings.php., 2008.
20. Aircraft Surface Coatings. NASA Contractor Report 165928, 1982.
21. Blackford, R., Performance demands on aerospace paints relative to environmental legislation. *Pigment and Resin Technology,* 1999. 28(6): p. 331–335.
22. Fisher, F.A., J.A. Plumer, and R.A. Perala, *Lightning Protection of Aircraft.* Lightning Technologies, Pittsfield, MA, 1990.
23. Rupke, E.J.P., *Lightning Direct Effects Handbook.* Lightning Technologies, Pittsfield, MA, 2002.
24. Feraboli, P. and M. Miller, Damage resistance and tolerance of carbon/epoxy composite coupons subjected to simulated lightning strike. *Composites Part A,* 2009. 40(6–7): p. 954–967.
25. Gardiner, G. Lightning strike protection for composite structure. *High-Performance Composites,* July 2006.
26. Gates, D., Building the 787 | When Lightning Strikes. http://seattletimes.nwsource.com/news/business/links/boeinglightni_a.pdf, 2006.
27. Wang, B. et al., Investigation of Lightning and EMI Shielding Properties of SWNT buckypaper Nanocomposites. 2005, DTIC Accession Number: ADA430333.

28. Jilbert, G.H. and J.E. Field, Synergistic effects of rain and sand erosion. *Wear*, 2000. 243(1–2): p. 6–17.
29. Meng, H.C. and K.C. Ludema, Wear models and predictive equations: their form and content. *Wear*, 1995. 181–183(Part 2): p. 443–457.
30. Warren (Andy) Thomas, S.C. Hong, Chin-Jye (Mike) Yu, Edwin L. Rosenzweig, Enhanced Erosion Protection for Rotor Blades. Paper presented at the American Helicopter Society 65th Annual Forum, Grapevine, TX, May 27–29, 2009.
31. Krevelen, D.W., *Properties of Polymers: Their Correlation with Chemical Structure; Their Numerical Estimation and Prediction from Additive Group Contributions.* Elsevier Sci. Pub, Third revised edition, Amsterdam, The Netherlands, 1990.
32. Slikkerveer, P.J., M.H.A. Van Dongen, and F.J. Touwslager, Erosion of elastomeric protective coatings. *Wear*, 1999. 236(1–2): p. 189–198.
33. Slikkerveer, P.J. et al., Erosion and damage by sharp particles. *Wear*, 1998. 217(2): p. 237–250.
34. Brandstadter, A. et al., Solid-particle erosion of bismaleimide polymers. *Wear*, 1991. 147(1): p. 155–164.
35. Walley, S.M., J.E. Field, and P. Yennadhiou, Single solid particle impact erosion damage on polypropylene. *Wear*, 1984. 100: p. 263–280.
36. Hutchings, I.M., D.W.T. Deuchar, and A.H. Muhr, Erosion of unfilled elastomers by solid particle impact. *Journal of Materials Science*, 1987. 22(11): p. 4071–4076.
37. Arnold, J.C. and I.M. Hutchings, Flux rate effects in the erosive wear of elastomers. *Journal of Materials Science*, 1989. 24(3): p. 833–839.
38. Li, J. and I.M. Hutchings, Resistance of cast polyurethane elastomers to solid particle erosion. *Wear*, 1990. 135(2): p. 293–303.
39. Lancaster, J.K., Material-specific wear mechanisms: relevance to wear modelling. *Wear*, 1990. 141(1): p. 159–183.
40. Barkoula, N.M. and J. Karger-Kocsis, Review Processes and influencing parameters of the solid particle erosion of polymers and their composites. *Journal of Materials Science*, 2002. 37(18): p. 3807–3820.
41. Barkoula, N.M., J. Gremmels, and J. Karger-Kocsis, Dependence of solid particle erosion on the cross-link density in an epoxy resin modified by hygrothermally decomposed polyurethane. *Wear*, 2001. 247(1): p. 100–108.
42. Stachowiak, G.W. and A.W. Batchelor, *Engineering Tribology.* Amsterdam; Elsevier, 1993 (Tribology series; 24).
43. Cook, S.S., Erosion by water-hammer. *Proceedings of the Royal Society of London. Series A, Containing Papers of a Mathematical and Physical Character*, 1928. 119(783): p. 481–488.
44. Rouse, H. and S. Ince, *History of Hydraulics.* 1957: Iowa Institute of Hydraulic Research, State University of Iowa.
45. Field, J.E., The physics of liquid impact, shock wave interactions with cavities, and the implications to shock wave lithotripsy. *Physics in Medicine and Biology*, 1991. 36: p. 1475–1484.
46. Bowden, F.P. and J.E. Field, The brittle fracture of solids by liquid impact, by solid impact, and by shock. *Proceedings of the Royal Society of London. Series A, Mathematical and Physical Sciences*, 1964. 282(1390): p. 331–352.
47. Lesser, M., Thirty years of liquid impact research: A tutorial review. *Wear*, 1995. 186: p. 28–34.

48. Hammond, D.A., M.F. Amateau, and R.A. Queeney, Cavitation erosion performance of fiber reinforced composites. *Journal of Composite Materials*, 1993. 27(16): p. 1522.

49. Weigel, W.D., Advanced Rotor Blade Erosion Protection System. Final Report, 1996.

50. 3M Products and Solutions for the Aerospace Industry, http://multimedia. mmm.com/mws/mediawebserver.dyn?CCCCCClJVORCXf1Cpf1CCcRGdnzb bbbA-.

51. Oka, Y.I. et al., Control and evaluation of particle impact conditions in a sand erosion test facility. *Wear*, 2001. 250(1–12): p. 736–743.

52. Tewari, U.S. et al., Solid particle erosion of carbon fiber- and glass fiber-epoxy composites. *Composites Science and Technology*. 63(3–4): p. 549–557.

53. Wood, R.J.K. and D.W. Wheeler, Design and performance of a high velocity air-sand jet impingement erosion facility. *Wear*, 1998. 220(2): p. 95–112.

54. Westmark, C. and G.W. Lawless, A discussion of rain erosion testing at the United States Air Force rain erosion test facility. *Wear*, 1995. 186: p. 384–387.

55. Goodwin, J.E., W. Sage, and G.P. Tilly, Study of erosion by solid particles. *Proc. Inst. Mech. Eng*, 1969. 184: p. 279–292.

56. Ranaudo, R.J. et al. Performance Degradation of a Typical Twin Engine Commuter Type Aircraft in Measured Natural Icing Conditions. 1984.

57. *Aircraft Icing Handbook*. Civil Aviation Authority, Lower Hutt, Zew Zealand, 2000.

58. Armitage, N.P., J.C.P. Gabriel, and G. Grüner. Quasi-Langmuir–Blodgett thin film deposition of carbon nanotubes. *Journal of Applied Physics*, 2004. 95: p. 3228.

59. Simonite, T. Nanotube coating promises ice-free windscreens. NewScientist-Tech, http://www.newscientist.com/article/dn10850-nanotube-coating-prom-ises-icefree-windscreens.html, December 2006.

60. Yoon, Y.H. et al. Transparent film heater using single-walled carbon nanotubes. *Advanced Materials*, 2007. 19(23): p. 4284–4287.

61. http://nationalsecurity.battelle.org/clients/inno_defense.aspx?id=15; March 14, 2007.

62. Ferrick, M.G. et al. Double Lap Shear Testing of Coating Modified Ice Adhesion to Liquid Oxygen Feed Line Bracket, Space Shuttle External Tank. 2006, Storming Media.

63. Bhamidipati, M., Smart anti-ice coatings, Cape Cod Research, Command: NAVSEA Topic: N04-084. 2000: p. 1–6.

64. Examples of Aircraft Corrosion. http://www.corrosionsource.com/technical-library/corrdoctors/Modules/Aircraft/Frames.htm.

65. Miller, D. Corrosion control on aging aircraft: What is being done. *Materials Performance*, October 1990: p. 10–11.

66. Findlay, S.J. and N.D. Harrison. Why aircraft fail. *Materials Today*, 2002. 5(11): p. 18–25.

67. Huttunen-Saarivirta, E. et al. Corrosion behaviour of aircraft materials in acetate and formate based antiskid chemicals. *Corrosion Engineering, Science and Technology*, 2008. 43(1): p. 64–80.

68. Legg, K. *The Emerging Future of Surface Technology in Defense*. Rowan Technology Group, 2009. SUR/FIN.

69. Sørensen, P.A. et al. Anticorrosive coatings: A review. *Journal of Coatings Technology and Research*, 2009. 6(2): p. 135–176.

70. Huttunen-Saarivirta, E. et al. Corrosion behaviour of aircraft coating systems in acetate-and formate-based de-icing chemicals. *Materials and Corrosion*, 2009. 60(3): p. 173–191.
71. Kjernsmo, D., Kleven, and K, Scheie, J. *Corrosion Protection*. Bording A/S, Copenhagen 2003.
72. Lambourne, R. and T.A. Strivens. *Paint and Surface Coatings—Theory and Practice*. Woodhead, Cambridge, 1999.
73. Fu, M.H. et al. *Polyurethane Elastomer and Its Application*. Chemical Industry Impress, Beijing, 1994.
74. Howarth, G.A. Polyurethanes, polyurethane dispersions and polyureas: Past, present and future. *Surface Coatings International Part B: Coatings Transactions*, 2003. 86(2): p. 111–118.
75. Chattopadhyay, D.K. and K. Raju. Structural engineering of polyurethane coatings for high performance applications. *Progress in Polymer Science*, 2007. 32(3): p. 352–418.
76. Liu, W., S.V. Hoa, and M. Pugh. Organoclay-modified high performance epoxy nanocomposites. *Composites Science and Technology*, 2005. 65(2): p. 307–316.
77. Yao, K.J. et al. Polymer/layered clay nanocomposites: 2 polyurethane nanocomposites. *Polymer*, 2002. 43(3): p. 1017–1020.
78. Ahmadi, S.J., Y. Huang, and W. Li, Fabrication and physical properties of EPDM–organoclay nanocomposites. *Composites Science and Technology*, 2005. 65(7–8): p. 1069–1076.
79. Liu, T. et al. Preparation and characterization of nylon 11/organoclay nanocomposites. *Polymer*, 2003. 44(12): p. 3529–3535.
80. Lam, C. et al. Effect of ultrasound sonication in nanoclay clusters of nanoclay/epoxy composites. *Materials Letters*, 2005. 59(11): p. 1369–1372.
81. Lau, K. et al. Thermal and mechanical properties of single-walled carbon nanotube bundle-reinforced epoxy nanocomposites: the role of solvent for nanotube dispersion. *Composites Science and Technology*, 2005. 65(5): p. 719–725.
82. Delozier, D.M. et al. Preparation and characterization of space durable polymer nanocomposite films. *Composites Science and Technology*, 2005. 65(5): p. 749–755.
83. Wang, Y. et al. Study on the preparation and characterization of ultra-high molecular weight polyethylene–carbon nanotubes composite fiber. *Composites Science and Technology*, 2005. 65(5): p. 793–797.
84. Bright, I. et al. Carbon nanotubes for integration into nanocomposite materials. *Microelectronic Engineering*, 2006. 83(4–9): p. 1542–1546.
85. Zhou, R. et al. Mechanical properties and erosion wear resistance of polyurethane matrix composites. *Wear*, 2005. 259(1–6F): p. 676–683.
86. Fornes, T.D. et al. Nylon 6 nanocomposites: the effect of matrix molecular weight. *Polymer*, 2001. 42(25): p. 09929–09940.
87. Fukushima, Y. and S. Inagaki. Synthesis of an intercalated compound of montmorillonite and 6-polyamide. *Journal of Inclusion Phenomena and Macrocyclic Chemistry*, 1987. 5(4): p. 473–482.
88. Hare, C.H. *Protective Coatings: Fundamentals of Chemistry and Composition*. C. H. Hare, Technology Publishing Co., Pittsburgh, 1994, 514.
89. Funke, W., Towards environmentally acceptable corrosion protection by organic coatings: problems and realization. *JCT, Journal of Coatings Technology*, 1983. 55(705): p. 31–38.

90. Wang, D. and C.A. Wilkie, A stibonium-modified clay and its polystyrene nano-composite. *Polymer Degradation and Stability*, 2003. 82(2): p. 309–315.
91. LeBaron, P.C., Z. Wang, and T.J. Pinnavaia, Polymer-layered silicate nanocomposites: An overview. *Applied Clay Science*, 1999. 15(1–2): p. 11–29.
92. Yano, K., A. Usuki, and A. Okada. Synthesis and properties of polyimide-clay hybrid films. *Journal of Polymer Science Part A: Polymer Chemistry*, 1997. 35(11): p. 2289–2294.
93. Daniel, L. and M. Chipara, Carbon Nanotubes and Their Polymer-Based Composites in Space Environment. AIAA-2009-6769, AIAA SPACE 2009 Conference & Exposition Pasadena, CA, September 14–17, 2009.
94. Mawhinney, D.B. et al. Surface defect site density on single walled carbon nano-tubes by titration. *Chemical Physics Letters*, 2000. 324(1–3): p. 213–216.
95. Kannan, B. and B. Marko. Chemically functionalized carbon nanotubes. *Small*, 2005. 1(2): p. 180–192.
96. Zhao, J. et al. Electronic properties of carbon nanotubes with covalent sidewall functionalization. *The Journal of the Physical Chemistry B*, 2004. 108(14): p. 4227–4230.
97. Chen, J. et al. Solution properties of single-walled carbon nanotubes. *Science*, 1998. 282(5386): p. 95.
98. Hamon, M.A. et al. Dissolution of single-walled carbon nanotubes. *Advanced Materials*, 1999. 11(10): p. 834–840.
99. Monthioux, M. et al. Sensitivity of single-wall carbon nanotubes to chemical processing: an electron microscopy investigation. *Carbon*, 2001. 39(8): p. 1251–1272.
100. Hu, H. et al. Determination of the acidic sites of purified single-walled carbon nanotubes by acid–base titration. *Chemical Physics Letters*, 2001. 345(1–2): p. 25–28.
101. Grossiord, N. et al. Toolbox for dispersing carbon nanotubes into polymers to get conductive nanocomposites. *Chemistry of Materials*, 2006. 18(5): p. 1089–1099.
102. Andreas, H. Functionalization of single-walled carbon nanotubes. *Angewandte Chemie International Edition*, 2002. 41(11): p. 1853–1859.
103. Jung, Y.C., N.G. Sahoo, and J.W. Cho. Polymeric nanocomposites of polyure-thane block copolymers and functionalized multi-walled carbon nanotubes as crosslinkers. *Macromolecular Rapid Communications*, 2006. 27(2): p. 126–131.
104. Sun, Y.P. et al. Functionalized carbon nanotubes: properties and applications. *Accounts of Chemical Research*, 2002. 35(12): p. 1096–1104.
105. Homenick, C.M., G. Lawson, and A. Adronov, Polymer grafting of carbon nanotubes using living free-radical polymerization. *Polymer Reviews*, 2007. 47(2): p. 265–290.
106. Gao, C. et al. Polyurea-functionalized multiwalled carbon nanotubes: synthesis, morphology, and Raman spectroscopy. *The Journal of Physical Chemistry B*, 2005. 109(24): p. 11925–11932.
107. Xiong, J. et al. The thermal and mechanical properties of a polyurethane/multi-walled carbon nanotube composite. *Carbon*, 2006. 44(13): p. 2701–2707.
108. Dai, H. et al. Synthesis and characterization of carbide nanorods. *Nature*, 1995. 375: p. 769–772.
109. Han, W. et al. Synthesis of gallium nitride nanorods through a carbon nanotube-confined reaction. *Science*, 1997. 277(5330): p. 1287.
110. Bezryadin, A., C.N. Lau, and M. Tinkham, Quantum suppression of supercon-ductivity in ultrathin nanowires. *Nature*, 2000. 404: p. 971–974.

111. Wildgoose, G.G., C.E. Banks, and R.G. Compton, Metal nanoparticles and related materials supported on carbon nanotubes: Methods and applications. *Small*, 2006. 2(2): p. 182.

112. Gruner, G. Carbon nanonets spark new electronics. *Scientific American Magazine*, 2007. 296(5): p. 76–83.

113. Rinzler, A.G. et al. Large-scale purification of single-wall carbon nanotubes: Process, product, and characterization. *Applied Physics A: Materials Science and Processing*, 1998. 67(1): p. 29–37.

114. New NCC company to explore nanomaterials. *High Performance Composites*, Composites Technology Article Archive, 2009. http://www.compositesworld. com/news/news-ncc-company-to-explore-nanomaterials.

115. Gruner, G. Carbon nanotube films for transparent and plastic electronics. *Journal of Materials Chemistry*, 2006. 16(35): p. 3533–3539.

116. Akil, H.M. et al. Effect of various coupling agents on properties of alumina-filled PP composites. *Journal of Reinforced Plastics and Composites*, 2006. 25(7): p. 745.

117. Bauer, F. et al. Preparation of scratch-and abrasion-resistant polymeric nanocomposites by monomer grafting onto nanoparticles, 4 Application of MALDI-TOF mass spectrometry to the characterization of surface modified nanoparticles. *Macromolecular Chemistry and Physics*, 2003. 204(3): p. 375–383.

118. Glasel, H.J. et al. Preparation of scratch and abrasion resistant polymeric nanocomposites by monomer grafting onto nanoparticles. 2. Characterization of radiation-cured polymeric nanocomposites. *Macromolecular Chemistry and Physics*, 2000. 201(18): p. 2765–2770.

119. Bauer, F. et al. Trialkoxysilane grafting onto nanoparticles for the preparation of clear coat polyacrylate systems with excellent scratch performance. *Progress in Organic Coatings*, 2003. 47(2): p. 147–153.

120. Kuo, M.C. et al. PEEK composites reinforced by nano-sized SiO_2 and $Al2O_3$ particulates. *Materials Chemistry and Physics*, 2005. 90(1): p. 185–195.

121. Hosseinpour, D. et al. The effect of interfacial interaction contribution to the mechanical properties of automotive topcoats. *Progress in Organic Coatings*, 2005. 54(3): p. 182–187.

122. Wigg, P.J.C., P.W. May, and D. Smith. Stiffness measurements of diamond fiber reinforced plastic composites. *Diamond and Related Materials*, 2003. 12(10–11): p. 1766–1770.

123. Shenderova, O.A., V.V. Zhirnov, and D.W. Brenner. Carbon nanostructures. *Critical Reviews in Solid State and Materials Sciences*, 2002. 27(3–4): p. 227–356.

124. Gogotsi, Y. et al. Conversion of silicon carbide to crystalline diamond-structured carbon at ambient pressure. *Nature*, 2001. 411(6835): p. 283–286.

125. Aleksenskii, A.E. et al. The structure of diamond nanoclusters. *Physics of the Solid State*, 1999. 41(4): p. 668–671.

126. Krueger, A., The structure and reactivity of nanoscale diamond. *Journal of Materials Chemistry*, 2008. 18(13): p. 1485–1492.

127. Khabashesku, V.N., J.L. Margrave, and E.V. Barrera. Functionalized carbon nanotubes and nanodiamonds for engineering and biomedical applications. *Diamond and Related Materials*, 2005. 14(3–7): p. 859–866.

128. Lasseter, T.L. et al. Covalently modified silicon and diamond surfaces: resistance to nonspecific protein adsorption and optimization for biosensing. *Journal of the American Chemical Society*, 2004. 126(33): p. 10220–10221.

129. Shenderova, O.A., V.V. Zhirnov, and D.W. Brenner, Carbon nanostructures. *Critical Reviews in Solid State and Materials Sciences*, 2002. 27(3–4): p. 227.
130. Liu, Y. et al. Functionalization of nanoscale diamond powder: fluoro-, alkyl-, amino-, and amino acid-nanodiamond derivatives. *Chemistry Materials*, 2004. 16(20): p. 3924–3930.
131. Spitsyn, B.V. et al. Inroad to modification of detonation nanodiamond. *Diamond and Related Materials*. 15(2–3): p. 296–299.
132. Lisichkin, G.V. et al., Photochemical chlorination of nanodiamond and interaction of its modified surface with C-nucleophiles. *Russian Chemical Bulletin*, 2006. 55(12): p. 2212–2219.
133. Butenko, Y.V. et al. Photoemission study of onionlike carbons produced by annealing nanodiamonds. *Physical Review B*, 2005. 71(7): p. 075420.
134. Krueger, A. et al. Deagglomeration and functionalisation of detonation diamond. *Physica Status Solidi (a)*, 2007. 204(9): p. 2881–2887.
135. Tsubota, T. et al. Chemical modification of hydrogenated diamond surface using benzoyl peroxides. *Physical Chemistry Chemical Physics*, 2002. 4(5): p. 806–811.
136. Nakamura, T. et al. Chemical modification of diamond powder using photolysis of perfluoroazooctane. *Chemical Communications*, 2003. 2003(7): p. 900–901.
137. Ando, T. et al. Chemical modification of diamond surfaces using a chlorinated surface as an intermediate state. *Diamond and Related Materials*, 1996. 5(10): p. 1136–1142.
138. Ando, T. et al. Direct interaction of elemental fluorine with diamond surfaces. *Diamond and Related Materials*, 1996. 5(9): p. 1021–1025.
139. Li, L., J.L. Davidson, and C.M. Lukehart. Surface functionalization of nanodiamond particles via atom transfer radical polymerization. *Carbon*, 2006. 44(11): p. 2308–2315.
140. Maitra, U. et al. Mechanical properties of nanodiamond-reinforced polymer-matrix composites. *Solid State Communications*, 2009. 149(39–40): p. 1693–1697.
141. Moniruzzaman, M. and K.I. Winey. Polymer nanocomposites containing carbon nanotubes. *Macromolecules*, 2006. 39(16): p. 5194–5205.
142. Tian, M. et al. Structure and properties of fibrillar silicate/SBR composites by direct blend process. *Journal of Materials Science*, 2003. 38(24): p. 4917–4924.
143. *Painting and Marking of Army Aircraft*. Technical Manual, December 30, 1998. TM 55-1500-345-23.
144. Sheet, Q., J. Lyons, and F. Ko. Melt electrospinning of polymers: A review. *Polymer News*, 2005. 30: p. 1–9.
145. Ajayan, P.M. et al. Single-walled carbon nanotube-polymer composites: strength and weakness. *Advanced Materials*, 2000. 12(10): p. 750–753.
146. Cheremisinoff, N.P. *Elastomer Technology Handbook*. CRC Press, Boca Raton, FL, 1993.
147. Xia, H. and M. Song. Preparation and characterisation of polyurethane grafted single-walled carbon nanotubes and derived polyurethane nanocomposites. *Journal of Materials Chemistry*, 2006. 16(19): p. 1843–1851.
148. Ramasubramaniam, R., J. Chen, and H. Liu. Homogeneous carbon nanotube/polymer composites for electrical applications. *Applied Physics Letters*, 2003. 83: p. 2928.
149. Gooch, J.W. and J.K. Daher. *Electromagnetic Shielding and Corrosion Protection for Aerospace Vehicles*. Springer Science+ Business Media, LLC, Berlin, 2007.

150. Yang, Y., M.C. Gupta, and K.L. Dudley. Towards cost-efficient EMI shielding materials using carbon nanostructure-based nanocomposites. *Nanotechnology*, 2007. 18: p. 345701.
151. Gou, J. et al. Carbon nanofiber paper for lightning strike protection of composite materials. *Composites Part B: Engineering*. 41(2): p. 192–198.
152. Petrovi, Z.S. et al. Effect of silica nanoparticles on morphology of segmented polyurethanes. *Polymer*, 2004. 45(12): p. 4285–4295.
153. Song, M. et al. High performance nanocomposites of polyurethane elastomer and organically modified layered silicate. *Journal of Applied Polymer Science*, 2003. 90(12): p. 3239–3243.
154. Chen, X. et al. Surface and interface characterization of polyester-based polyurethane/nano-silica composites. *Surface and Interface Analysis*, 2003. 35(4): p. 369–374.
155. Cho, S.H., S.R. White, and P.V. Braun. *Self-Healing Polymer Coatings*. Advanced Materials, 2009. 21: p. 645–649.
156. Parkin, I.P. and R.G. Palgrave. Self-cleaning coatings. *Journal of Materials Chemistry*, 2005. 15(17): p. 1689–1695.
157. http://www.paintpro.net/Articles/PP705/PP705_ProductProfiles.cfm.
158. Armanios, D.E. Parochialism in Eu economic policy: case study between the Boeing Company and the Airbus Company. *International Journal of Technology, Policy and Management*, 2006. 6(1): p. 66–85.
159. Turion-Kahlmann, M. Akzo Nobel supplies Airbus A380 with advanced coatings system. *Anti-Corrosion Methods and Materials*, 53, 2006.

9

Surface Modification of Carbon Nanotubes (CNTs) for Composites

Joong Hee Lee and Nam Hoon Kim

Chonbuk National University

N. Satheesh Kumar

National University of Malaysia (UKM)

Basavarajaiah Siddaramaiah

Sri Jayachamarajendra College of Engineering

CONTENTS

9.1 Introduction

Carbon nanotubes (CNTs) were first discovered in 1991 [1]. A CNT is defined as a sheet of carbon atoms rolled up into a tube with a diameter in the order of tens of nanometers. Researchers all over the world have attempted to obtain purified single or multiwalled carbon nanotubes. Multiwalled carbon nanotubes (MWNTs) are comprised of several concentric cylinders of graphite sheets, with an approximately 3.4 Å spacing between layers. MWNTs usually have a diameter in the order of 2–100 nm (typically 2–10 nm in internal diameter), which is slightly larger than that of a single layer cylindrical graphite sheet of single-walled carbon nanotubes (SWNTs), which is approximately 0.2–2 nm in diameter. The length of the CNTs can vary from micrometers to centimeters, giving them a very high aspect ratio (length to diameter).

The one dimensional, all-carbon structure of a CNT is responsible for its superior mechanical, electrical, and thermal properties [2–5]. The high tensile strength (in the range 20–100 GPa), low density (for light weight materials), and high aspect ratio (with nanoscale diameter) of CNTs have made CNTs attractive as reinforcing agents at very low concentrations of polymer composites [6]. These unique properties of CNTs have opened up the development of multifunctional materials, such as conducting polymers with improved mechanical performance [7].

CNTs have been examined with greater emphasis, improving the mechanical and electronic properties of polymer composite materials. CNTs are held together as bundles through van der Waals interactions. Therefore, CNTs exhibit inferior properties to the exfoliated individual CNTs. CNTs alone cannot be used to form structural components without a supporting medium, such as a matrix. The effective utilization of CNTs in a polymer matrix to fabricate composites for a variety of applications requires (1) improvement in interfacial bonding between CNTs and polymer chains, and (2) proper dispersion of CNTs in a polymer matrix. Interfacial bonding between the CNTs and host polymer dictates the final properties of the CNT-reinforced polymer composite. The expected interfacial bonding stress between CNTs and the polymer can be as much as 500 MPa [8]. The short response time and low input voltage are the main advantages of CNTs compared to other electrochemically active materials [9].

The utilization of both SWNTs and MWNTs as reinforcing agents for thermoset (polyimide, epoxy, phenolic, etc.) [10–14] and thermoplastic polymers (polyether ether ketone [PEEK], polymethyl methacrylate [PMMA], nylon-12, etc.) [15–17] has been reported. Tailoring the nature of the CNT wall is important for enhancing the interfacial interactions between the polymer chains and CNTs. The surface modification of CNTs improves the interfacial interactions between CNTs and polymer chains. Although there are few reports on the utilization of CNTs without surface modification, surface modified

CNTs exhibit improved dispersion and interfacial bonding between the polymer and nanotubes. The two main approaches for the surface modification of CNTs are (1) covalent attachment [18–21] of functional groups to the walls of the nanotubes, and (2) noncovalent attachment controlled by thermodynamic criteria [22]. The mechanical and electrical properties of CNTs are not expected to change after noncovalent surface modification [23]. Covalent modification occurs through the attachment of chemical groups to the nanotube. This chapter discusses the reported research on the different surface modification routes used for CNTs.

9.2 Surface Modification of CNTs

While a large knowledge base exists for incorporating micron-sized carbon-based fibers, there are few reports on the incorporation of CNTs. A major drawback in the utilization of CNTs as bulk materials is the difficulty in integrating them homogeneously into a matrix because of their pronounced agglomeration tendency. It is likely that chemical functionalization will facilitate dispersion, stabilize the CNTs, and prevent agglomeration, which can lead to the formation of defects. For many applications, it is essential to tailor the chemical nature of CNT walls to take advantage of their unique properties. There are two main approaches for the surface modification of CNTs. Noncovalent attachment, which for some polymer chains is called "wrapping," is controlled by thermodynamic criteria [24] and can alter the nature of the CNT surface, rendering it more compatible with the polymer matrix. The advantage of noncovalent attachment is that the perfect structure of the nanotube is not altered. Therefore, its mechanical properties are not expected to change. The disadvantage of noncovalent attachment is the weak force of attraction between the wrapping molecule and nanotube. This can lead to poor load transfer when the CNTs are applied in composites. The main approach for modifying CNTs is the chemical covalent attachment of functional groups. Acid oxidation is used most commonly in this category. The CNTs are opened at the ends, and terminal carbons are converted to carboxylic acids by oxidization in concentrated sulfuric or nitric or mixed acid [25]. Considerable work has also been carried out on amination [26,27], fluoration [28,29], and long alkyl chain grafting [30] through chemical reactions. Several other methods, such as polymer wrapping [24], electrochemical [31], mechanical–chemical [32], and plasma treatment have been reported as routes for the functionalization of CNTs. The covalent attachment of functional groups to the surface of CNTs is expected to improve the efficiency of load transfer. However, it should be noted that these functional groups might introduce defects to the walls of the perfect structure of the CNTs, which will reduce the strength of the reinforcing component. Hence, there is a trade-off between the strength of the interface and that of the nanotube filler. The functional groups attached can be small molecules [33,20] or polymer chains. Chemical functionalization is a particularly attractive target because it can improve the solubility [34], processability, and allow the unique properties of SWNTs to be coupled with other materials. The ability to dissolve and separate discrete CNT molecules from tight bundles opens a new avenue in the field of nanotechnology. In addition these capabilites would aid in the purification of the molecules, eventually allowing easier manipulation. This section focuses on recent developments in the surface modification of CNTs with polymers.

9.2.1 Chemical Modification

CNTs are assembled as ropes or bundles. The as-grown CNTs contain some residual catalyst, bucky onions, spheroidal fullerenes, amorphous carbon, polyhedron graphite nanoparticles, and other forms of impurities. Therefore, purification, "cutting," or disentangling, and activation treatments are needed before chemical functionalization. The utilization of CNTs in polymer–CNT composite applications requires a good dispersion of CNTs.

Covalent modification and the surface chemistry of CNTs are very important factors for nanotube processing and applications [20]. Recently, many efforts toward polymer–composite reinforcement have focused on integrating chemically modified CNTs containing different functional groups into a polymer matrix. Covalent functionalization can be realized either by the modification of surface-bound carboxylic acid groups on the nanotubes or by the direct addition of reagents to the sidewalls of nanotubes. Considering the above, in situ polymerization is one of the main approaches for preparing polymer-grafted nanotubes. Therefore, studies on the preparation of polymer grafted CNTs frequently overlap the work on in situ polymerization processing. This subsection discusses only the recent advances in the production of polymer-grafted CNTs, which have been used to improve the mechanical properties of polymer composites.

To facilitate CNTs for different applications, many approaches have been used to functionalize them, either covalently or noncovalently. Among the many functional groups, the carboxyl group is quite attractive because it can be used for further covalent and noncovalent functionalization of CNTs [25,29,32,36–42]. More recently, a number of research groups focused on the functionalization of CNTs with polymers because the long polymer chain helps the tubes dissolve in common solvents, even with a low degree of functionalization. Covalent sidewall functionalization also helps to disentangle the CNT bundles and disperse the CNTs into high-aspect-ratio, small bundles or individual tubes which increases the tensile modulus and strength of the resulting CNT–polymer composites [43]. Figure 9.1 shows various types of covalent sidewall functionalization of CNTs.

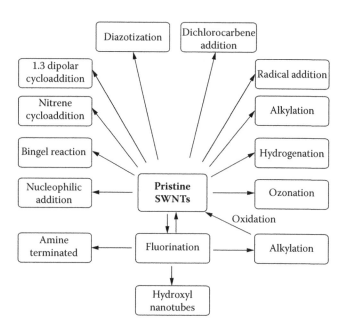

FIGURE 9.1
Schematic diagram of the covalent sidewall functionalization of CNTs. (Redrawn after Banarjee et al., *Adv Mater* 2005; 1:17 © 2009, John Wiley & Sons.)

The two-step functionalization of nanotubes through the introduction of carboxyl groups by oxidation followed by the formation of amide or ester linkages allows stable chemical modification. However, it has little influence on the electrical and mechanical properties of the nanotubes. In comparison, addition reactions enable the direct coupling of functional groups to the p-conjugated carbon framework of the tubes. The required reactive species (atoms, radicals, carbenes, or nitrenes) are generally made available through thermally activated reactions. Small-diameter tubes are preferred on the account of their higher chemical reactivity due to the increased curvature [44]. A series of addition reactions have been well documented, the most important of which are listed in Figure 9.2. The initial experiments showed the addition of three functional groups per 100 carbon atoms to the sidewall through an addition reaction [45]. However, the procedures developed at the later stages showed a higher degree of functionalization up to 10% [46]. In principle, the addition reaction can be initiated exclusively on an intact sidewall, or in parallel at the defect sites from where the reaction can proceed further. For most reactions shown in Figure 9.2, more study will be needed to determine the extent to which these two possibilities contribute. One exception is nanotube fluorination, in which direct addition to the defect-free sidewall appears viable [47]. Nonetheless, the addition of fluorine has a noticeable activation barrier because the reaction requires slightly elevated temperatures (>150°C). Analogous to nanotube functionalization

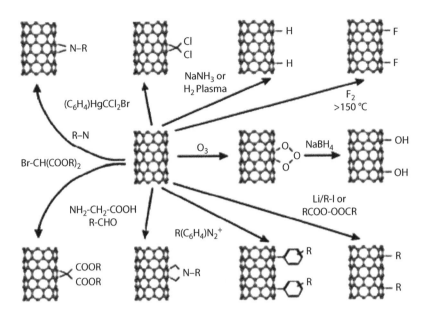

FIGURE 9.2
Overview of the possible addition reactions for functionalization of the nanotube sidewall.
(Reprinted with permission from Balasubramanian, K. and Burghard, M. *Small* 2005; 2:180. ©
2009, Wiley–VCH Verlag GmbH & Co.)

with carboxyl groups, the direct covalent attachment of functional moieties
to the sidewalls strongly enhances the solubility of the nanotubes. The good
solubility of CNTs modified with organic groups has been exploited for their
effective purification [48]. In this procedure, small particles are first separated from the solution by chromatography or filtration and the covalently
attached groups are then removed by thermal annealing (>25°C).

9.2.2 Substitution Reactions

The production of one-dimensional (1D) nanomaterials utilizing CNTs has
received considerable attention. The formation of nanowires and nanotubes
has been achieved successfully using CNT as a template [49–51]. Some nanotubes can be formed on the surfaces of CNTs without a chemical reaction
with the template, while other products are formed via a direct substitution
reaction between the CNTs and starting materials [52]. Either the CNT template maintains the original structure, or completely changes to a new compound with the skeleton of the CNT template maintaining the final products
in the 1D structure. The possibility of nanotubes made from noncarbon elements, such as boron nitride (BN), B-, and N-doped carbon nanotubes, has
been investigated [53–56]. Substitution reactions in CNTs by B, N, and other
species contribute to the generation of interesting materials for electronic
applications [57]. It was reported that a bamboo sheet-like morphology can

be obtained by the synthesis of N-doped multiwalled CNTs [58]. The potential doping of CNTs with different types of atoms might provide a mechanism for controlling their electronic properties. CNTs doped with B and/or N can be used to engineer the band gaps of quasi-one-dimensional materials for electronics applications [59]. The substitution reaction within a nanotube may involve a three step process: (1) the creation of a vacancy or a defect in the perfect structures of CNTs, (2) incorporation of external substituting atoms at a newly created vacancy or defect site, and (3) equilibration or redistribution via the diffusion of one or many substituted atoms into energetically favorable morphologies or configurations in the systems. Carbon nanotubes have been used as a template for the synthesis of one-dimensional silicon carbide (SiC) and other nanostructures [51,60,61]. Li et al. [62] examined substitution reactions between multiwalled CNTs and silicon monoxide vapor by transmission electron microscopy (TEM). The authors observed a reaction inside and on the external surfaces of multiwalled nanotubes, which led to the formation of silicon carbide nanowires with a core–shell structure. The investigation by Deppak et al. [57] on nitrogen substitution reactions in a graphene sheet and carbon nanotubes showed the formation of a vacancy in a curved graphene sheet or a CNT, which caused a curvature-dependent local reconstruction of the surface. The formation of crystalline nano rods with a typical diameter of 6 and 30 mm was reported by Han et al. [63] with boron-doped CNTs prepared by a partial substitution reaction. X-ray diffraction (Figure 9.3) revealed boron-doped CNTs to be more graphitic and crystalline than undoped CNTs, as evidenced by the higher shift in the angle of incidence (20). Li et al. [64] reported the synthesis and growth mechanism of silicon carbide nanowires (SiCNWs) on a CNT template via a direct substitution reaction. They showed that high purity SiCNWs with a typical diameter and length of approximately 40 nm and > 10 μm respectively, could be synthesized on a CNT template. The synthesis of alumina nanotubes using CNTs as a template at 1473 K was reported by Zhang et al. [65]. They showed that the synthesized alumina nanotubes and nanowires were polycrystalline and single crystals, respectively. The effect of temperature on the BN substitution of SWNTs was studied by Hasi et al. [66], who reported that no reaction took place below 1150°C. Between 1150°C and 1270°C, there was a weak shift in the heat-treated tubes indicating some substitution reaction had occurred.

9.2.3 Electrochemical Modification

The electrochemical modification (ECM) of CNTs is a clean and nondestructive method that results in electrochemical functionalization via the electrogeneration of radicals in the vicinity of the SWNT sidewall. ECM can be used for the selective grafting of individual metallic nanotubes. ECM has been used to produce covalently or noncovalently bonded organic moieties on individual contacted SWNTs through appropriate control of the electrochemical conditions. As a result, electrochemistry has developed into an elegant tool for

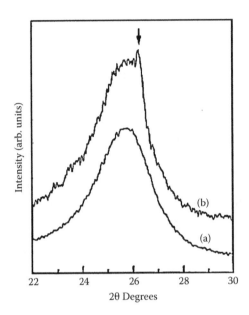

FIGURE 9.3
X-ray diffraction of (a) undoped carbon nanotube and (b) boron-doped carbon nanotube. (Reprinted with permission from Han et al., *Chem Phys Lett* 1999; 299:368–373 © 2009, Elsevier.)

functionalizing CNTs in a selective and controlled manner. To this end, a constant potential (potentiostatic) or constant current (galvanostatic) is applied to a CNT electrode immersed in a solution containing a suitable reagent whereby highly reactive (radical) species can be generated by electron transfer between the CNTs and reagent. Many organic radical species tend to react with the starting reagent or self-polymerize, resulting in a polymer coating on the tubes. Depending on the reagent used, the polymeric layer may or may not be bonded in a covalent manner to the nanotube sidewall. In addition to being simple, clean, and efficient, ECM appears to be versatile because it allows for accurate control over the extent of film deposition through the choice of suitable electrochemical conditions, that is, the duration and magnitude of the applied potential. Moreover, the surface properties of the coated nanotubes can be tailored from highly hydrophilic to predominantly hydrophobic using reagents containing the appropriate substituents,. ECM has proven to be a suitable tool for modifying entangled SWNTs networks and individual SWNTs.

Phenyl residues have been grafted covalently onto bulk samples of SWNTs by the electrochemical coupling of aromatic diazonium salts [32]. In one of the first experiments performed in this direction, phenyl rings bearing various substituents were linked covalently to a SWNT. The Raman spectra of the SWNT showed a noticeable change after modification, most importantly an increase in the relative intensity of the D-line at ~1290 cm^{-1}, which is sensitive to disorder within the sp^2-bonded carbon framework [67]. In addition, there was a significant decrease in the intensity of the radial breathing mode (RBM), which suggests a strong disturbance in the hexagonal lattice of the sidewalls. The extent of modification has been found to vary according to the type of aryl diazonium salt used. After ECM, up to one in 20 carbon atoms was found to possess an aromatic residue corresponding to a 5% degree of functionalization. The coupling mechanism has been interpreted as involving the transfer of an electron from the CNTs. This reduces the diazonium salt to a highly reactive phenyl radical, which then attaches covalently to the sidewall. More recently, Raman spectroscopy demonstrated the formation of aryl chains on the sidewalls of the nanotubes [68]. Although the first electrochemical couplings were achieved using HIPCO-produced SWNTs, whose reactivity is enhanced by their small diameters, down to 0.6 nm, [69] large-diameter nanotubes were also found to be amenable to covalent ECM under specific conditions [70].

The strong capability of ECM becomes particularly apparent when applied to individual nanotubes. This has been demonstrated by the covalent attachment of substituted phenyl groups to individual SWNTs via the diazonium coupling route [71]. This method yields homogeneous tube coatings with a thickness as low as 3–4 nm. The reported layer thickness increased with increasing duration of ECM, the magnitude of the applied potential, and the concentration of diazonium salt. Detailed studies on the effect of grafted coatings on the electronic and structural properties still remain a challenge when performed at the single nanotube level. The combination of electrical transport measurements and confocal Raman microscopy is useful for assessing the aforementioned properties [72]. Diazonium coupling causes an increase in both the electrical resistance and relative Raman D-line intensity for a metallic SWNT. Combined electrical and Raman studies also allow differentiation between the changes in semiconducting tubes that arise from doping due to covalent modification of their carbon framework. Although the tube clearly shows an increase in resistance after the coating, the D-band in the Raman spectrum is almost unaffected. This suggested that the radical cations created do not attach covalently to the nanotube wall, which means the observed increase in resistance is due to n-type doping through the oxidation of the amine, leading to a lower hole concentration in p-type tubes.

According to the literature procedures, the acid-functionalization of MWNT sidewalls (MWNTs-R$_1$-COOH) is achieved by a free-radical addition reaction of alkyl groups terminated with a carboxylic acid group (Scheme 9.1) [72]. Purified MWNTs (500 mg) were placed in a dry 500 mL flask containing 200 mL of dry DMF and sonicated (40 kHz) for 30 min to disperse the MWNTs. The suspension was then kept between 80°C and 90°C for 10 days while 1.0 g of peroxide was added each day. After the reaction was completed, the suspension was cooled and filtered through a polytetrafluoroethylene (PTFE) membrane with a pore size of 0.2 μm. The functionalized MWNTs (MWNTs-R$_1$-COOH) were collected on the membrane. During filtration, a large amount of ethanol was

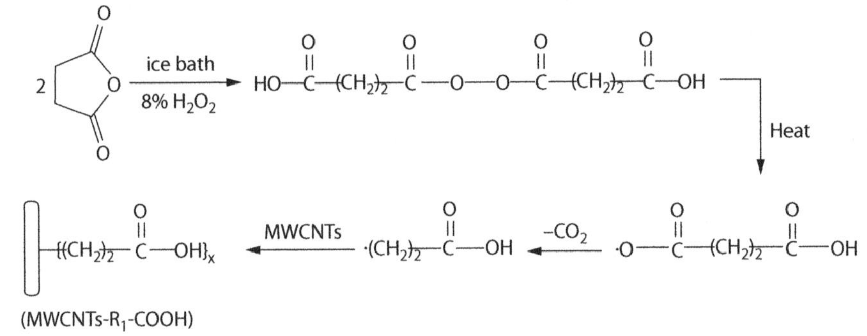

SCHEME 9.1
Preparation of sidewall acid–functionalized MWNTs. (Reprinted with permission from Peng et al., *J Am Chem Soc* 2003; 125:151–174 © 2009, ACS Publication.)

used to wash off the unreacted reagents and by-products. Finally, the solids of MWNTs-R_1-COOH were obtained after vacuum drying overnight at 70°C.

The amino-functionalized MWNTs (MWNTs-R_2-NH_2) were prepared using the following procedures (Scheme 9.2): MWNTs-R_1-COOH (150 mg) was placed in a dry 100 mL flask filled with 50 mL of $SOCl_2$, and the mixture was heated under reflux for 24 h. The residual $SOCl_2$ was removed by reduced-pressure distillation. A large amount of petroleum ether was then used to remove the unreacted reagents. The solid precipitate was filtered through a PTFE membrane with a pore size of 0.2 µm. The acyl chloride-functionalized MWNTs (MWNTs-R_1-COCl) were collected on the membrane. The prepared MWNTSL-R_1-COCl was placed immediately in a dry 100 mL flask, filled with 50 mL of anhydrous DMF and an excess of 1, 6-diaminohexane was added. The mixture was stirred at room temperature for 12 h, and the MWNTSLs-R_2-NH_2 sample was obtained after repeated filtration, washing, and vacuum drying overnight at 70°C.

9.2.4 Photochemical Modification

Carbon nanotubes exhibit unusual optoelectronic properties, as evidenced by a report on the ignition of dry and fluffy SWNTs in air by a camera flash [73]. The observed ignition was attributed to a rapid increase in temperature

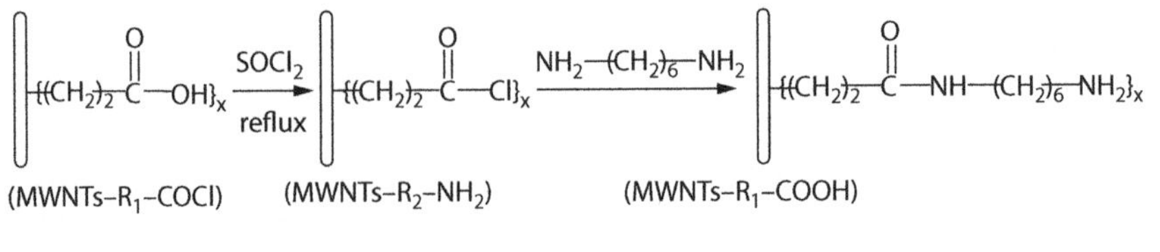

SCHEME 9.2
Preparation of the amido derivatives of MWNTs. (Reprinted with permission from Gabriel et al., *Carbon* 2006; 44:1891 © 2009, Elsevier.)

within the nanotubes upon exposure to the flash (the estimated temperature was 1500°C within a short time). Many electronic devices require electrical conductors that are optically transparent to visible light. Transparent electrical conductors function by transmitting electrical power to the user interfaces, such as touch screens or send a signal to a pixel in a LCD display. Transparent conductors are an essential component in many optoelectronic devices, including flat panel displays, touch screens, electroluminescent lamps, solar panels, "smart" windows, and OLED lighting systems.

In contrast to chemical functionalization routes based upon thermally activated chemistry or electrochemistry, photochemical approaches show slow progress. Photoirradiation is used to generate reactive species, such as nitrenes, during sidewall addition reactions [74]. Photoactivation exclusively employs azido compounds as the nitrene precursor. "True" photochemical modification of SWNTs was achieved by sidewall osmylation [75,76]. A pronounced increase in electrical resistance was observed after exposing the SWNTs to osmium tetroxide (OsO_4) under UV light irradiation. The observed changes were attributed to the photo-induced cyclo addition of OsO_4 to the partial carbon–carbon double bonds, which resulted in a decrease in π electron density in the nanotubes. Interestingly, OsO_4 addition is reversible in the absence of humidity, that is, the cyclo adduct can be cleaved easily by photo irradiation, whereby the original resistance is restored.

In a later study, the UV light-induced osmylation of SWNTs was also examined in organic solvents [76]. Under these conditions, OsO_4 appears to react preferentially with the metallic tubes in the sample. This selectivity may be due to the availability of the electronic states at the Fermi level in the metallic tubes, which facilitate the formation of an intermediate charge transfer complex. In contrast to the reactions performed by exposure to gaseous OsO_4, the major product in an organic medium is SWNTs coated with OsO_2 particles, which results in the formation of extended tube aggregates. This difference might occur because the cyclo adduct is readily hydrolyzed in the presence of traces of water.

Paul [77] in his patent explained the possibility of obtaining an electrically conducting coating involving ultraviolet light with sufficient intensity to functionalize the carbon nanotube sidewall groups and a photoreactive chemical, such as osmium tetraoxide, in the presence of oxygen. It is believed that the sidewall groups of CNTs may be functionalized by cyclo addition, such as osmyl ester or quinine-type functionality. Nakamura et al. [78] reported a useful method for sidewall modification of single-walled carbon nanotubes (SWNTs) with thiol- and amino-containing substituents via radical processes of the photolysis of 1,2-dithiane and acetonitrile. The authors irradiated a suspension of purified SWNT in acetonitrile using a Xe excimer lamp for 3 h at room temperature under an argon atmosphere with constant stirring. The recorded x-ray photoelectron spectra (XPS) (Figure 9.4) before and after irradiation indicated a new peak at 400.5 eV for nitrogen 1s along with the sp^2 carbon peak at 284.6 eV after irradiation.

In another study, Nakamura et al. [79] reported the photochemical modification of single-walled carbon nanotubes with amino functionality. Water-soluble

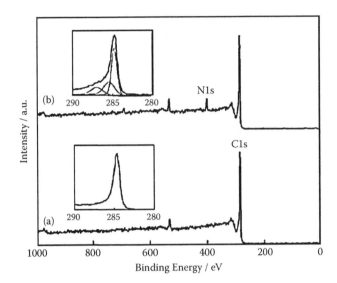

FIGURE 9.4
X-ray photo electron (XPS) spectra of SWNTs (a) before and after irradiation in presence of acetonitrile. (Reprinted with permission from Nakumara et al., *New Diamond and Frontier Carbon Technology.* 2007; 17(6). © 2009, MYU Tokyo.)

DNA-coated multiwalled CNTs were prepared by Moghaddam et al. [80] using azide photochemistry and DNA oligonucleotides. They reported that the DNA-coated vertically aligned nanotubes can offer the architecture for a highly loaded three-dimensional DNA chip. Dai and Lee [81] examined the asymmetric end functionalization of CNTs using a photochemical approach.

9.2.5 In Situ Polymerization

The method for producing CNT-filled composites using an in situ polymerization process consists of the following: combining CNTs with a monomer in the presence or absence of a polymerization catalyst; activating the polymerization catalyst if needed; and exposing the mixture to polymerization conditions resulting in a polymer mixture with highly dispersed CNTs, thereby forming a carbon nanotube-filled composite [82]. Yao et al. [83] developed a novel carbon nanotube–poly (methyl methacrylate) (CNT/PMMA) composite electrode as a sensitive amperometric detector of microchip capillary electrophoresis (CE). The authors fabricated a composite electrode using the in situ polymerization of a mixture of CNTs and prepolymerized methyl methacrylate in the micro channel of a piece of fused silica capillary under heat. Some conjugated or conducting polymers were attached to the surfaces by in situ polymerization to improve the processability of CNTs, as well as the electrical, magnetic, and optical properties. Composite materials based on the coupling of conducting polymers and CNTs have been reported to exhibit the properties of the individual components with a synergistic effect [84].

The development of nanostructures afforded by conducting polymers, such as polyaniline (PAni) and polypyrrole (PPy), has proceeded independently and includes the preparation of nanotubes [85–87] nanofibers [88–91] and the coating of various substrates with a thin polymer film [92,93]. The combination of both types of materials on a nano level is an obvious challenge. This approach is illustrated by the coating of CNTs with conducting polymers. The deposition process follows a similar principle as the coating of polymer fibers with diameters in the micrometer range [94], which is a well-established technique for preparing conducting textiles [95]. Both SWNTs [96–98] and MWNTs [99–106] have been coated with PAni or PPy in situ during the polymerization of their respective monomers. Conducting polymers have been deposited electrochemically in some cases [96,97,101–103, 107] but the chemical polymerization of aniline salts using ammonium peroxydisulfate as an oxidant is the most popular way of preparing PAni coatings [108–111].

Among the various conducting polymers, PAni has potential use in the synthesis of polymer–CNT composites due to its environmental stability, good processability, and reversible control of conductivity both by protonation and charge-transfer doping [112,113]. Considerable progress has been made in the design and fabrication of PAni–CNT composites with examples including PAni–MWNT composites that show site-selective interactions between the quinoid ring of the polymer and MWNTs [114], doped PAni–MWNT composites with or without protonic acid synthesized via in situ polymerization [115–117], coaxial nanowires of a PAni–MWNT composite prepared by an electrochemical reaction [118], and the fabrication of a PAni–SWNTs composite [119,120]. These results show site-selective interactions between the quinoid ring of PAni and MWNTs, which open the way for charge transfer processes and improved electric properties of PAni–MWNT composites.

Several methods have been used to prepare these polymer–CNT composites. These include the deposition of PAni onto CNT-modified electrodes [121] or onto aligned CNTs [122], as well as the use of self-assembly [123] or in situ polymerization [124] to prepare PAni–CNT composites. PPy–CNT composite nanowires are prepared by a template-directed electropolymerization of PPy in the presence of a CNT dopant [125]. PPy–CNT films are synthesized on macro-sized electrodes by electropolymerization with anionic CNTs acting as the dopant [126]. Despite these efforts, there is still a need for effective strategies for reliably and predictably incorporating CNTs into polymeric materials. Surface coatings are recognized as advanced and intriguing methods for dispersing MWNTs homogeneously throughout a matrix without destroying their integrity [127,128]. Many deposition approaches are used widely for the surface coating of nanoscale objects, such as self-assembly [123], electrochemical [122,123], miniemulsion [127], and inverse microemulsion polymerization [135]. Inverse microemulsion polymerization was recently reported to be a useful strategy for covering CNTs by precipitating

inorganic nanoparticles onto a nanotube surface [129]. It is a simple, reproducible procedure, in which the thickness and adherence of the coating can
be controlled fairly easily by the monomer concentration and reaction conditions. An inverse microemulsion consisting of an oil, surfactant, and water
molecule is a thermodynamically stable and isotropic transparent solution.
In an inverse microemulsion system, microdrops of an aqueous phase are
trapped within the assemblies of the surfactant molecules dispersed in a
continuous oil phase. The domain size of the dispersed phase is usually
small (a few nanometers), and a range of functional reactants can be introduced deliberately into the nanometer-sized aqueous phase confined within
the inversed micelles. These microdrops provide a large interfacial area and
reduce the interfacial energy values almost to zero [130]. Therefore, water
droplets act as a binder between the MWNTs and nanoparticles. Moreover,
their confinement in microemulsions produces nanoparticulates that attach
to the MWNTs, making them more homogeneous.

The procedure of in situ emulsion polymerization of MWNTs includes 6 g
of SDBS added to 5 mL of a 0.1 M HCl, followed by the addition of 3 mL of
butanol and 60 mL of hexane [131]. The mixture was then stirred until the
system became transparent. Subsequently, 2.0 mg of MWNTs was added to
the homogeneous solution, which was then sonicated for 6 h without any
traces of carbon deposits and cooled to 0°C. Upon stirring, a precooled solution of deionized water (5 mL) containing 1.32 g of ammonium persulfate
(APS) was added dropwise to the suspension, which was stirred for a further
30 min. A pyrrole monomer (0.20 mL) was added to the well-stirred reaction
mixture. The reaction mixture was sonicated for 10 min and allowed to stand
at 0° C under nitrogen with constant stirring for 24 h. The polymerization
reaction was quenched by pouring the mixture into acetone, resulting in the
precipitation of the MWNT–PPy composites. The precipitate was filtered and
washed with 0.1 M HCl and acetone until the washing solution became clear
and the SDBS was removed thoroughly. The powder was dried in a vacuum
for 48 h at room temperature before characterization.

Tang and Xu [132] synthesized poly (phenyl acetylene)-wrapped carbon nanotubes (PPA-CNTs), which were soluble in organic solvents, such as tetrahydrofuran, toluene, chloroform, and 1,4-dioxane. The fullerene tips and graphene
sheets of CNTs might undergo nonlinear optical (NLO) absorption processes,
and the cylindrical bodies of the CNTs with a high aspect ratio can also function
as light scattering centers. Both NLO absorption and light scattering of CNTs
protect the PPA chains from photodegradation under severe laser irradiation.
Therefore, PPACNTs show a strong photostabilization effect. Interestingly, PPA-
CNTs can be processed macroscopically, and shearing of their solutions readily
aligns the CNTs along the direction of the applied mechanical force. Fan et al. [133]
synthesized conducting polypyrrole-coated carbon nanotubes (PPy-CNTs), and
found that the magnetization of PPy-CNTs is the sum of the two components, PPy
and CNTs. Star et al. [134] synthesized poly (metaphenylenevinylene)-wrapped,
single-walled carbon nanotubes (PmPV-SWNTs). The UV–Vis absorption spectra

confirmed the p–p interactions between the SWNT and fully conjugated PmPV backbone. These results suggest that the photoexcited PmPV has a dipole moment that alters the local electric field at the surfaces of the SWNTs. Xiao and Zhou [135] deposited PPy or poly (3-methylthiophene) (PMeT) on the surfaces of MWNTs by in situ polymerization. The Faraday effect of the conducting polymer enhances the performance of super-capacitors with MWNTs deposited with a conducting polymer.

Jia et al. synthesized poly (methyl methacrylate) (PMMA)–MWNT composites by in situ radical polymerization [136]. In this study, in situ polymerization was performed using the radical initiator 2,2′-azobisisobutyronitrile (AIBN). The authors believed that, p-bonds in the CNTs were initiated by AIBN. Therefore, CNTs can participate in PMMA polymerization to form a strong interface between the MWNTs and PMMA matrix. Velasco-Santos et al. [137] and Putz et al. [138] also used AIBN to initiate the in situ radical polymerization to incorporate functionalized MWNTs and SWNTs into the PMMA matrices. Wu et al. [139] converted the hydroxyl groups on the surface of the CNTs to hydroxymethyl groups ($-CH_2OH$) by the formalization reaction with formaldehyde, as shown in Scheme 9.3. Zhou [140] also converted the carboxylic acid and hydroxyl groups on the surfaces of the CNTs to vinyl groups ($-CH-CH_2$) by a reaction with 3-isopropenyl-α, α-dimethylbenzyl isocyanate.

Chen et al. [25] activated the CNT surface by plasma modification. Acetaldehyde and ethylene diamine vapors were plasma polymerized onto the surfaces of the CNTs, thereby introducing active aldehyde (–CHO) and amino ($-NH_2$) groups. Mickelson et al. added the fluorine functional groups to the side walls of the SWNT [29]. These fluorinated SWNTs can be dissolved in alcohol and reacted with other species, particularly strong nucleophiles, such as alkyllithium reagents. Mawhinney et al. [141] reported that, carboxylic acids, anhydrides, quinines, and esters can also be introduced to CNTs by ozone oxidation. Bahr et al. [31] functionalized SWNTs successfully by the electrochemical reduction of aryl diazonium compounds, resulting in the formation of a free radical that could attach to the SWNT surface. Georgakilas et al. [48] reported a method for functionalizing SWNTs with the 1,3-dipolar cyclo addition of azomethine ylides. Chen et al. [142] and Holzinger et al. [41]

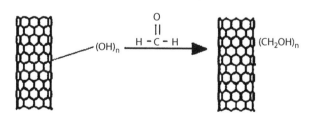

SCHEME 9.3
Reaction of CNTs with formaldehyde. (Redrawn after Wu et al., *Chem J Chinese Univ* 1995; 16:1641© 2009.)

functionalized SWNTs by direct addition based on nucleophilic carbenes cyclo addition of nitrenes and radicals. These functionalizing methods were reviewed by Bahr and Tour [44].

MWNTs have been wrapped successfully by polyethyleneimine (PI) through in situ polymerization in *N, N*-dimethylformamide (DMF) [143]. The following process was used. First, 4,4-diaminodiphenyl ether (ODA) was adsorbed on the surface of the MWNTs in DMF via a π–π electron interaction. PI acid was then formed by the addition of pyromelitic dian-hydride (PMDA). The growing PI acid polymer chain would wedge away the MWNTs due to π–π electron interactions between the MWNTs and the benzene ring of PI acid. Finally, after evaporating the DMF, PI was attached to the surfaces of the MWNTs, and the MWNTs–PI composite was obtained. Thermogravimetric analysis and differential scanning calorimetry sug-gested an improvement in thermal stability of the functionalized MWNT–PI composites. The MWNTs–PI composites might be good candidates as light-weight materials for electronic devices in the aerospace and microelectron-ics industries on account of their excellent electronic, optical, and thermal properties.

Composites from polyurethane and CNT have attracted considerable attention [144–148]. Recently, Xia et al. reported the poly (propylene gly-col) grafted MWNT–polyurethane (PU) system [149]. They concluded that grafted MWNTs can improve the rheological behavior of the polyol–MWNT dispersion and have a better reinforcing effect on PU.

Xianhong et al. [149] functionalized MWNTs by the in situ polycondensa-tion of polyurethane (PU) using a two-step method as described in the fol-lowing text.

PU-functionalized MWNTs (MWNTs-PU) were prepared via in situ poly-condensation. First, 100 mL of anhydrous DMF and 100 mg of MWNTs-R_2-NH_2 were placed in a 250 mL three-necked flask, and sonicated at 80°C for 30 min under a N_2 atmosphere until the suspension became homogeneous. TDI was then dissolved in the aforementioned solution with constant stir-ring at 80°C for 2 h. Subsequently, PBAG (at a molar ratio of half that of TDI) was added and prepolymerized for 24 h to form a prepolymer. 1,4 butanediol [BDO] (at a molar ratio of half that of TDI) was added to the prepolymer with constant stirring at 80°C for 24 h to ensure a complete reaction. Finally, DMF was added to the flask to dilute the reaction mixture. After vacuum filtration, the sample was washed thoroughly with DMF. The MWNTs-PU sample was obtained after filtering and washing with THF, followed by drying over-night in a vacuum oven (70°C).

The SWNTs were modified with polyethylene (PE) and prepared through in situ Ziegler–Natta polymerization [150]. The SWNTs modified by in situ polymerization of PE were mixed with commercially obtained PE by melt blending. The mechanical properties of the resulting composites were supe-rior to those of the SWNTs without modification. The in situ polymerized PE layer played an important role as an interfacial modifier.

9.2.6 Atom Transfer Radical Polymerization (ATRP)

Although conventional radical polymerization methods are good for growing homopolymers or random copolymers under mild conditions, controlled radical methods are more desirable for achieving interesting functionality and architectures with controlled molecular weights and low polydispersities [151–161]. Atom transfer radical polymerization (ATRP) has been suggested to be the most successful route for controlled radical polymerization and allows the development of polymers with interesting functionality and architectures. ATRP has been employed to modify a range of silica surfaces [162–164] as well as the pores of two-dimensional hexagonal-structured silica (SBA-15) [165] through the growth of polymer chains. ATRP has also been used to modify the surfaces of gold [166,167], magnetic nanoparticles [168,169], and CdS nanoparticles [170]. Recently, ATRP was used to functionalize CNTs through the development of polymer brushes on the surfaces of nanotubes [171].

Shanmugharaj et al. [172] prepared poly (styrene-co-acrylonitrile) grafted MWNTs using a surface-initiated atom transfer radical (ATR) polymerization method. The reaction scheme involves fixation of the ATRP initiator to the carboxylic acid groups and polymerization. The 2-bromoisobutyl bromide group is an excellent ATRP initiator for the living radical polymerization of styrene monomers. DSC showed that the polymers grafted onto MWNTs had higher glass-transition temperatures. The polymer-grafted MWNTs exhibited relatively good dispersibility in an organic solvent such as tetrahydrofuran.

Baskaran et al. employed MWNT-COOH to attach the ATRP initiator, hydroxyethyl-2-bromoisobutyrate and initiator-modified MWNTs were used for the in situ ATRP of styrene and MMA from the MWNTs [171]. SWNTs were functionalized along their sidewalls with phenol groups using the 1,3-dipolar cycloaddition reaction. These phenols could be derivatized further with 2-bromoisobutyl bromide resulting in the attachment of ATRP initiators to the sidewalls of the CNTs. These initiators were active in the polymerization of MMA and *tert*-butyl acrylate (t-BA) from the surfaces of the CNTs (Figure 9.1) [173].

9.2.7 Grafting

The pronounced tendency of CNTs to agglomerate is a disadvantage in the preparation of composites with polymers. Therefore, the surface modification of CNTs has been investigated widely to address this problem. The covalent attachment of polymers—that is, the grafting of polymers—onto CNTs is quite effective because polymers grafted onto the surface prevent the aggregation of CNTs. Accordingly, modification of the CNT surface by the grafting of polymers has been examined widely [174,175]. The methodologies used to achieve the grafting of polymers onto CNTs include (1) grafting-onto method, (2) polymer reaction, (3) grafting-from method, and (4) stepwise growth [176].

Yoon et al. [177] reported the surface-initiated ring opening polymerization, "grafting-from" approach, to form a shortened SWNT/poly (*p*-dioxanone) composite. The authors were interested in the functionalization of a variety of substrates with biocompatible/biodegradable aliphatic polyesters using surface-initiated, ring-opening polymerization for potential applications in biomedical areas [178–181].

9.2.7.1 Preparation of MWNT-Graft-Poly TDI

To 250 mL DMF, 5 g of MWNT-OH, 35 g of poly TDI (toluene diisocyanate), and 0.15 g Dabco-33LV were added. The mixture was then transferred to a reaction vessel and ultrasonicated at 300 W for 30 min at room temperature to ensure good dispersion and contact between the MWNT-OH and organic reagents. The mixture was then stirred for 24 h at 100°C. The MWNT-graft-poly TDI obtained was filtered and washed three times with DMF and subjected to Soxhlet extraction with DMF for 48 h to remove the unreacted polyTDI.

9.2.7.2 Preparation of MWNT-Graft-PU

A 200 mL solution of DMF containing 3.4 g of MWNT-graft-poly TDI, 17 g of polyether polyol, and 0.1 g of Dabco-33LV was transferred to a reaction vessel and ultrasonicated at 300 W for 30 min at room temperature. The mixture was then stirred for 24 h at 100°C. The MWNT-graft-PU obtained was filtered and washed three times with DMF and subjected to Soxhlet extraction with DMF for 48 h to remove the unreacted polyether polyol.

9.2.7.3 Preparation of MWNT-Graft-PU/Polyether Polyol Dispersion

Typically, 2 g MWNT-graft-PU and 198 g polyether polyol were placed into a jar and subjected to ball milling at 80°C for 48 h. A well-dispersed 1% MWNT-graft-PU/polyether polyol dispersion was obtained. As a control experiment, an ungrafted MWNT/polyether polyol dispersion was prepared using this method.

9.2.7.4 Preparation of MWNT-Graft-PU/PU Composites

MWNT-graft-PU/PU composites with a hard segment content of 26% were prepared using the following procedure: A certain amount of MWNT-graft-PU/polyether polyol dispersion was blended with 1, 4-butanediol, polymeric MDI, Dabco-33LV and BYK-535 at room temperature for 2 min and degassed under vacuum for 3 min to remove any bubbles. A viscous pre-polymer was poured into a metal mold and cured at 50°C for 24 h and 80°C for 1 week to obtain the PU/MWNT-graft-PU composites. Blank PU and PU/ MWNT composites were also prepared as control samples.

Two main steps were used to obtain the MWNT graft PU. First, MWNT-OH was reacted with polyTDI (compound a), which is a type of high active substance with NCO-groups, to obtain the MWNT-graft-polyTDI. The MWNT-graft-polyTDI was reacted with trifunctional poly (propylene) polyol (compound b) to produce the MWNT-graft-PU. For both reactions, N, N-dimethyl formamide (DMF) was used as the solvent. The mixture was first subjected to ultrasonic irradiation to ensure good dispersion and contact between the MWNT and organic reagents. Compounds a and b were added at a fivefold excess compared to MWNT-OH and MWNT-graft-poly TDI.

Transmission electron microscopy (TEM) is a useful instrument to study the morphology of the grafted polymer on the CNTs. After grafting, many spots with a size of several nanometers on the sidewalls of the MWNTs could be observed on the grafted polymer. In most cases, the grafted polymer was wrapped on the surface of the CNTs [182–185]. It is believed that these grafted polymer spots with chemical bonding will play an important role in improving the interaction between the CNTs and solvent or polymer matrix.

Bin et al. examined the water soluble SWNTs–poly (m-amino benzene sulfonic acid) [PABS] graft copolymer [186]. The SWNTs-PABS graft copolymer is quite soluble in water and exhibits an order of magnitude higher electrical conductivity than the neat PABS.

Significant work has been carried out to attach macromolecules to the tips and convex walls of CNTs via special reactions, including etherification, amidization, radical coupling, and other reactions [25]. However, the macromolecules must possess suitable reactive functionality and radicals. Kamigaito et al. [187] discussed the role of metals on the living radical polymerization of monomers. Recently, the grafting-from approach was used to functionalize MWNTSLs via the dense in situ grafting of polymer chains onto the convex walls of CNTs. Many linear polymers, such as polystyrene (PS) [188], poly(sodium 4-styrenesulfonate) [189], poly(methyl methacrylate) (PMMA) [190], polyimide (PI) [191], poly(2-vinylpyridine) [192], poly(ethylene glycol) (PEG) [193], poly(vinyl alcohol) (PVA) and its related copolymer poly(vinyl acetate-co-vinyl alcohol) [186], and poly (m-aminobenzenesulfonic acid) [194], as well as dendrons [195] dendrimers [196], and hyper branched polymers [197] have been bonded successfully to CNTs. Although significant work has been carried out on the grafting of polymers to CNTs, there are no reports on the grafting of copolymers, such as poly (styrene-co-acrylonitrile) (SAN), to the surface. SAN has significant importance in many areas, such as automotive, packaging, and electronic and medical applications. The successful grafting of styrene-co-acrylonitrile onto the surface of CNTs can provide a means of fabricating polymer composites with a fine dispersion of functionalized CNTs which can be useful in electronic applications.

9.2.8 Oxygen Plasma Treatment

An oxygen plasma treatment is used for surface modification of the CNTs. It is a useful modification technique because it allows the fixation of chemical groups to the surface without altering the properties of CNTs [198]. The main advantage of this technique is the surface modification of CNTs within a very short treatment time. An oxygen plasma treatment is generally carried out in a plasma reactor equipped with a microwave generator [199] and is more applicable to MWNTs rather SWNTs. A plasma treatment of SWNTs may destroy the sp² structure of the single graphene layers, which can result in a loss of their excellent mechanical properties [199].

Greton et al. [200] concluded that an oxygen plasma treatment is the best of the different plasma treatments for MWNTs they evaluated due to the selectivity of the stable carbon-hydrogen bonds and lower treatment time than the chemical oxidation route. They observed both nitrogen and oxygen by XPS when an Ar/NH_3 plasma treatment was used. The authors successfully coated MWNTs with a thin and cross-linked polymer layer when a methyl methacrylate plasma treatment was adopted. Furthermore, nanocomposites were fabricated using three different types of plasma-treated MWNTs with epoxy resin. The MWNT/epoxy nanocomposites fabricated showed a higher elastic modulus when the plasma treated MWNTs were used.

Chirila et al. [201] examined the effect of different oxygen plasma treatment conditions, such as the plasma power, chamber pressure, plasma frequency (radio frequency-[RF] or microwave-[MW]) and treatment time on the quantity of functional groups (carboxylic, carbonilic, hydroxylic, phenolic) introduced to the vapor-grown carbon nanotubes (VGCNT) fibers' surfaces. The MW-plasma treated fibers showed better wettability, which helped increase fiber-polymer adhesion in the composites.

Kim et al. [202] compared the effect of the plasma treatment on the morphological behavior of epoxy–CNT composites with that of an acid and amine treatment using field emission scanning electronic microscopy. They reported that fiber pullout (Figure 9.5) was much lower after the plasma treatment than after the other treatments. This was attributed to the formation of oxygen groups, which contributes to the good dispersion and interfacial bonding. Shi et al. reported a technique for forming a thin film on multiwalled and aligned CNTs through a plasma polymerization method [203]. They observed the deposition of a thin polymer film (2~7 nm) and improved dispersibility of the nanotubes. Feltan et al. [204] examined the effect of an oxygen radio frequency plasma treatment on the modification of MWNTs. The high-resolution transmission electron microscopy (HRTEM) study suggested that the plasma-treated CNTs improved the uniformity of the distribution of surface defects compared to gold-decorated pristine CNTs. Xu et al. [205], who performed a simple and nonpolluting large scale oxygen plasma treatment work, suggest that the position of the sample outside the barrel with a direct discharge not only allows the rapid grafting of polar functional

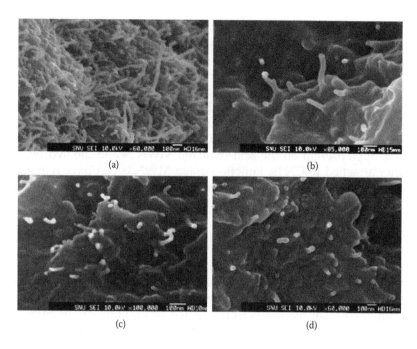

FIGURE 9.5

Field emission scanning electron microphotograph of (a) untreated (b) acid treated (c) amine treated and (d) plasma treated epoxy/MWCNT composites. (Reprinted with permission from Jin Ah Kim et al., *Carbon* 2006; 44:1898–1905 © 2009, Elsevier.)

groups but can also easily damage the MWNTs after a longer time, particularly at high power.

9.2.9 Modifications with Water Soluble Polymers

Recently, an exploration of the potential biological and medical applications of MWNTs has attracted considerable attention [206–211]. However, the insolubility of MWNTs in water limits its applications in these fields. Hence, modification of the solubility of MWNTs in water or the preparation of water-soluble MWNTs is in demand. The modification of MWNTs with water soluble molecules to graft, warp, and cover is performed to improve the solubility of MWNTs in water [20,212–214]. Among these methods, covalently grafting water-soluble polymers onto the MWNT surface is more effective [215–217] because such polymers bound covalently to the MWNT surface play an effective role as a compatibilizer in aqueous solutions. Functionalization is generally realized by the esterification or amidation of the carboxylic acid present on the surface of the MWNTs purified with strong oxidative acids and a polymer containing hydroxyl or amino groups. Lin et al. [193,218] prepared poly(vinyl alcohol)-functionalized MWNTs via an esterification reaction of polymer-bound hydroxyl groups with nanotube-bound carboxylic acid as well as amino polymer-functionalized MWNTs by an amidation reaction of nanotube-bound carboxylic acid with the polymers-bound amino groups. These functionalized MWNT samples were quite soluble in water. Therefore, poly (acrylic acid) (PAA) is an excellent water-soluble polymer with outstanding performance.

Liu et al. [219] covalently functionalized MWNTs with poly(acrylic acid) by grafting a poly (acryloyl chloride) to the nanotube surface via an esterification reaction of an acyl chloride-bound polymer with the hydroxyl functional groups present on the acid-oxidized MWNTs and the hydrolysis of polymer attached to CNTs. Functionalized MWNTs can have remarkable solubility in water, which may assist in an examination of the potential biological and medical applications of MWNTs with the excellent performance of PAA.

To functionalize MWNTs, a solution of poly (acrylic acid) (PAC) was added to a suspension of acid-oxidized MWNTs (100 mg) and sodium bicarbonate (75 mg) in anhydrous dioxane (50 mL). The mixture was stirred vigorously at 50°C for 48 h [220]. Schlenk technology was employed to separate the PAC-functionalized MWNTs from the resulting suspension. After vacuum filtration, the remaining solids were washed several times with anhydrous tetrahydrofuran (THF) at room temperature. The collected black solid was dispersed in anhydrous THF by sonication for 20 min. This suspension was then filtered and washed with anhydrous THF. After repeating the dispersion, filtration and wash procedures five times, the obtained filtrate was dropped into a standard $AgNO_3$ solution without a white deposit. This suggests that the un-reacted and physically absorbed PAC in the functionalized samples had been removed completely. The remaining PAC-functionalized MWNTs were obtained after vacuum drying. All the dispersion, washing and drying operations were performed under nitrogen. PAC-functionalized MWNT materials were obtained after dispersing the PAC-functionalized MWNTs in deionized water for 8 h, filtering and washing repeatedly with deionized water, followed by extraction with 1,4-dioxane and drying in a vacuum.

Since poly (ethylene oxide) (PEO) is a representative hydrophilic polymer with good solubility in many solvents and polymers, MWNTs complexed with PEO have been attractive fillers in a variety of polymer composites. Goh et al. [220] suggested the MWNT–poly (methyl methacrylate) (PMMA) composite, where amine-terminated PEO ($PEONH_2$) acted as a compatibilizer. The terminal amine groups of PEO-NH_2 interacted ionically with the carboxylic acid groups on the MWNTs, while the PEO chain formed miscible entanglements with the PMMA matrix. Adding the modified MWNTs to the PMMA matrix increased the storage modulus of the composites. Sun et al. employed diamine terminated PEO to dissolve the pristine and purified SWNTs [221]. Haddon et al. [222] and Sano et al. [37] also reported a PEO grafted SWNT. Basiuk et al. [223] suggested the possibility of the direct functionalization of MWNTs with hydrocarbons possessing amine functional groups. Scheme 9.1 shows the complexation of MWNT-COOH and MWNT-COCl with PEO-NH$_2$ [224]. Jung Sook et al. [225] reported the functionalization of MWNTs with monoamine terminated PEO. The resulting MWNT–PEO complexes exhibited excellent dispersibility in solvents, including water, ethanol and DMF. They also found no significant difference between MWNT-COOH and MWNT-COCl for the complexation reaction indicating that MWNT-COOH and MWNT-COCl have similar reactivity against PEO-NH$_2$.

Riggs et al. [226] functionalized SWNTs and MWNTs with poly (propionylethylenimine-co-ethylenimine) (PPEI-EI). The carboxylated CNTs were heated under reflux in thionyl chloride (SOCl$_2$) to afford the acyl chloride functionality before a reaction with PPEI-EI. Both the PPEI-EI–functionalized SWNTs and MWNTs were luminescent and soluble in a range of solvents including water. Zhang et al. [227] dissolved MWNTs in chitosan solutions that contained reactive amino and hydroxyl functional groups. The noncovalent interaction between the MWNTs and chitosan offers a straightforward method for the manipulation, purification, and modification of MWNTs. Soon Huat Tan et al. [228] functionalized MWNTs with water-soluble poly (2-ethyl-2-oxazoline) (PEOX) using a polymer wrapping process. SEM showed that the diameter of the MWNTs was increased by approximately 10 nm after functionalization with PEOX. They reported that the functionalized MWNTs produced an excellent and stable dispersion in water and other organic solvents, such as ethanol and DMF.

9.2.10 Modification with Biomolecules

The combination of biological molecules with CNTs is quite important in the process of developing new nanoscale devices for future biological, medical, and electronic applications. The carboxylic groups generated on CNTs have been used to covalently bind biomaterials, such as sugar moieties, [229] oligonucleotides, [230–233], peptide nucleic acids (DNA), [234] and proteins [235] via the carbodiimide coupling of the respective amino functionalized biomolecules or employing hetero bifunctional coupling reagents. For example, ferritin and bovine serum albumin (BSA) have been bound covalently to nitrogen-doped MWNTs using a two-step process of diimide-activated amidation (Figure 9.6) [236]. First, carboxylated MWNTs were activated by N-ethyl-N'-(3-dimethylaminopropyl) carbodiimide hydrochloride (EDC) forming a stable active ester in the presence of N-hydroxysuccinimide (NHS). Subsequently, the active ester was reacted with the amine groups of ferritin or BSA to form amide bonds between the MWNTs and proteins. This two-step process avoided the intermolecular conjugation of the proteins, and allowed the uniform attachment of proteins to the MWNTs. This approach provides a universal and efficient method for attaching biomolecules to CNTs under ambient conditions.

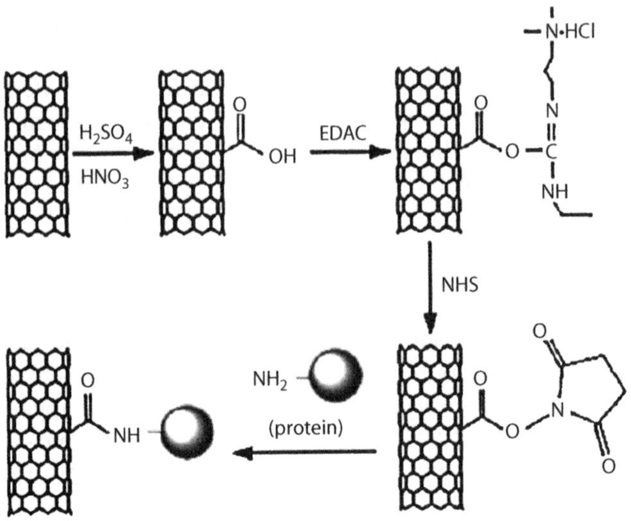

FIGURE 9.6
Attachment of proteins to carbon nanotubes via a two-step process of diimide-activated amidation. (Reprinted with permission from Liu, *Eur Polym J* 2005; 41: 2693–2703 © 2009, Elsevier.)

9.2.11 Modification with Engineering Polymers

Aromatic polyimides (PI) are used widely as matrix polymers for advanced applications in the aerospace and microelectronics industries on account of their good chemical and mechanical properties as well as their excellent thermal stability and radiation resistance [237–239]. Nanocomposites of polyimides with CNTs are promising materials for many applications, particularly as lightweight materials in space applications. Recently, some studies on CNTs–polyimide composites have been undertaken. Ge et al. [239] reported the chemical grafting of polyetherimide onto CNTs. Hill et al. [240] reported that the nanotube content in CNTs/PI composites can reach 25 wt% using CNTs functionalized with derivatized polyimide. Recently, some studies [265–270] on CNTs–PI composites were reported. Ounaies et al. [241] examined the electrical properties of PI composites reinforced with SWNTs. Ge et al. [239] chemically grafted polyetherimide to the CNTs. Jiang et al. [242] evaluated the electrical and mechanical properties of PI/CNTs composites formed by in situ polymerization.

9.3 Conclusions

There is a tremendous shift from developing traditional materials, such as metals, ceramics, polymers, composites, etc., to a more revolutionary trend of developing functionalized nanomaterials. The commercial availability of

CNTs in large quantities has prompted the development of CNT–polymer composites. Compared to other electrochemically active materials, CNTs are advantageous owing to their short response time, high degree of deformation, and low voltages that need to be applied. The recently developed chemical and electrochemical functionalization schemes have expanded the applications of CNTs significantly. However, there have been limited attempts to understand the structure of functionalized nanotubes on the atomic scale. The modification of CNTs with polymers is a particularly attractive target because it can improve the solubility, processability, ease of characterization, and unique properties that can be coupled with other types of materials. Research to improve the quality of nanotubes with particular emphasis on their uniformity may enhance the development of nanomaterial technology.

References

1. Iijima, S. Helical microtubules of graphite carbon. *Nature* 1991; 354:56–58.
2. Saito, R., Dresselhaus, G., and Dresselhaus, M.S. *Physical Properties of Carbon Nanotubes*. Imperial College Press: London, 1998.
3. Ouyang, M., Huang, J.L., and Lieber, C.M. *Annu Rev Phys Chem* 2002; 53:201.
4. Harris, P.J.F. *Carbon Nanotubes and Related Structures—New Materials for the Twenty-First Century*. Cambridge University Press: Cambridge, 2001.
5. Dai, L. *Carbon Nanotechnology: Recent Developments in Chemistry, Physics, Materials Science and Device Applications*. Elsevier: Amsterdam, 2006.
6. Dennig, P. and Irvin, G. *J Reinforced Plastics Comp* 2006; 25(2):175–188.
7. Valentini, L., Armentano, I., Santilli, P., Kenny, J.M., Lozzi, L., and Santucci, S. Electrical transport properties of conjugated polymer onto self-assembled aligned carbon nanotubes. *Diamond Related Mater* 2003; 2(9):1524–1531.
8. Lau, K.T. and Hui, D. *Comp Part B: Eng* 2002; 33:263–277.
9. Friedrich H. Development of carbon nanotube (CNT) actuators: dispersion, alignment, electrolytes. Fraunhofer ISC Annual Report 2005; 42–45.
10. Erik, C.L., Thostenson, T., and Chou, T.W. *Comp Sci Technol* 2005; 65:491–516.
11. Moniruzzaman, M., Du, F., Romero, N., and Winey K.I. Increased flexural modulus and strength in SWNT/epoxy composites by a new fabrication method. *Polymer* 2006; 47:293–298.
12. Cheol, P., Joycelyn, H., and Peter, L. Evidence of piezoelectricity in SWNT-polyimide and SWNT-PZT-polyimide composites. *J Thermoplastic Comp Mater* 2008; 21:393–409.
13. Cheol, P., Roy, E.C., Emilie, J.S., Joycelyn, S.H., Neal, E., and Edward, K. Adhesion study of polyimide to single-wall carbon nanotube bundles by energy-filtered transmission electron microscopy. *Nanotechnology* 2003; 4:11–14.
14. Nyan-Hwa, T., Meng-Kao, Y., and Tai-Hao, P. Experimental study and theoretical analysis on the mechanical properties of SWNTs/phenolic composites. *Comp Part B: Eng* 2008; 39(26):932.

15. Kooi, S.E., Schlecht, U., Burghard, M., and Kern, K. *Angew Chem Int Ed* 2002; 41:1353–1355.
16. Li, S., Hui, Z., Zhong, Z., and Sishen, X. Processing and performance improvements of SWNT paper reinforced PEEK nanocomposites. *Comp Part A: Appl Sci Manuf* 2007; 38(2):388–392.
17. Jianfeng, D., Qing, W., Weixue, L., Zhiqiang, W., and Guangji, X. Properties of well aligned SWNT modified poly (methyl methacrylate) nanocomposites. *Mater Lett* 2007; 61(1):27–29.
18. Reto, H., Fangming, D., John, E.F., and Karen, I.W. Interfacial in situ polymerization of single wall carbon nanotube/nylon 6,6 nanocomposites. *Polymer* 2006; 47(7):2381–2388.
19. Special issue on carbon nanotubes. *Acc Chem Res* 2002; 35:997–1113.
20. Tasis, D., Tagmatarchis, N., Georgakilas, V., and Prato, M. *Chem Eur J* 2003; 9:4000–8.
21. Peng, L. *European Polym J* 2005; 41:2693–2703.
22. O'Connell, M.J., Boul, P., Ericson, L.M., Huffman, C., Wang, Y., and Haroz, E. *Chem Phys Lett* 2001; 342:265–71.
23. Carrillo, A., Swartz, J.A., Gamba, J.M., Kane, R.S., Chakrapani, N., Wei, B., and Ajayan, P. *Nano Lett* 2003; 3:1437–1440.
24. O Connell, M.J., Boul, P., Ericson, L.M., Huffman, C., Wang, Y., and Haroz, E. *Chem Phys Lett* 2001; 342:265–71.
25. Chen, J., Hamon, M.A., Hu, H., Chen, Y.S., Rao, A.M., Eklund, P.C., and Haddon, R.C. *Science* 1998; 282:95.
26. Valentini, L., Macan, J., Armentano, I., Francesco, M., and Kenny, J.M. *Carbon* 2006; 44:2196.
27. Gojny, F.H., Nastalczyk, J., Roslaniec, Z., and Schulte, K. *Chem Phys Lett* 2003; 370:820.
28. An, K.H., Heo, J.G., Jeon, K.G., Bae, D.J., Jo, C., Yang, C.W., Park, C.Y., Lee, Y.H., Lee, Y.S., and Chung, Y.S. *Appl Phys Lett* 2002; 80:4235.
29. Mickelson, E.T., Huffman, C.B., Rinzler, A.G., Smalley, R.E., Hauge, R.H., and Margrave, J.L. *Chem Phys Lett* 1998; 296:188.
30. Chen, Q.D., Dai, L.M., Gao, M., Huang, S.M., and Mao, A. *J Phys Chem* 2001; 105:618.
31. Bahr, J.L., Yang, J.P., Kosynkin, D.V., Bronikowski, M.J., Smalley, R.E., and Tour, J.M. *J Am Chem Soc* 2001; 123:6536.
32. Kónya, Z., Vesselenyi, I., Niesz, K., Kukovecz, A., Demortier, A., Fonseca, A., Delhalle, J., Mekhalif, Z., Magy, J.B., Koós, A.A., Osvath, Z., Kocsonya, A., Biro, L.P., and Kiricsi, I. *Chem Phys Lett* 2002; 360:429.
33. Kooi, S.E., Schlecht, U., Burghard, M., and Kern, K. *Angew Chem Int Ed* 2002; 41:1353–1355.
34. Bahr, J.L., Mickelson, E.T., Bronikowski, M.J., Smalley, R.E., and Tour, J.M. *Chem Commun* 2001; 193–194.
35. Banerjee, S., Hemraj-Benny, T., and Wong, S.S. *Adv Mater* 2005; 17:17.
36. Sun, Y.P., Fu, K.F., Lin, Y., and Huang, W.J. *Acc Chem Res* 2002; 35:1096.
37. Sano, M., Kamino, A., Okamura, J., and Shinkai, S. *Langmuir* 2001; 17:5125–5128.
38. Niyogi, S., Hamon, M.A., Hu, H., Zhao, B., Bhowmik, P., and Sen, R. *Acc Chem Res* 2002; 35:1105.
39. Pompeo, F. and Resasco, D.E. *Nano Lett* 2002; 2:369.

40. Dyke, C.A. and Tour, J.M. *J Am Chem Soc* 2003; 125:1156.
41. Holzinger, M., Vostrowsky, O., Hirsch, A., Hennrich, F., Kappes, M., Weiss, R., and Jellen, F. *Angew Chem Int Ed* 2001; 40:4002.
42. Georgakilas, V., Kordatos, K., Prato, M., Guldi, D.M., Holzinger, M., and Hirsch, A. *J Am Chem Soc* 2002; 124:760.
43. Banarjee, S., Hemraj-Benny, T., and Wong, S.S. *Adv Mater* 2005; 1:17.
44. Bahr, J.L. and Tour, J.M. *J Mater Chem* 2002; 12:1952.
45. Hirsch, A. *Angew Chem* 2002; 114:1933.
46. Dyke, C.A. and Tour, J.M. *Chem Eur J* 2004; 10:813.
47. Balasubramanian, K. and Burghard, M. *Small* 2005; 2:180.
48. Georgakilas, V., Voulgaris, D., Zquez, E.V., Prato, M., Guldi, D.M., Kukovecz, A., and Kuzmany, H. *J Am Chem Soc* 2002; 124:143.
49. Zhu, Y.Q., Hsu, W.K., Kroto, H.W., and Walton, D.R.M. *Chem Commun* 2001; 2184.
50. Zhu, Y.Q., Hsu, W.K., Kroto, H.W., Walton, D.R.M. *J Phys Chem B* 2002; 106(31):7623.
51. Han, W., Redlich, P., Ernst, F., and Ruhle, M. *Appl Phys Lett* 1999; 75:1875.
52. Rubio, A., Corkill, J.L., and Cohen, M.L. *Phys Rev* 1994; 49:5081.
53. Blase, X., Rubio, A., Louie, S.G., and Cohen, M.L. *Euro Phys Lett* 1994; 28:335.
54. Gleize, P., Schouler, M.C., Gadelle, P., and Caillet, M. *J Mater Sci* 1994; 29:1575.
55. Gleize, P., Herreyre, S., Gadelle, P., Mermoux, M., Cheynet, M.C., and Abello, L. *J Mater Sci Lett* 1994; 13:1413.
56. Chopra, N.G., Luyken, R.J., Cherry, K., Crespi, V.H., Cohen, M.L., Louie, S.G., and Zettl, A. *Science* 1995; 269:966.
57. Deepak, S., Madhu, M., Daraio, C., and Jin, S.V. Mediated mechanism of nitrogen substitution in carbon nanotubes. *Phys Rev B* 2004; 69:153–41.
58. Terrones, M. and Terrones, H. *Appl Phys Lett* 1999; 75:3932.
59. Czerw, R. *Nano Lett* 2001; 1:457.
60. Han, W.Q., Fan, S.S., Li, Q.Q., and Hu, Y.D. *Science* 1997; 277:1287.
61. Dai, H.J., Wong, E.W., Lu, Y.Z., Fan, S., and Lieber, C.M. *Nature* 1995; 375:769.
62. Li, C.P., Gerald, J.F., Zou, J., and Chen, Y. Transmission electron microscopy investigation of substitution reactions from carbon nanotube template to silicon carbide nanowires. *New J Phys* 2007; 9:137.
63. Han, W., Bando, Y., Kurashima, K., and Tadao, Sato. Boron-doped carbon nanotubes prepared through a substitution reaction. *Chem Phys Lett* 1999; 299:368–373.
64. Li, C.P., John, F.G., Jin, Z., and Ying, C. Synthesis of Silicon Carbide Nanowires on Carbon Nanotube Template. IEEE Xplore ICONN 2006.
65. Zhang, Y., Jun, L., Rongrui, H., Qi, Z., Zhang, X., and Jing, Zhu. Synthesis of alumina nanotubes using carbon nanotubes as templates. *Chem Phy Lett* 2002; 360(5–6):579–584.
66. Hasi, F., Simon, F., Hulman, M., and Kuzmany, H. CP685, Molecular Nanostructure: XVII Intl Winter School/Euro Conference on Electronic Properties of Novel Materials. American Institute of Physics 2003.
67. Maultzsch, J., Reich, S., Thomsen, C., Webster, C., Czerw, R., Carroll, D.L., Vieira, S.M.C., Birkett, P.R., and Rego, C.A.. *Appl Phys Lett* 2002; 81:2647.
68. Marcoux, P.R., Hapiot, P., Batail, P., and Pinson, P. *New J Chem* 2004; 28:302.
69. Nikolaev, P., Bronikowski, M.J., Bradley, R.K., Rohmund, F., Colbert, D.T., Smith, K.A., and Smalley, R.E. *Chem Phys Lett* 1999; 313:91.
70. Balasubramanian, K., Friedrich, M., Jiang, C., Fan, Y., Mews, A., Burghard, M., and Kern, K. *Adv Mater* 2003; 15:1515.

71. Knez, M., Sumser, M., Bittner, A.M., Wege, C., Jeske, H., Kooi, S., Burghard, M., and Kern, K. *J Electroanal Chem* 2002; 522:70.
72. Peng, H., Alemany, L.B., Margrave, J.L., and Khabashesku, V.N. *J Am Chem Soc* 2003; 125:151–174.
73. Ajayan, P.M., Terrones, M., De la Guardia, A., Huc, V., Grobert, N., Wei, B.Q., Lezec, H., Ramanath, G., and Ebbesen, T.W. *Science* 2002; 296:705.
74. Moghaddam, M.J., Taylor, S., Gao, M., Huang, S.M., Dai, L.M., and McCall, M.J. *Nano Lett* 2004; 4:89.
75. Cui, B., Burghard, M., and Kern, K. *Nano Lett* 2003; 3:613.
76. Banerjee, S. and Wong, S.S. *J Am Chem Soc* 2004; 126:2073.
77. Paul, G.J. Patterning carbon nanotube coatings by selective chemical modification. Patent WO/20060/78286.
78. Nakamura, T., Tsuguyori, O., Masatou, I., Akie, O., and Tomomi. K. *New Diamond Frontier Carbon Technol.* 2007; 17(6).
79. Nakamura, T., Ohana, T., Ishihara, M., Hasegawa, M., and Koga, Y. Photochemical modification of single-walled carbon nanotubes. *Diamond Related Mater* 2008; 17:559–562.
80. Moghaddam, M.J., Sarah, T., Mei, G., Shaoming, H., Liming, D., and Maxine, J.M. Highly efficient binding of DNA on the sidewalls and tips of carbon nanotubes using photochemistry. *Nano Lett* 2004; 4:89–93.
81. Dai, L. and Lee, K.M. Asymmetric end-functionalization of carbon nanotube U.S. Patent 2006/0257556 A1, February 16, 2006.
82. Barazza, H.J., Balzano, L., Pompeo, F., Rueda, O.L., O'Rear, E.A., and Resasco, D.E. Carbon nanotube-filled composites prepared by in-situ polymerization. U.S. Patent 7153903.
83. Yao, X., Huixia, W., Joseph, W., Song, Q., and Gang, C. *Chem A Eur J* 2006; 13:846.
84. Hughes, M., Chen, G.Z., Shaffer, M.S., Fray, D.J., and Windle, A.H. *Chem Mater* 2002; 14:1610.
85. Zhang, L.J. and Wan, M.X. *Nanotechnology* 2002; 13:750.
86. Zhang, Z., Wei, Z., Zhang, L., and Wan, M. *Acta Mater* 2005; 53:1373.
87. Konyushenko, E.N., Stejskal, J., Sedenková, I., Trchová, M., Sapurina, I., and Cieslar, M. *Polym Int* 2006; 55:31.
88. Huang, J. and Kaner, R.B. *Angew Chem Int Ed* 2004; 43:5817.
89. Nickels, P., Dittmer, W.U., Beyer, S., Kotthaous, J.P., and Simmel, F.C. *Nanotechnology* 2004; 15:1524.
90. Ma, Y., Zhang, J., and He, H. *J Am Chem Soc* 2004; 126:7097.
91. Kim, B.K., Kim, Y.H., Won, K., Chang, H., Choi, Y., and Kong, K.J. *Nanotechnology* 2005; 16:1177.
92. Geng, Y., Li, J., Sun, Z., Jing, X., and Wang, F. *Synth Met* 1998; 96:1.
93. Stejskal, J. and Sapurina, I. *Pure Appl Chem* 2005; 77:815.
94. Tzou, K. and Gregory, R.V. *Synth Met* 1992; 47:267.
95. Martin, C.R. *Handbook of Conducting Polymers.* In: Skotheim, T.A., Elsenbaumer, R.L., Reynolds, J.R., Editors. 2nd ed. New York: Marcel Dekker, 1998. pp. 409–21.
96. Vivekchand, S.R.C., Sudheendra, L., Sandeep, M., Govindaraj, A., and Rao, C.N.R. *J Nanosci Nanotechnol* 2002; 2:631.
97. Karim, R.M., Lee, C.J., Park, Y.T., and Lee, M.S. *Synth Met* 2005; 151:131.
98. Ham, H.T., Choi, Y.S., Jeong, N., and Chung, I.J. *Polymer* 2005; 46:6308.
99. Wu, M., Snook, G.A., Gupta, V., Shaffer, M., Fray, D.J., and Chen, G.Z. *J Mater Chem* 2005; 15:2297.

100. Choi, H.J., Park, S.J., Kim, S.T., and Jhon, M.S. *Diam Relat Mater* 2005; 14:766.
101. Guo, D.J. and Li, H.L. *J Solid State Electrochem* 2005; 9:445.
102. Cheng, G., Zhao, J., Tu, Y., He, P., and Fang, Y. *Anal Chim Acta* 2005; 533:11.
103. Qu, F., Yang, M., Jiang, J., Shen, G., and Yu, R. *Anal Biochem* 2005; 344:108.
104. Zhang, X., Zhang, J., and Liu, Z. *Appl Phys A* 2005; 80:1813.
105. Wu, T.-M., Lin, Y.W., and Liao, C.S. *Carbon* 2005; 43:734.
106. Sainz, R., Benito, A.M., Martı́nez, M.T., Galindo, J.F., Sotres, J., and Baró, A.M. *Adv Mater* 2005; 17:278.
107. Han, G., Yuan, J., Shi, G., and Wei, F. *Thin Solid Films* 2005; 474:64.
108. Cochet, M., Maser, W.K., Benito, A.M., Callejas, A.M., Martı́nez, M.T., and Benoit, J.-M. *Chem Commun* 2001; 1450.
109. Deng, M.G., Yang, B.C., and Hu, Y.D. *J Mater Sci* 2005; 40:5021.
110. Yu, Y., Che, B., Si, Z., Li, L., Chen, W., and Xue, G. *Synth Met* 2005; 150:271.
111. Philip, B., Xie, J., Abraham, J.K., and Varadan, V.K. *Polym Bull* 2005; 53:127.
112. Skotheim, T.A., Elsenbaumer, R.L., and Reynolds, J.R. *Handbook of Conducting Polymers*. New York: Marcel Dekker, 1997.
113. Premamoy, G., Samir, K.S., and Amit, C. Characterization of poly (vinyl pyrrolidone) modified polyaniline prepared in stable aqueous medium. *Eur Polym J* 1999; 35:699–710.
114. Cochet, M., Maser, W.K., Benitor, A., Callejas, A., Martinez, M.T., Benoit, J.M., Schreiber, J., and Chauvet, O. *Chem Commun* 2001; 1450.
115. Deng, J.G., Ding, X.B., Zhang, W.C., and Peng, Y.X. Carbon nanotube–polyaniline hybrid materials. *Eur Polym J* 2002; 38:2497–501.
116. Zengin, H., Zhou, W.S., Jin, J.Y., Czerw, R., Smith, D.W., Echegoyen, L., Carroll, D.L., Foulger, S.H., and Ballato, J. *J Adv Mater* 2002; 14:1480.
117. Wei, Z.X., Wan, M.X., Lin, T., and Dai, L.M. Polyaniline nanotubes doped with sulfonated carbon nanotubes made via a self-assembly process. *Adv Mater* 2003; 15:136–139.
118. Gao, M., Huang, S.M., Dai, L.M., and Wallace, G. Aligned coaxial nanowires of carbon nanotubes sheathed with conducting polymers. *Angew Chem Int* 2000; 39:3664–7.
119. Blanchet, G.B., Fincher, C.R., and Gao, F. Polyaniline nanotube composites: a high-resolution printable conductor. *Appl Phys Lett* 2003; 82:1290–2.
120. Li, X.H., Wu, B., Huang, J.E., and Zhang, J. Fabrication and characterization of well-dispersed single-walled carbon nanotube/polyaniline composites. *Carbon* 2003; 41:1670–3.
121. Downs, C., Nugent, J., Ajayan, P.M., Duquette, D.J., and Santhanam, K.S.V. *Adv Mater* 1999; 11:1028.
122. Do Nascimento, G.M., Corio, P., Novickis, R.W., Temperini, M.L.A., and Dresselhaus, M.S.J. *J Polym Sci* Part A: *Polym Chem* 2005; 43:815.
123. Chen, G.Z., Shaffer, M.S.P., Coleby, D., Dixon, G., Zhou, W.Z., Fray, D.J., and Windle, A.H. *Adv Mater* 2000; 12:522.
124. Weiss, Z., Mandler, D., Shustak, G., and Domb, A.J. *J Polym Sci Part A: Polym Chem* 2004; 42:1658.
125. Caruso, F. *Adv Mater* 2001; 13:11.
126. Caruso, R.A. and Antonietti, M. *Chem Mater* 2001; 13:3272.
127. Barraza, H.J., Pompeo, F., O'Rear, E.A., and Resasco, D.E. *Nano Lett* 2002; 2:797.
128. Santra, S., Tapec, R., Theodoropoulou, N., Dobson, J., Hebard, A., and Tan, W.H. *Langmuir* 2001; 17:2900.

129. Sun, J., Gao, L., and Iwasa, M. *Chem Commun* 2004; 832.
130. Hentze, H.P. and Kaler, E.W. *Curr Opin Colloid Interface Sci* 2003; 8:164.
131. Yan, F. and Xue, G. *J Mater Chem* 1999; 9:3035.
132. Tang, B.Z. and Xu, H.Y. *Macromolecules* 1999; 32:2569.
133. Fan, J.H., Wan, M.X., Zhu, D.B., Chang, B.H., Pan, Z.W., and Xie, S.S. *J Appl Polym Sci* 1999; 74:2605.
134. Star, A., Stoddart, J.F., Steuerman, D., Diehl, M., Boukai, A., Wong, E.W., Yang, X., Chung, S.W., Choi, H., and Heath, J.R. *Angew Chem Int Ed* 2001; 40:1721.
135. Zhou, X. and Xiao, Q.F. *Electrochem Acta* 2003; 48:575.
136. Jia, Z., Wang, Z., Xu, C., Liang, J., Wei, B., Wu, D., and Zhu, S. *Mater Sci Eng A* 1999; 271:395.
137. Velasco-Santos, C., Martinez-Hernandez, A.L., Fisher, F.T., Ruoff, R., and Castano, V.M. *Chem Mater* 2003; 15:4470.
138. Putz, K.W., Mitchell, C.A., Krishnamoorti, R., and Green, P.F. *J Polym Sci Part B: Polym Phys* 2004; 42:2286.
139. Wu, B.Y., Liu, A.H., Zhou, X.P., and Jiang, Z.D. *Chem J Chinese Univ* 1995; 16:1641.
140. Zhou, W. Bachelor Thesis, Huazhong University of Science and Technology. Wuhan; China. 2003.
141. Mawhinney, D.B., Naumenko, V., Kuznetsova, A., and Yates, J.T. *J Am Chem Soc* 2000; 122:2383.
142. Chen, Y., Haddon, R.C., Fang, S., Rao, A.M., Lee, W.H., Dickey, E.C., Grulke, E.A., Pendergrass, J.C., Chavan, A., Haley, B.E., and Smalley, R.E. *J Mater Res* 1998; 13:2423.
143. Zhi, Y., Xiaohua, C., Chuansheng, C., Wenhua, L., Hua, Z., Longshan, X., and Bin, Y. *Polym Comp* 2007; 36–41.
144. Hesheng, X., Mo, S., Jie, J., and Lei, C. *Macromol Chem Phys* 2006; 207:1945–1952.
145. Koerner, H., Price, G., Pearce, N.A., Alexander, M., and Vaia, R.A. *Nat Mater* 2004; 3:115.
146. Koerner, H., Liu, W., Alexander, M., Mirau, P., Dowty, H., and Vaia, R.A. *Polymer* 2005; 46:4405.
147. Cho, J.W., Kim, J.W., Jung, Y.C., and Goo, N.S. *Macromol Rapid Commun* 2005; 26:412.
148. Sen, R., Zhao, B., Perea, D., Itkis, M.E., Hu, H., Love, J., Bekyarova, E., and Haddon, R.C. *Nano Lett* 2004; 4:459.
149. Xianhong, C., Xiaojin, C., Ming, L., Wenbin, Z., Xiaohua, C., and Zhenhua, C. *Macromol Chem Phys* 2007; 208:,964–972.
150. Xin, T., Chang, L., Hui-Ming, C., Haichao, Z., Feng, Y., and Xuequan, Z. *J Appl Polym Sci* 2004; 92:3697–3700.
151. Mun˜ oz-Bonilla, A., Madruga, E.L., and Fernández-García, M. *J Polym Sci Part A: Polym Chem* 2005; 43:71–77.
152. Ishizu, K. and Kakinuma, H. *J Polym Sci Part A: Polym Chem* 2005; 43:63–70.
153. Cao, C., Zou, J., Dong, J.Y., Hu, Y., and Chung, T.C. *J Polym Sci Part A: Polym Chem* 2005; 43:429–437.
154. Meng, J.Q., Du, F.S., Liu, Y.S., and Li, Z.C. *J Polym Sci Part A: Polym Chem* 2005; 43:752–762.
155. Wang, X., Zhang, H., Shi, M., Wang, X., and Zhou, O. *J Polym Sci Part A: Polym Chem* 2005; 43:733–741.
156. Jakubowski, W., Lutz, J.F., Slomkowski, S., and Matyjaszewski, K. *J Polym Sci Part A: Polym Chem* 2005; 43:1498–1510.

157. Jankova, K., Bednarek, M., and Hvilsted, S. *J Polym Sci Part A: Polym Chem* 2005; 43:3748–3759.
158. Muñoz-Bonilla, A., Cerrada, M.L., and Fernández-García, M. *J Polym Sci Part A: Polym Chem* 2005; 43:4828–4837.
159. Abraham, S., Ha, C.S., and Kim, I. *J Polym Sci Part A: Polym Chem* 2005; 43:6367–6378.
160. París, R. and Luis de la Fuente, J. *J Polym Sci Part A: Polym Chem* 2005; 43:6247–6261.
161. Brar, A.S. and Kaur, S. *J Polym Sci Part A: Polym Chem* 2006; 44:1745–1757.
162. Pyun, J., Kowalewski, T., and Matyjaszewski, K. *Macromol Rapid Commun* 2003; 24:1043–1059.
163. Ramakrishnan, A., Dhamodharan, R., and Ruhe, J. *J Polym Sci Part A: Polym Chem* 2006; 44:1758–1769.
164. Chen, R., Feng, W., Zhu, S., Botton, G., Ong, B., and Wu, Y. *J Polym Sci Part A: Polym Chem* 2006; 44:1252–1262.
165. Kruk, M., Dufour, B., Celer, E.B., Kowalewski, T., Jaroniec, M., and Matyjaszewski, K. *J Phys Chem B* 2005; 109:9216–9225.
166. Kim, J.B., Bruening, M.L., and Baker, G.L. *J Am Chem Soc* 2000; 122:7616–7617.
167. Kotal, A., Mandal, T.K., and Walt, D.R. *J Polym Sci Part A: Polym Chem* 2005; 43:3631–3642.
168. Vestal, C.R. and Zhang, Z.J. *J Am Chem Soc* 2002; 124:14312–14313.
169. Gravano, S.M., Dumas, R., Liu, K., and Patten, T.E. *J Polym Sci Part A: Polym Chem* 2005; 43:3675–3688.
170. Cui, T., Zhang, Jm., Wang, J., Cui, F., Chen, W., Xu, F., Wang, Z., Zhang, K., and Yang, B. *Adv Funct Mater* 2005; 15:481–486.
171. Baskaran, D., Mays, J.M., and Bratcher, M.S. *Angew Chem Int Ed* 2004; 43:2138–2142.
172. Shanmugharaj, A.M., Bae, J.H., Nayak, R.R., and Ryui, S.H. *J Polym Sci Part A: Polym Chem* 2007; 45:460–470.
173. Yao, Z.L., Braidy, N., Botton, G.A., and Adronov, A. *J Am Chem Soc* 2003; 125:16015–16024.
174. Tsubokawa, N. *Prog Polym Sci* 1992; 17:417.
175. Tsubokawa, N. *Bull Chem Soc Jpn* 2002; 75:2115.
176. Tsubokawa, N. Preparation and properties of polymer-grafted carbon nanotubes and nanofibers. *Polymer J* 2005; 37(9):637–655.
177. Yoon, K.R., Wan-Joong, K., and Choi, L.S. *Macromol Chem Phys* 2004; 205:1218–1221.
178. Yoon, K.R., Koh, Y.J., and Choi, L.S. *Macromol Rapid Commun* 2003; 24:207.
179. Yoon, K.R., Chi, Y.S., Lee, K.B., Lee, J.K., Kim, D.J., Koh, Y.J., Joo, S.W., Yun, W.S., and Choi, L.S. *J Mater Chem* 2003; 13:2910.
180. Yoon, K.R., Lee, K.B., Chi, Y.S., Yun, W.S., Joo, S.W., and Choi, L.S. *Adv Mater* 2003; 15:2063.
181. Yoon, K.R., Kim, Y., and Choi, L.S. *J Polym Res* [in press].
182. Kong, H., Gao, C., and Yan, D.Y. *J Am Chem Soc* 2004; 126:412.
183. Qin, S.H., Qin, D.Q., Ford, W.T., Resasco, D.E., and Herrera, J.E. *Macromolecules* 2004; 37:752.
184. Hong, C.Y., You, Y.Z., Wu, D.C., Liu, Y., and Pan, C.Y. *Macromolecules* 2005; 38:2606.
185. Shaffer, M.S.P. and Koziol, K. *Chem Commun* 2002; 18:2074.
186. Bin, Zhao., Hui, Hu., and Haddon, R.C. *Adv Func Mater* 2004; 14(1):71–76.

187. Kamigaito, M., Ando, T., and Sawamoto, M. *Chem Rev* 2001; 101:3689–3746.
188. Liu, Y.Q. and Yao, Z.L. *Macromolecules* 2005; 38: 1172–1179.
189. Koshio, A., Yudasaka, M., Zhang, M., and Iijima, S. *Nano Lett* 2001; 1:361–363.
190. Park, C., Ounaies, Z., Watson, K.A., Crooks, R.E., Smith, J., Lowther, S.E., Connell, J.W., Siochi, E.J., Harrison, J.S., and St. Clair, T.L. *Chem Phys Lett* 2002; 364:303–308.
191. Lou, X.D., Detrembleur, C., Pagnoulle, C., Jerome, R., Bocharova, V., Kiriy, A., and Stamm, M. *Adv Mater* 2004; 16:2123–2127.
192. Kahn, M.G.C., Banerjee, S., and Wong, S.S. *Nano Lett* 2002; 2:1215–1218.
193. Lin, Y., Zhou, B., Fernando, K.A.S., Allard, L.F., and Sun, Y.P. *Macromolecules* 2003; 36:7199–7204.
194. Sun, Y.P., Huang, W., and Carroll, D.L. *Chem Mater* 2001:13:2864–2869.
195. Sano, M., Kamino, A., and Shinkai, S. *Angew Chem Int Ed* 2001; 40:4661–4663.
196. Cao, L., Yang, W., Yang, J., Wang, C., and Fu, S. *Chem Lett* 2004; 33:490–491.
197. Bubert, H., Haiber, S., Brandl, W., Marginean, G., Heintze, M., and Bruser, V. Characterization of the uppermost layer of plasma-treated carbon nanotubes. *Diam Relat Mater* 2003; 12:811–5.
198. Chirala, V., Marginean, G., Brandl, W., and Iclanzan, T. Vapour grown carbon nanofibers-polypropylene composites and their properties. In *Carbon Nanotubes*. Edited by Popov, V.N. and Lambin, P. Springer; Heidelberg. 2006. p. 227.
199. Miyagawa, H., Misra, M., and Mohanty, A.K. *J Nano Sci Nanotechnol* 2005; 5:1593.
200. Greton, Y., Depleux, S., Benoit, R., Salvestar, J.P., Sinturel, C., Beguin, F., Bonnamy, S., Besarmot, G., and Boufend, L. *Mol Crystals Liquid Crystals* 2002; 387:359.
201. Chirila, V., Marginean, G., and Brandl, W. *Surf Coatings Technol* 2005; 200:1238.
202. Jin, Ah Kim, Dong, Gi Seong., Tae, Jin Kang, and Jae, Ryoun Youn. *Carbon* 2006; 44:1898–1905.
203. Shi, D., Jie, L., He, P., Wang, L.M., Van Ooij, W.J., Schulz, M., Liu, Y.J., and Mast, D.B. *Appl Phys Lett* [In press].
204. Felten, A., Ghijsen, J., Pireaux, J.J., Johnson, R.L., Whelan, C.M., Liang, D., Tendeloo, G., and Bittencourt, C. *J. Phys D: Appl Phys* 2007; 40:7379–7382.
205. Xu, T., Yang, J., Liu, J., and Fu, Q. *Appl Surf Sci* 2007; 253(22):8945–8951.
206. Bianco, A., Kostarelos, K., Partidos, C.D., and Prato, M. *Chem Commun* 2005; 571.
207. Pantarotto, D., Singh, R., McCarthy, D., Erhardt, M., Briand, J.P., Prato, M., Kostarelos, K., and Bianco, A. *Angew Chem Int Ed* 2004; 43:5242.
208. Chen, X., Lee, G.S., Zettl, A., and Bertozzi, C.R. *Angew Chem Int Ed* 2004; 43:6111.
209. Bianco, A. and Prato, M. *Adv Mater* 2003; 15:1765.
210. Davis, J.J., Coleman, K.S., Azamian, B.R., Bagshaw, C.B., and Green, M.L.H. *Chem Eur J* 2003; 9:3732.
211. Gao, H.J., Kong, Y., and Cui, D.X. *Nano Lett* 2003; 3:471.
212. O'Connell, M.J., Bachilo, S.M., Huffman, C.B., Moore, V.C., Strano, M.S., Haroz, E.H., Rialon, K.L., Boul, P.J., Noon, W.H., Kittrell, C., Ma, J.P., Hauge, R.H., Weisman, R.B., and Smalley, R.E. *Science* 2002; 297:593.
213. Islam, M.F., Rojas, E., Bergey, D.M., Johnson, A.T., and Yodh, A.G. *Nano Lett* 2003; 3:269.
214. Kim, O.K., Je, J.T., Baldwin, J.W., Kooi, S., Pehrsson, P.E., and Buckley, L.J. *J Am Chem Soc* 2003; 125:4426.
215. Dyke, C.A. and Tour, J.M. *Nano Lett* 2003; 3:1215.
216. Zhao, B., Hu, H., and Haddon, R.C. *Adv Funct Mater* 2004; 14:71.
217. Qin, S.H., Qin, D.Q., Ford, W.T., Herrera, J.E., Resasco, D.E., Bachilo, S.E., and Weisman, R.B. *Macromolecules* 2004; 37:3965.

218. Lin, Y., Rao, A.M., Sadanadan, B., Kenik, E.A., and Sun, Y.P. *J Phys Chem B* 2002; 106:1294.
219. Yan-Xin, LIU., Zhong-Jie, D.U., Yan, L.I., Zhang, Chen., and Hang-Quan, L.I. *Chinese J Chem* 2006; 24: 563–568.
220. Goh, S.H., Wang, M., and Pramoda, K.P. *Carbon* 2006; 44:613.
221. Huang, W., Fernando, S., Allard, L.F., and Sun, Y.P. *Nano Lett* 2003; 3:565.
222. Haddon, R.C., Zhao, B., Hu, H., Yu, A., and Perea, D. *J Am Chem Soc* 2005; 127:8197.
223. Basiuk, E.V., Monroy-Pelaez, M., Puente-Lee, L., and Basiuk, V.A. *Nano Lett* 2004; 4:863.
224. Gabriel, G., Sauthier, G., Fraxedas, J., Moreno-Manas, M., Martinez, M.T., Miravitlles, C., and Casabo, J. *Carbon* 2006; 44:1891.
225. Jung Sook, An, Byeong-Uk, Nam, Soon Huat, Tan, and Sung Chul, Hong. *Macromol Symp* 2007; 249:276–282.
226. Riggs, E., Guo, Z., Carroll, D.L., and Sun, Y.P. *J Am Chem Soc* 2000; 122:5879.
227. Zhang, M., Smith, A., and Gorski, W. *Anal Chem* 2004; 76:5045.
228. Soon Huat, Tan, Jeung Choon, Goak, Lee, N., Jeong-Yeol, Kim., and Hong, S.C. *Macromol Symp* 2007; 270–275.
229. Nguyen, C.V., Delzeit, L., Cassel, A.M., Li, J., Han, J., and Meyyappan, M. *Nano Lett* 2002; 2:1079–1081.
230. Baker, S.E., Cai, W., Lasseter, T.L., Weidkamp, K.P., and Hamers, R.J. *Nano Lett* 2002; 2:1413–1417.
231. Dwyer, C., Guthold, M., Falvo, M., Washburn, S., Superfine, R., and Erie, D. *Nanotechnology* 2002; 13:601–604.
232. Hazani, M., Naaman, R., Hennrich, F., and Kappes, M.M. *Nano Lett* 2003; 3:153–155.
233. Williams, K.A., Veenhuizen, P.T.M., de la Torre, B.G., Eritjia, R., and Dekker, C. *Nature* 2002; 420:761.
234. Huang, W., Taylor, S., Fu, K., Lin, Y., Zhang, D., Hanks, T.W., Rao, A.M., and Sun, Y.P. *Nano Lett* 2002; 2:311–314.
235. Jiang, K., Schadler, L.S., Siegel, R.W., Zhang, X., Zhang, H., and Terrones, M. *J Mater Chem* 2004; 14:37–39.
236. Liu, P. *Eur Polym J* 2005; 41: 2693–2703.
237. Ge, J.J., Li, C.Y., Xue, G., Mann, I.K., Zhang, D., Harris, F.W., Cheng, S.Z.D., Hong, S.C., Zhuang, X., and Shen, Y.R. Rubbing-induced molecular reorientation on an alignment surface of an aromatic polyimide containing cyanobiphenyl side chains. *J Am Chem Soc* 2001; 123:5768.
238. Lim, H., Cho, W.J., Ha, C.S., Ando, S., Kim, Y.K., Park, C.H., and Lee, K. Electrospinning of continuous carbon nanotube-filled nanofiber yarns. *Adv Mater* 2003; 15:1161.
239. Ge, J.J., Zhang, D., Li, Q., Hou, H.Q., Graham, M.J., Dai, L.M., Harris, F.W., and Cheng, S.Z.D. Multiwalled carbon nanotubes with chemically grafted polyetherimides. *J Am Chem Soc* 2005; 127:9984.
240. Hill, D., Lin, Y., Qu, L.W., Kitaygorodskiy, A., Connell, J.W., Allar, L.F., and Sun, Y.P. Functionalization of carbon nanotubes with derivatized polyimide. *Macromolecules* 2005; 38:7670.
241. Ounaies, Z., Park, C., Wise, K.E., Siochi, E.J., and Harrison, J.S. *Compos Sci Technol* 2003; 63:1637.
242. Jiang, X.W., Bin, Y.Z., and Matsuo, M. *Polymer* 2005; 46(18):7418.

10

Ocean Engineering Application of Nanocomposites

Yansheng Yin and Xueting Chang

Shanghai Maritime University, Shanghai, China

CONTENTS

10.1 Introduction

Basic and applied research in the field of nanomaterials and nanocomposites materials science enables great progress in the development of new kinds of ocean engineering materials and coatings applied in marine environments.

Nanocomposite materials are used increasingly in marine structures due to their many advantages. However, the unique and hostile marine environment, with the presence of seawater and moisture, temperature extremes, time-dependent three-dimensional loading due to wave slamming, hydrostatic pressure, and other factors gives rise to significant challenges for designers of composite marine structures [1]. Additional requirements for naval structures include the ability to withstand highly dynamic loading due to weapons impact or to air or underwater explosions.

In the last century, the research programs mostly supported studies on the mechanical behavior of marine structural materials, providing the scientific basis for the effective design and utilization of affordable marine structures. Although the protection of marine structures has been extensively examined, the understanding of their interaction with seawater and microbiology is far from complete. Damage of marine structures still occurs from time to time, with two general failure modes evident [2]. The first mode is structural failure, caused by wave forces acting on and damaging the structure itself.

423

The second mode is foundation failure, caused by corrosion and erosion of the seawater and microbiology adhesion in the vicinity of the structure, resulting in collapse of the structure as a whole [3–5].

Today, the major focus of marine composites is on nanocomposite sandwich structures and nanomembranes, especially on nanocomposite coatings.

Nanocomposite research in the ocean structure application field deals with deformation, damage initiation, damage growth, failure and corrosion in nanoparticles, such as nanofiber and nanoclay, and reinforced composites and nanocomposite sandwich structures subjected to static, cyclic, and dynamic multiaxial loading conditions in severe ocean environments; the areas of emphasis include constitutive behavior, failure modes and mechanisms, failure criteria, environmental effects, dynamic response, and structural effects. Additionally, concepts for enhancing the mechanical properties of marine composites through the introduction of nanoparticles should also be explored.

10.2 Oceanic Environmental Effects on Composites

During the past decade, safety analysis of marine and offshore engineering products such as ships, marine cranes, and offshore topsides has attracted a great deal of public attention. This is because a marine or offshore engineering product is usually a large, expensive and complex structure, and a serious failure could cause disastrous consequences. The research of marine and offshore materials is to decrease the probabilities of serious system failures, which could cause death or injury, damage, or loss of property [6]. As we know, the conjoint parts of different kinds of materials applied in ocean aqueous environments may cause potential difference corrosion. The combination of aggressive liquid media and the possibility of high burdens of seawater associated with microbiology and strain corrosion are representative of some of the most severe operating conditions faced by today's marine materials. Substantial direct and indirect costs are typically associated with failure of components. Economic factors have therefore ensured that the volume of study of marine materials has increased in recent years [7]. Submersibles that are capable of reaching great depths have of course been around for some time. The pressure hulls of these vessels are typically constructed of high strength steels or alloys of aluminum and titanium. This gives high weight–to–displacement ratios that make them unsuitable for use in an autonomous vehicle with limited energy-carrying capability that has long endurance requirements [8]. New materials such as solid glass, ceramics, ceramic composites, and metal–matrix composites have great potential but are considered technically unfeasible at present. This leaves nanoglass and nanocarbon fiber-reinforced polymers as the only current options that will provide a pressure hull with a low enough weight-to-displacement ratio to allow the required payload to be carried. Most authors appear to favor the theoretically stronger nanocarbon fiber composites (see, for example, Q. M. Jia [9] and K-T Lau [10]).

10.3 Marine Nanocomposite Sandwich Structures

Foam core composite sandwich structures are candidate materials for naval structures. These materials offer many advantages, such as corrosion resistance and high stiffness-to-weight ratio. The latter feature can lead to a most substantial weight savings in marine structural applications. However, it is well known that polymers and polymeric composites absorb fluids when exposed to ambient liquid environments, and that fluid absorption is accompanied by expanded strains and may degrade material properties [11].

As commented by P. M. Schubel and A. C. Marshall [12,13], the use of composite sandwich structures is increasing in the aerospace and marine industry, as well as in other areas where a lightweight structure material with elevated flexural stiffness is required. The design concept behind these structures is the increase in flexural stiffness by placing two stiff, strong, and thin face sheets separated by a lightweight thicker flexible core. By employing this strategy, the moment of inertia is enhanced and consequently the flexural properties increase while keeping the weight practically unaffected. According to Vinson [14], as face sheets carry most of the load, an optimal design for sandwich composite structures must take into consideration the face sheet response to static and dynamic loads. Poly (methacrylimide) (PMI) foams have high specific compressive strength properties, up to 67 MPa [13]. Unfortunately, as discussed by Lim and his coworkers [15], most of the previous studies have focused on the strength capacity of sandwich structures under quasi-static loading. However, Abrate [16] called attention to the response of sandwich composite structures to impact loadings. According to him, this class of structures is susceptible to damage caused by foreign object impact.

In Emmanuel Ayorinde's study [17], the development of reliable low-cost nondestructive examination (NDE) methods for marine composites was summarized, especially composite sandwich structures. It has been shown that low-to-medium strain rates have strong impacts on the damage history of composite structures. Small quantities of nano-particles have been found to make significant differences to the material properties and NDE responses of basic sandwich and laminate composites. Particulars of enhancement with silicon carbide and titanium dioxide nano-sized materials have also been given.

Desmond et al. used carbon nanofibers (CNFs) to increase the compressive properties of poly (arylene ether sulfone) (PAES) foams [18]. The polymer composite pellets were produced by melt blending the PAES resin with CNFs in a single screw extruder. The pellets were saturated and foamed with water and CO_2 in a one-step batch process method. The foams had cell nucleation densities between 109 and 1010 cells/cm^3, two orders of magnitude higher than unreinforced PAES foam, suggesting that the CNFs acted as heterogeneous nucleating agents. The CNF–PAES foam exhibited improved compressive properties compared to unreinforced PAES foam produced from a similar method. Both the specific compressive modulus and strength increased by over 1.5 times that of unreinforced PAES foam.

Due to the wide application of foam-cored composite sandwich structures in seawater environments under long-term exposure, special attention was focused on seawater-induced damage in foam materials, weight gains, and expansion strains, as well as on possible degradation in the properties of foam materials due to such extended exposure. Xiaoming Li [19] found that after 2400–3600 h immersion in seawater, damage in the form of cell wall swelling and breakage, as well as pitting, appeared over the outer regions of the samples. Upon continued immersion, some of the aforementioned pits enlarged in size and penetrated the sample to depths of up to 2.8 mm, as is shown in Figure 10.1

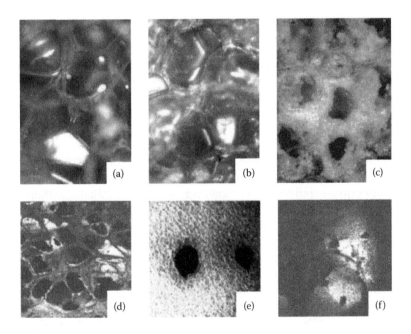

FIGURE 10.1
Undamaged PVC foams and seawater-induced damage in the same foam materials. (From X.M. Li and Y.J. Weitsman. 2004. *Composites B* 35: 451–459.) (a,b) Undamaged dry PVC foams. (c) Seawater causes swelling of foam cell walls. (d) Confocal microscope photograph shows damage (pits and breakage of cell walls) inside the core, 1.78 mm below the surface. (e) Cavities were observed to form on the surfaces of the foam materials after extended immersion in seawater. (f) A confocal microscope photograph shows pits forming at the bottom of a cavity (2.8 mm below the surface).

Siriruk et al. [20] investigated seawater effects on the interfacial mechanical response between foam and facing due to sustained seawater exposure using carefully controlled laboratory conditions. Results indicated that the delamination crack propagates close to the interface in the wet case, while it stays within the foam in the dry case. A good agreement between data and predictions indicated a reduction in fracture toughness by 30% due to sustained exposure to seawater.

Polyester resins are widely employed for pleasure boat construction. In order to satisfy new environmental legislation on styrene emissions, resin suppliers have proposed modified formulations, but these show lower failure strains under tensile loading. Maurin et al. examined the influence of wet aging on mechanical properties of low styrene polyester, together with standard polyester and vinyl ester resins, and their glass-reinforced composites. Nano-indentation was used to study local changes in the resin after aging. The low styrene resins appear to lose strength more slowly than the standard resins, but after 9 months in seawater at 40°C similar property losses are noted [21]. Meanwhile, a set of sandwich composites plates made of fiberglass–nano-modified epoxy face sheets and polystyrene foams was prepared by Antonio F. Ávila. The results show that the addition of 5 wt% of nanoclay led to more efficient energy absorption. The failure modes were also analyzed, and they seem to be affected by the nanoclay addition to face sheets [22].

10.4 Oceanic-Applied Nanomembrane

Development and modification of membranes for desired functions such as better filtration performance and fouling control have been pursued by ocean researchers and industries worldwide since the 1980s [23,24]. Membrane separation is advantageous and widely used in seawater treatment and purification operations, but fouling control and cost reduction are very challenging, especially in MBR (membrane bioreactor) and related microfiltration, ultrafiltration, and nanofiltration–reverse osmosis processes [25–27]. To contribute to meeting the challenges, especially to removal of micropollutants by microfiltration at low separation pressure, some adsorptive components and photocatalytic components, such as nno-TiO_2, were incorporated into the PVA modification of terylene MF membranes to make them more hydrophilic, lower in cost, and more adsorptive. Further gains were better fouling control and micropollutant removal [28].

Reverse osmosis (RO) membranes have been widely used in water and wastewater treatment and seawater desalination. Most commercially available RO membranes are of thin film composite (TFC) type, where a dense polyamide (PA) thin layer several hundred nanometers thick is formed on a microporous polysulfone (PS) supported by an interfacial polymerization process [29]. The physiochemical properties of this critical PA layer essentially determine the performance (flux and rejection) of TFC membranes. For example, membrane zeta potential has been used to correlate the transport of some trace organic solutes through RO and nanofiltration (NF) membranes [30]. In addition, the flux performance and fouling behavior of a membrane may also be affected by its zeta potential [31]. Other properties, such as the chemical composition and morphology of the polyamide layer, are also important to RO membrane performance [32]. However, the major drawback that limits the application of RO is the decrease in permeate flux due to membrane fouling. One of the tactics to overcome the decline in flux is to operate the membrane process below critical flux. Consequently, understanding of the physiochemical properties of the polyamide layer become critical in order to control membrane fouling and trace organics rejection. The performance of polyamide (PA) composite reverse osmosis (RO) membranes is essentially determined by a dense skin layer several hundred nanometers thick [33]. Improved understanding of this critical thin layer will likely advance our understanding and control of membrane fouling and trace organics rejection. Vineet K. Gupta [34] developed a novel protocol to characterize nanofiltration (NF) and RO membranes that employ the Spiegler–Kedem membrane-transport model based on irreversible thermodynamics, coupled with a film theory description of the concentration polarization. The novel aspect of this characterization protocol is extracting the concentration polarization boundary layer thickness as well the three membrane transport parameters from permeation data. This methodology was applied to four dilute aqueous salt systems for a Dow Filmtech NF-255 membrane and one dilute aqueous salt system for a Dow Filmtech BS-30 membrane that were studied using a well-mixed flat sheet membrane permeation cell. This design correlation formalism in principle can be used to characterize any NF or RO membrane irrespective of the type of membrane contactor employed.

After all, biofouling is one the most critical problems in seawater desalination plants, and science has not yet found effective ways to control it. Typical side effects of microbiological fouling are [35]: (1) a reduction of the performance of membranes due to formation of biofilm on their surfaces, (2) the secondary pollution of the purified water by bacterial cells and products of their metabolism, (3) an increase in power consumption because of the higher pressure requirements to overcome biofilm resistance and the flux decline. In attempts to reduce membrane biofouling, Yang et al. proposed an innovative biofouling control approach by surface modification of the RO membrane and spacer with nanosilver coating. Silver compounds and ions are historically recognized for their effective antimicrobial activity. Nanosilver particles have been applied as a biocide in many aspects of disinfection, including healthcare products and water treatment. A chemical reduction method was used for directly coating nanosilver particles on a membrane sheet and spacer. The silver-coating membranes and spacers, along with an unmodified membrane sheet, were tested in the membrane cell and compared on the basis of their antifouling performance [36]. In the silver-coated spacer test, there was almost no multiplication of cells detected on the membrane during the whole testing period. Besides, the cells adhering to the membrane seemed to lose their activity quickly. According to the RO performance and microbial growth morphology, the nanosilver coating technology is valuable for use in biofouling control in seawater desalination.

V. Kochkodan deposited TiO_2 particles on membrane surfaces with following ultraviolet (UV) irradiation at 365 nm. It was shown that due to photobactericidal effect the fluxes of surface modified membranes were 1.7–2.3 times higher compared with those for control membrane samples (without TiO_2 deposition and UV treatment) [35].

For employing direct microscopic observation to elucidate mechanisms governing bacterial cell deposition on clean and organic-fouled nanofiltration (NF) and RO membranes, bovine serum albumin (BSA) and alginic acid (AA) were used as models for protein and polysaccharide-rich organic matter in secondary wastewater effluents by Arun Subramani [37]. An extended DLVO analysis of bacterial cells and membrane surface properties suggested that bacterial deposition correlated most strongly with the Lewis acid–base free energy of adhesion and root mean square (RMS) roughness, whereas van der Waals and electrostatic free energies were weakly correlated. Chung et al. suggested a polymeric membrane reservoir, which can be directly applied into seawater as shown in Figure 10.2. Different types of nanoporous polymeric matrices were selected and investigated for the polymeric membrane reservoir.

From this study [38], ease of water diffusion to the reservoir was the most important factor for the reservoir system to be applied to lithium recovery from seawater; a membrane reservoir prepared by Kimtex® showed the best result. The proposed system has the advantage of direct application in the sea without a pressurized flow system.

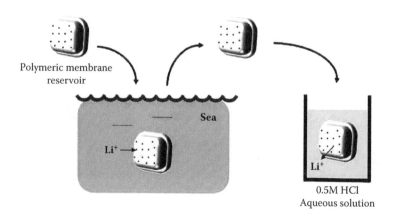

FIGURE 10.2
Conceptual illustration of nano lithium recovery from seawater using polymeric membrane reservoir. (From T. Liu et al. *Electrochim. Acta* 52: 3709–3713.)

Chong et al. [39] described the determination of critical flux and the cake-enhanced osmotic pressure (CEOP) effect of colloidal silica fouling in reverse osmosis (RO), using a sodium chloride tracer response technique. The tracer technique gave the transient values of concentration polarization (CP) and cake resistance (Rf). It was found that the critical flux concept was applicable to colloidal silica fouling in the RO process. It was observed that the number of particles convected to and finally deposited on the membrane surface, known as the fractional deposition constant, was strongly influenced by the crossflow velocity. The CEOP effect, as deduced from the transient CP data, was found to be more severe at high flux and low cross-flow velocity conditions, due to formation of a thicker cake layer and flux-induced polarization. Even for the small (20 nm) colloids used in this study, the CEOP effect can readily exceed the Rf effect due to fouling.

It is well known that a surface with large water contact angle (≥150°) is called a *super-hydrophobic surface*. By increasing the surface roughness, the surface water repellency can be dramatically enhanced. Such surface is of great importance for many industrial applications, and may present a solution to the long-standing problems of environmental contamination and corrosion of metals and metal alloys. Tao Liu and Yansheng Yin [40] have pretreated the super-hydrophobic nanofilm by a *n*-tetradecanoic acid ($CH_3(CH_2)12COOH$) etch, formed on the fresh copper surface. The structure of the film is similar to that of a haulm or flower, and the seawater contact angle is larger than 150°, as is shown in Figure 10.3.

Moreover, the corrosion resistance of bare and modified samples in seawater was investigated by cyclic voltammogram (CV) and electrochemical impedance spectroscopy (EIS). Experimental results show that the corrosion rate of Cu with a super-hydrophobic surface decreases dramatically because of its special microstructure.

However, in Asim K. Ghosh's research on the impacts of organic solvent properties, reaction and curing conditions on polyamide composite reverse osmosis membrane separation performance, film structure, and interfacial properties, the direct experimental evidence showed that the rugose nanomorphology

(a)   (b)   (c)

FIGURE 10.3
(a) Water droplet on the modified surface immersed for 3 days. (b) Water droplet on the modified surface immersed for 5 days. (c) Water droplet on the modified surface immersed for 10 days.

and relative hydrophobicity of high performance MPD–TMC membranes might enhance concentration polarization and exacerbate surface fouling [41].

To embed molecular sieve nanoparticles throughout the polyamide thin film layer of an interfacial composite RO membrane, Byeong-Heon Jeong has reported on a new concept for formation of mixed matrix reverse osmosis membranes by interfacial polymerization of nanocomposite thin films in situ on porous polysulfone supports [42]. Nanocomposite films created for this study comprise NaA zeolite nanoparticles dispersed within 50–200 nm thick polyamide films. Hand-cast pure polyamide membranes exhibit surface morphologies characteristic of commercial polyamide RO membranes, whereas nanocomposite membranes have measurably smoother and more hydrophilic, negatively charged surfaces. At the highest nanoparticle loadings tested, hand-cast nanocomposite film morphology is visibly different, and pure water permeability is nearly double that of hand-cast polyamide membranes with equivalent solute rejections. Comparison of membranes formed using pore-filled and pore-opened zeolites suggest nanoparticle pores play an active role in water permeation and solute rejection. The best performing nanocomposite membranes exhibit permeability and rejection characteristics comparable to commercial RO membranes. As a concept, thin film nanocomposite membrane technology may offer new degrees of freedom in tailoring RO membrane separation performance and material properties.

10.5 Marine Nanocomposite Coating

Marine fouling is the major problem for the application of materials in marine environments. The man-made equipment and ocean coastal structures placed in the marine environment encounter fouling by marine organisms and biofilm formation on surfaces. The greatest disadvantage of the biofouling is corrosion and huge flow resistance. Fouling in one form or other can occur in most locations during all seasons, and the research of antifouling methods is essential to reduce the consequences for oceanic installations. The most common method of combating biofouling is using toxic antifouling coatings containing metal compounds such as copper and tri-*n*-butyltin (TBT) or biocides like silver or surfactants.

However, these coatings are not recommended for use by virtue of their potential damage to marine organisms leading environmental pollution problems. Some researchers focus on nanocomposite coatings because of the special physical and chemical properties, such as extremely small surface area and super-hydrophobic properties. Nanocomposite coatings are a class of materials that show improved properties at low loading levels of nanoparticles when compared to conventional filler contents in coatings. Improved properties include mechanical and optical permeability and wettability, among others that can be tailored by the choice of nanoparticle in the coating formulation. Marielle Wouters [43] evaluated the antifouling and fouling-release properties of sol–gel based coating modified by anisotropic particles. The obtained surface structures of the coatings were on their antifouling and fouling-release properties. The adherence and ease of removal by hydrodynamic shear forces were determined for marine bacteria. The coatings were found to inhibit settlement of the marine bacteria and showed promising results with respect to fouling-release properties in comparison with state-of-the-art fouling-release coatings and antifouling coatings.

In order to obtain the desired functionality of the composite, the nanoparticle needs to be modified in most cases prior to introduction into the binder material to obtain proper dispersion. By selecting the type of modifier for the clay particle, the compatibility with the coating formulation can be tuned, as can the final coating properties [44]. Recently, nanocomposite coating systems are produced by introduction of anisotropic nanoparticles. These nanoparticles (sepiolite) are ionically and/or covalently functionalized with one or more different organic groups, that is, providing the particle with more than one function. By the introduction of coating-compatible and coating-incompatible modifiers, a structured coating can be obtained.

The dissolution of soluble pigments from both tin-based and tin-free chemically active antifouling (AF) paints is a key process influencing the corrosion resistance properties of the composite coatings. Yebra et al. [45] developed a method to measure the dissolution rate of seawater soluble pigments for ZnO antifouling paints. Rougher and more porous exposed surfaces, with a larger number of defects in the lattice structure, are hypothesized to be responsible for faster seawater attacks by the pellets compared to the ZnO crystals. In any case, the ZnO dissolution rates are markedly lower than those associated with the seawater dissolution of cuprous oxide (Cu_2O) particles, which are also used in AF paints.

Y. Castro [46] and H.M. Hawthorne [47] discussed the preparation and characterization of nanoparticulate sol–gel composite coatings on different substrates. In the former study, incorporation of nanoparticles to the sol was applied to make it possible to increase the coating thickness, without increasing the sintering temperature. Commercial SiO_2 nanoparticles were suspended in an acid-catalyzed SiO_2 sol, whose stability is largely increased by adding tetramethylammonium hydroxide (TMAH) up to pH 6. The coatings as thick as 5 mm showed good corrosion resistance in marine water when the suspension stability was increased by 15 times compared with that of the starting acid suspension. In the later study, phosphate-bonded sol–gel composite alumina coatings were prepared on stainless steel substrates at processing temperatures of 300°C, 400°C, and 5008°C. It was found that the coating corrosion and electrochemical behavior was largely controlled by the degree of cracking and porosity in the coatings, which is minimum in those processed at the lowest temperature.

Dineshram et al. [48] tested the antifouling properties of titania, niobia, and silica coatings, derived from their respective nanoparticle dispersions or sols and fabricated on soda lime glass substrates in marine environments for antimacrofouling applications for marine optical instruments. The number of barnacles, *H. elegans,* and oyster settlement can be found in Figure 10.4.[a–c]

The present results show colonization of biofilms on the nanocoatings and clearly indicate biofilm formation, which in turn allows macrofouling organisms to deposit easily.

Nanoparticle-reinforced metal matrix composites (MMCs) generally exhibited wide engineering applications due to their enhanced hardness, better wear, and corrosion resistance when compared to pure metal or alloy. Research into the production of nanocomposite coatings applied in oceanic environments for anticorrosion, anti-fouling and reduced moving resistance has been done by numerous investigators. Gul et al. [49] prepared Ni/Al_2O_3 MMC coatings from a modified Watt's type electrolyte containing nano-α-Al_2O_3 particles by a direct current (DC) plating method to increase the surface hardness and wear resistance of the electrodeposited Ni. Al_2O_3 nanopowders with average particle size of 80 nm were co-deposited with nickel matrix on the steel substrates. The results showed that the wear resistance of the

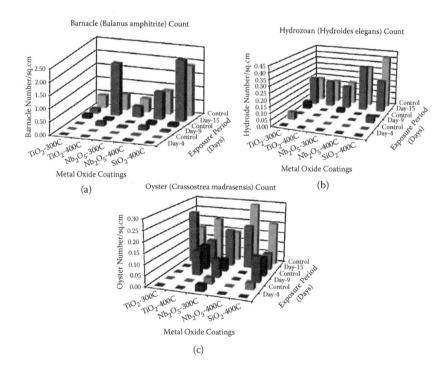

FIGURE 10.4

(a) Barnacle (*B. amphitrite*) count, (b) hydroides (*H. elegans*) count and (c) oyster (*C. madrasensis*) count after static immersion for 4, 9, and 15 days of different metal oxide coatings on glass substrate (front side) relative to controls without metal oxide coatings (back side). (From R. Dineshram et al. *Colloid. Surf. B* 74: 75–83.)

nanocomposites was approximately 2–3.5 times increased compared with unreinforced Ni deposited material. Xu et al. [50] researched a duplex surface treatment of AISI 316L stainless steel to enhance the erosion–corrosion resistance. The duplex surface treatment consisted of Ni/nano-SiC and Ni/nano-SiO_2 predeposited by brush plating and a subsequent surface alloying with Ni–Cr–Mo–Cu by a double glow of the substrate. The results suggested that the highly dispersive nano-SiO_2 particles were helpful to improve the erosion–corrosion resistance of composite alloying layer. But in F. Erler and his colleagues' works [51], the co-deposition of nanoscaled Al_2O_3 and TiO_2 particles on a conventional Watts nickel electrolyte showed decreasing corrosion stability, which indicates an attack along the interfaces of the nickel and particles.

References

1. D. Hui and Y.D.S. Rajapakse. 2004. Marine composites: foreword. *Composites: B* 35: 447–450.
2. D.S. Jeng, D.H. Cha, Y.S. Lin, and P.S. Hu. 2001. Wave-induced pore pressure around a composite breakwater. *Ocean Eng.* 28: 1413–1435.
3. J. Wen, K.L. Zhao, T.Y. Gu, and Issam I. Raad. 2009. A green biocide enhancer for the treatment of sulfate-reducing bacteria (SRB) biofilms on carbon steel surfaces using glutaraldehyde. *Int. Biodeter. Biodegrn.* 63: 1–5.
4. S. Cheng, J.T. Tian, S.G. Chen, Y.H. Lei, X.T. Chang, T. Liu, and Y.S. Yin. 2009. Development of nanostructures for medicine special issue microbially influenced corrosion of stainless steel by marine bacterium *Vibrio natriegens:* (I) Corrosion behavior. *Mater. Sci. Eng. C* 29: 751–755.
5. X.T. Chang, S.G. Chen, G.H. Gao, Y.S. Yin, S. Cheng, and T. Liu. 2009. Electrochemical behavior of microbiologically influenced corrosion on Fe_3Al in marine environment. *Acta Metallur. Sin.* 22: 313–320.
6. J. Wang, J.B. Yang, P. Sen, and T. Ruxton. 1996. Safety based design and maintenance optimisation of large marine engineering systems. *Appl. Ocean Res.* 18: 13–27.
7. A. Neville, T. Hodgkiess, and J.T. Dallas. 1995. A study of the erosion-corrosion behaviour of engineering steels for marine pumping applications. *Wear* 186–187: 497–507.
8. D. Graham. 1995. Composite pressure hulls for deep ocean submersibles. *Compos Struct.* 32: 331–343.
9. Q.M. Jia, M. Zheng, C.Z. Xu, and H.X. Chen. 2006. The mechanical properties and tribological behavior of epoxy resin composites modified by different shape nanofillers. *Polym. Adv. Technol.* 17: 168–173.
10. K.-T. Lau, C. Gu, and D. Hui. 2006. A critical review on nanotube and nanotube/nanoclay related polymer composite materials. *Composites B* 37: 425–436.
11. Y.J. Weitsman and M. Elahi. 2000. Effects of fluids on the deformation, strength and durability of polymeric composites—an overview. *Mech. Time-Depend Mater.* 4: 107–126.

12. P.M. Schubel, J.J. Luo, and S.M. Daniel. 2005. Low-velocity impact behavior of composite sandwich panels. *Composites A* 36: 1389–1396.
13. A.C. Marshall and S.M. Lee. 1990. *Core Composite and Sandwich Structures.* International Encyclopedia of Composites.
14. J.R. Vinson. 1999. *The Behavior of Sandwich Structures of Isotropic and Composite Materials.* 2nd ed. Technomic Publishing, Lancaster, PA.
15. T.S. Lim, C.P. Lee, and D.G. Lee 2004. Failure modes of foam core sandwich beams under static and impact loads. *J. Compos. Mat.* 38: 1639–1662.
16. S. Abrate. 2001. Modeling of impacts on composite structures. *Compos. Struct.* 51:129–38.
17. E. Ayorinde, R. Gibson, S. Kulkarni, F.Z. Deng, H. Mahfuz, S. Islam, and S. Jeelani. 2008. Reliable low-cost NDE of composite marine sandwich structures. *Composites B* 39: 226–241.
18. D.J. VanHouten and D.G. Baird. 2009. Generation and characterization of carbon nano-fiber–poly(arylene ether sulfone) nanocomposite foams. *Polymer* 50: 1868–1876.
19. X.M. Li and Y.J. Weitsman. 2004. Sea-water effects on foam-cored composite sandwich lay-ups. *Composites B* 35: 451–459.
20. A. Siriruk, D. Penumadu, and Y.J. Weitsman. 2009. Effect of sea environment on interfacial delamination behavior of polymeric sandwich structures. *Compos. Sci. Technol.* 69: 821–828.
21. R. Maurin, Y. Perrot, A. Bourmaud, P. Davies, and C. Baley. 2009. Seawater ageing of low styrene emission resins for marine composites: mechanical behaviour and nano-indentation studies. *Composites A* 40: 1024 –1032.
22. A.F. Avila, M.G.R Carvalho, E.C. Dias, and D.T.L. Cruz. 2009. Nano-structured sandwich composites response to low-velocity impact. *Compos. Struct.* 92: 745–751.
23. D.S. Wavhal and E.R. Fisher. 2003. Membrane surface modification by plasma induced polymerization of acrylamide for improved surface properties and reduced protein fouling. *Langmuir* 19: 79–85.
24. M. Norddin, A.F. Ismail, and D. Rana. 2008. Characterization and performance of proton exchange membranes for direct methanol fuel cell: blending of sulfonated poly (ether ether ketone) with charged surface modifying macromolecule. *J. Membr. Sci.* 323: 404–413.
25. Q. Shi, Y.L. Su, and Z.Y. Jian. 2008. Zwitterionic polyethersulfone ultrafiltration membrane with superior antifouling property. *J. Membr. Sci.* 319: 271–278.
26. S.S. Madaeni and N. Ghaemi. 2007. Characterization of self-cleaning RO membranes coated with TiO_2 particles under UV irradiation. *J. Membr. Sci.* 303: 221–233.
27. W. Bouguerra, A. Mnif, and B. Hamrouni. 2008. Boron removal by adsorption onto activated alumina and by reverse osmosis. *Desalination* 223: 31–37.
28. L.F. Liu, L. Xiao, and F.L. Yang. 2009. Terylene membrane modification with polyrotaxanes, TiO_2 and polyvinyl alcohol for better antifouling and adsorption property. *J. Membr. Sci* 333: 110–117.
29. R.J. Petersen. 1993. Composite reverse osmosis and nanofiltration membranes. *J. Membr. Sci.* 83: 81–150.
30. K. Kimura, G. Amy, J.E. Drewes, T. Heberer, T.U. Kim, and Y. Watanabe. 2003. Rejection of organic micropollutants (disinfection by-products, endocrine disrupting compounds, and pharmaceutically active compounds) by NF/RO membranes. *J. Membr. Sci.* 227: 113.

31. A.E. Childress and M. Elimelech. 2000. Relating nanofiltration membrane performance to membrane charge (electrokinetic) characteristics. *Environ. Sci. Technol.* 34: 3710–3716.
32. X.H. Zhu and M. Elimelech. 1997. Colloidal fouling of reverse osmosis membranes: measurements and fouling mechanisms. *Environ. Sci. Technol.* 31: 3654–3662.
33. C.Y.Y. Tang, Y.N. Kwon, and J.O. Leckie. 2007. Probing the nano- and micro-scales of reverse osmosis membranes—A comprehensive characterization of physiochemical properties of uncoated and coated membranes by XPS, TEM, ATR-FTIR, and streaming potential measurements. *J. Membr. Sci.* 287: 146–156.
34. V.K. Gupta, S.T. Hwang, W.B. Krantz, and A.R. Greenberg. 2007. Characterization of nanofiltration and reverse osmosis membrane performance for aqueous salt solutions using irreversible thermodynamics. *Desalination* 208: 1–18.
35. V. Kochkodan, S. Tsarenko, N. Potapchenko, V. Kosinova, and V. Goncharuk. 2008. Adhesion of microorganisms to polymer membranes: A photobactericidal effect of surface treatment with TiO_2. *Desalination* 220: 380–385.
36. H.L. Yang, J.C.T. Lin, and C. Huang. 2009. Application of nanosilver surface modification to RO membrane and spacer for mitigating biofouling in seawater desalination. *Water Res.* 43: 3777–3786.
37. A. Subramani, X.F. Huang, and E.M.V. Hoek. 2009. Direct observation of bacterial deposition onto clean and organic-fouled polyamide membranes. *J. Colloid Interf. Sci.* 336: 13–20.
38. K.S. Chung, J.C. Lee, W.K. Kim, S.B. Kim, and K.Y. Cho. 2008. Inorganic adsorbent containing polymeric membrane reservoir for the recovery of lithium from seawater. *J. Membr. Sci.* 325: 503–508.
39. T.H. Chong, F.S. Wong, and A.G. Fane. 2008. Implications of critical flux and cake-enhanced osmotic pressure (CEOP) on colloidal fouling in reverse osmosis: Experimental observations. *J. Membr. Sci.* 314: 101–111.
40. T. Liu, Y.S. Yin, and S.G. Chen. 2007. Super-hydrophobic surfaces improve corrosion resistance of copper in seawater. *Electrochim. Acta* 52: 3709–3713.
41. A.K. Ghosh, B.H. Jeong, X.F. Huang, and E.M.V. Hoek. 2008. Impacts of reaction and curing conditions on polyamide composite reverse osmosis membrane properties. *J. Membr. Sci.* 311: 34–45.
42. B.H. Jeong, E.M.V. Hoek, Y.S. Yan, A. Subramani, X.F. Huanga, G. Hurwitz, A.K. Ghosh, and A. Jawor. 2007. Interfacial polymerization of thin film nanocomposites: A new concept for reverse osmosis membranes. *J. Membr. Sci.* 294: 1–7.
43. M. Wouters, C. Rentrop, and P. Willemsen. 2009. Surface structuring and coating performance: Novel biocidefree nanocomposite coating with anti-fouling and fouling-release properties. *Prog. Org. Coating* 68: 4–11.
44. C.I.W. Calcagnoa, C.M. Mariani, S.R. Teixeira, and R.S. Mauler. 2006. The effect of organic modifier of the clay on morphology and crystallization properties of PET nanocomposites. *Polymer* 48: 966–974.
45. M.Y. Diego, K. Soren, E.W. Claus, and D.J. Kim. 2006. Dissolution rate measurements of sea water soluble pigments for antifouling paints: ZnO. *Org. Coatings* 56: 327–337.
46. Y. Castro, B. Ferrari, R. Moreno, and A. Duran. 2004. Coatings produced by electrophoretic deposition from nano-particulate silica sol–gel suspensions. *Surf. Coat. Technol.* 182: 199–203.

47. H.M. Hawthorne, A. Neville, T. Troczynski, X. Hub, M. Thammachartb, Y. Xiea, J. Fuc, and Q. Yangc. 2004. Characterization of chemically bonded composite sol–gel based alumina coatings on steel substrates. *Surf. Coat. Technol.* 176: 243–252.

48. R. Dineshram, R. Subasri, K.R.C. Somaraju, K. Jayaraj, L. Vedaprakash, Krupa Ratnam, S.V. Joshi, and R. Venkatesan. 2009. Biofouling studies on nanoparticle-based metal oxide coatings on glass coupons exposed to marine environment. *Colloid. Surf. B* 74: 75–83.

49. H. Gul, F. Kilic, S. Aslan, A. Alp, and H. Akbulut. 2009. Characteristics of electro-co-deposited Ni–Al$_2$O$_3$ nano-particle reinforced metal matrix composite (MMC) coatings. *Wear* 267: 976–990

50. J. Xu, C.Z. Zhuo, D.Z. Han, J. Tao, L.L. Liu, and S.Y. Jiang. 2009. Erosion–corrosion behavior of nano-particle-reinforced Ni matrix composite alloying layer by duplex surface treatment in aqueous slurry environment. *Corros. Sci.* 51: 1055–1068.

51. F. Erler, C. Jakob, H. Romanus, L. Spiess, B. Wielage, T. Lampke, and S. Steinhauser. 2003. Interface behaviour in nickel composite coatings with nano-particles of oxidic ceramic. *Electrochim. Acta* 48: 3063–3070.

Index